FUNDAMENTALS OF REAL ANALYSIS

PURE AND APPLIED MATHEMATICS

A Program of Monographs, Textbooks, and Lecture Notes

MONOGRAPHS AND TEXTBOOKS IN PURE AND APPLIED MATHEMATICS

1. *K. Yano,* Integral Formulas in Riemannian Geometry (1970)*(out of print)*
2. *S. Kobayashi,* Hyperbolic Manifolds and Holomorphic Mappings (1970) *(out of print)*
3. *V. S. Vladimirov,* Equations of Mathematical Physics (A. Jeffrey, editor; A. Littlewood, translator) (1970) *(out of print)*
4. *B. N. Pshenichnyi,* Necessary Conditions for an Extremum (L. Neustadt, translation editor; K. Makowski, translator) (1971)
5. *L. Narici, E. Beckenstein, and G. Bachman,* Functional Analysis and Valuation Theory (1971)
6. *D. S. Passman,* Infinite Group Rings (1971)
7. *L. Dornhoff,* Group Representation Theory (in two parts). Part A: Ordinary Representation Theory. Part B: Modular Representation Theory (1971, 1972)
8. *W. Boothby and G. L. Weiss (eds.),* Symmetric Spaces: Short Courses Presented at Washington University (1972)
9. *Y. Matsushima,* Differentiable Manifolds (E. T. Kobayashi, translator) (1972)
10. *L. E. Ward, Jr.,* Topology: An Outline for a First Course (1972) *(out of print)*
11. *A. Babakhanian,* Cohomological Methods in Group Theory (1972)
12. *R. Gilmer,* Multiplicative Ideal Theory (1972)
13. *J. Yeh,* Stochastic Processes and the Wiener Integral (1973) *(out of print)*
14. *J. Barros-Neto,* Introduction to the Theory of Distributions (1973) *(out of print)*
15. *R. Larsen,* Functional Analysis: An Introduction (1973) *(out of print)*
16. *K. Yano and S. Ishihara,* Tangent and Cotangent Bundles: Differential Geometry (1973) *(out of print)*
17. *C. Procesi,* Rings with Polynomial Identities (1973)
18. *R. Hermann,* Geometry, Physics, and Systems (1973)
19. *N. R. Wallach,* Harmonic Analysis on Homogeneous Spaces (1973) *(out of print)*
20. *J. Dieudonné,* Introduction to the Theory of Formal Groups (1973)
21. *I. Vaisman,* Cohomology and Differential Forms (1973)
22. *B.-Y. Chen,* Geometry of Submanifolds (1973)
23. *M. Marcus,* Finite Dimensional Multilinear Algebra (in two parts) (1973, 1975)
24. *R. Larsen,* Banach Algebras: An Introduction (1973)
25. *R. O. Kujala and A. L. Vitter (eds.),* Value Distribution Theory: Part A; Part B: Deficit and Bezout Estimates by Wilhelm Stoll (1973)
26. *K. B. Stolarsky,* Algebraic Numbers and Diophantine Approximation (1974)
27. *A. R. Magid,* The Separable Galois Theory of Commutative Rings (1974)
28. *B. R. McDonald,* Finite Rings with Identity (1974)
29. *J. Satake,* Linear Algebra (S. Koh, T. A. Akiba, and S. Ihara, translators) (1975)

30. *J. S. Golan,* Localization of Noncommutative Rings (1975)
31. *G. Klambauer,* Mathematical Analysis (1975)
32. *M. K. Agoston,* Algebraic Topology: A First Course (1976)
33. *K. R. Goodearl,* Ring Theory: Nonsingular Rings and Modules (1976)
34. *L. E. Mansfield,* Linear Algebra with Geometric Applications: Selected Topics (1976)
35. *N. J. Pullman,* Matrix Theory and Its Applications (1976)
36. *B. R. McDonald,* Geometric Algebra Over Local Rings (1976)
37. *C. W. Groetsch,* Generalized Inverses of Linear Operators: Representation and Approximation (1977)
38. *J. E. Kuczkowski and J. L. Gersting,* Abstract Algebra: A First Look (1977)
39. *C. O. Christenson and W. L. Voxman,* Aspects of Topology (1977)
40. *M. Nagata,* Field Theory (1977)
41. *R. L. Long,* Algebraic Number Theory (1977)
42. *W. F. Pfeffer,* Integrals and Measures (1977)
43. *R. L. Wheeden and A. Zygmund,* Measure and Integral: An Introduction to Real Analysis (1977)
44. *J. H. Curtiss,* Introduction to Functions of a Complex Variable (1978)
45. *K. Hrbacek and T. Jech,* Introduction to Set Theory (1978)
46. *W. S. Massey,* Homology and Cohomology Theory (1978)
47. *M. Marcus,* Introduction to Modern Algebra (1978)
48. *E. C. Young,* Vector and Tensor Analysis (1978)
49. *S. B. Nadler, Jr.,* Hyperspaces of Sets (1978)
50. *S. K. Segal,* Topics in Group Rings (1978)
51. *A. C. M. van Rooij,* Non-Archimedean Functional Analysis (1978)
54. *L. Corwin and R. Szczarba,* Calculus in Vector Spaces (1979)
53. *C. Sadosky,* Interpolation of Operators and Singular Integrals: An Introduction to Harmonic Analysis (1979)
54. *J. Cronin,* Differential Equations: Introduction and Quantitative Theory (1980)
55. *C. W. Groetsch,* Elements of Applicable Functional Analysis (1980)
56. *I. Vaisman,* Foundations of Three-Dimensional Euclidean Geometry (1980)
57. *H. I. Freedman,* Deterministic Mathematical Models in Population Ecology (1980)
58. *S. B. Chae,* Lebesgue Integration (1980)
59. *C. S. Rees, S. M. Shah, and C. V. Stanojević,* Theory and Applications of Fourier Analysis (1981)
60. *L. Nachbin,* Introduction to Functional Analysis: Banach Spaces and Differential Calculus (R. M. Aron, translator) (1981)
61. *G. Orzech and M. Orzech,* Plane Algebraic Curves: An Introduction Via Valuations (1981)
62. *R. Johnsonbaugh and W. E. Pfaffenberger,* Foundations of Mathematical Analysis (1981)
63. *W. L. Voxman and R. H. Goetschel,* Advanced Calculus: An Introduction to Modern Analysis (1981)
64. *L. J. Corwin and R. H. Szcarba,* Multivariable Calculus (1982)
65. *V. I. Istrătescu,* Introduction to Linear Operator Theory (1981)
66. *R. D. Järvinen,* Finite and Infinite Dimensional Linear Spaces: A Comparative Study in Algebraic and Analytic Settings (1981)

67. *J. K. Beem and P. E. Ehrlich,* Global Lorentzian Geometry (1981)
68. *D. L. Armacost,* The Structure of Locally Compact Abelian Groups (1981)
69. *J. W. Brewer and M. K. Smith, eds.,* Emmy Noether: A Tribute to Her Life and Work (1981)
70. *K. H. Kim,* Boolean Matrix Theory and Applications (1982)
71. *T. W. Wieting,* The Mathematical Theory of Chromatic Plane Ornaments (1982)
72. *D. B. Gauld,* Differential Topology: An Introduction (1982)
73. *R. L. Faber,* Foundations of Euclidean and Non-Euclidean Geometry (1983)
74. *M. Carmeli,* Statistical Theory and Random Matrices (1983)
75. *J. H. Carruth, J. A. Hildebrant, and R. J. Koch,* The Theory of Topological Semigroups (1983)
76. *R. L. Faber,* Differential Geometry and Relativity Theory: An Introduction (1983)
77. *S. Barnett,* Polynomials and Linear Control Systems (1983)
78. *G. Karpilovsky,* Commutative Group Algebras (1983)
79. *F. Van Oystaeyen and A. Verschoren,* Relative Invariants of Rings: The Commutative Theory (1983)
80. *I. Vaisman,* A First Course in Differential Geometry (1984)
81. *G. W. Swan,* Applications of Optimal Control Theory in Biomedicine (1984)
82. *T. Petrie and J. D. Randall,* Transformation Groups on Manifolds (1984)
83. *K. Goebel and S. Reich,* Uniform Convexity, Hyperbolic Geometry, and Nonexpansive Mappings (1984)
84. *T. Albu and C. Năstăsescu,* Relative Finiteness in Module Theory (1984)
85. *K. Hrbacek and T. Jech,* Introduction to Set Theory, Second Edition, Revised and Expanded (1984)
86. *F. Van Oystaeyen and A. Verschoren,* Relative Invariants of Rings: The Noncommutative Theory (1984)
87. *B. R. McDonald,* Linear Algebra Over Commutative Rings (1984)
88. *M. Namba,* Geometry of Projective Algebraic Curves (1984)
89. *G. F. Webb,* Theory of Nonlinear Age-Dependent Population Dynamics (1985)
90. *M. R. Bremner, R. V. Moody, and J. Patera,* Tables of Dominant Weight Multiplicities for Representations of Simple Lie Algebras (1985)
91. *A. E. Fekete,* Real Linear Algebra (1985)
92. *S. B. Chae,* Holomorphy and Calculus in Normed Spaces (1985)
93. *A. J. Jerri,* Introduction to Integral Equations with Applications (1985)
94. *G. Karpilovsky,* Projective Representations of Finite Groups (1985)
95. *L. Narici and E. Beckenstein,* Topological Vector Spaces (1985)
96. *J. Weeks,* The Shape of Space: How to Visualize Surfaces and Three-Dimensional Manifolds (1985)
97. *P. R. Gribik and K. O. Kortanek,* Extremal Methods of Operations Research (1985)
98. *J.-A. Chao and W. A. Woyczynski, eds.,* Probability Theory and Harmonic Analysis (1986)
99. *G. D. Crown, M. H. Fenrick, and R. J. Valenza,* Abstract Algebra (1986)
100. *J. H. Carruth, J. A. Hildebrant, and R. J. Koch,* The Theory of Topological Semigroups, Volume 2 (1986)

101. *R. S. Doran and V. A. Belfi*, Characterizations of C*-Algebras: The Gelfand-Naimark Theorems (1986)
102. *M. W. Jeter* Mathematical Programming: An Introduction to Optimization (1986)
103. *M. Altman*, A Unified Theory of Nonlinear Operator and Evolution Equations with Applications: A New Approach to Nonlinear Partial Differential Equations (1986)
104. *A. Verschoren*, Relative Invariants of Sheaves (1987)
105. *R. A. Usmani*, Applied Linear Algebra (1987)
106. *P. Blass and J. Lang*, Zariski Surfaces and Differential Equations in Characteristic $p > 0$ (1987)
107. *J. A. Reneke, R. E. Fennell, and R. B. Minton*. Structured Hereditary Systems (1987)
108. *H. Busemann and B. B. Phadke*, Spaces with Distinguished Geodesics (1987)
109. *R. Harte*, Invertibility and Singularity for Bounded Linear Operators (1988).
110. *G. S. Ladde, V. Lakshmikantham, and B. G. Zhang*, Oscillation Theory of Differential Equations with Deviating Arguments (1987)
111. *L. Dudkin, I. Rabinovich, and I. Vakhutinsky*, Iterative Aggregation Theory: Mathematical Methods of Coordinating Detailed and Aggregate Problems in Large Control Systems (1987)
112. *T. Okubo*, Differential Geometry (1987)
113. *D. L. Stancl and M. L. Stancl*, Real Analysis with Point-Set Topology (1987)
114. *T. C. Gard*, Introduction to Stochastic Differential Equations (1988)
115. *S. S. Abhyankar*, Enumerative Combinatorics of Young Tableaux (1988)
116. *H. Strade and R. Farnsteiner*, Modular Lie Algebras and Their Representations (1988)
117. *J. A. Huckaba*, Commutative Rings with Zero Divisors (1988)
118. *W. D. Wallis*, Combinatorial Designs (1988)
119. *W. Więsław*, Topological Fields (1988)
120. *G. Karpilovsky*, Field Theory: Classical Foundations and Multiplicative Groups (1988)
121. *S. Caenepeel and F. Van Oystaeyen*, Brauer Groups and the Cohomology of Graded Rings (1989)
122. *W. Kozlowski*, Modular Function Spaces (1988)
123. *E. Lowen-Colebunders*, Function Classes of Cauchy Continuous Maps (1989)
124. *M. Pavel*, Fundamentals of Pattern Recognition (1989)
125. *V. Lakshmikantham, S. Leela, and A. A. Martynyuk*, Stability Analysis of Nonlinear Systems (1989)
126. *R. Sivaramakrishnan*, The Classical Theory of Arithmetic Functions (1989)
127. *N. A. Watson*, Parabolic Equations on an Infinite Strip (1989)
128. *K. J. Hastings*, Introduction to the Mathematics of Operations Research (1989)
129. *B. Fine*, Algebraic Theory of the Bianchi Groups (1989)
130. *D. N. Dikranjan, I. R. Prodanov, and L. N. Stoyanov*, Topological Groups: Characters, Dualities, and Minimal Group Topologies (1989)

131. *J. C. Morgan II,* Point Set Theory (1990)
132. *P. Biler and A. Witkowski*, Problems in Mathematical Analysis (1990)
133. *H. J. Sussmann,* Nonlinear Controllability and Optimal Control (1990)
134. *J.-P. Florens, M. Mouchart, and J. M. Rolin,* Elements of Bayesian Statistics (1990)
135. *N. Shell*, Topological Fields and Near Valuations (1990)
136. *B. F. Doolin and C. F. Martin*, Introduction to Differential Geometry for Engineers (1990)
137. *S. S. Holland, Jr.*, Applied Analysis by the Hilbert Space Method (1990)
138. *J. Okniński*, Semigroup Algebras (1990)
139. *K. Zhu*, Operator Theory in Function Spaces (1990)
140. *G. B. Price,* An Introduction to Multicomplex Spaces and Functions (1991)
141. *R. B. Darst,* Introduction to Linear Programming: Applications and Extensions (1991)
142. *P. L. Sachdev,* Nonlinear Ordinary Differential Equations and Their Applications (1991)
143. *T. Husain*, Orthogonal Schauder Bases (1991)
144. *J. Foran*, Fundamentals of Real Analysis (1991)

Other Volumes in Preparation

FUNDAMENTALS OF REAL ANALYSIS

James Foran

University of Missouri
Kansas City, Missouri

Marcel Dekker, Inc. New York · Basel · Hong Kong

Library of Congress Cataloging-in-Publication Data

Foran, James,
Fundamentals of real analysis / James Foran.
p. cm. -- (Monographs and textbooks in pure and applied mathematics ; 143)
Includes bibliographical references and index.
ISBN 0-8247-8453-7
1. Functions of real variables. 2. Mathematical analysis. I. Title. II. Series.
QA331.5.F67 1991
515'.8--dc20 90-23801
CIP

This book is printed on acid-free paper.

MARCEL DEKKER, INC.
270 Madison Avenue, New York, New York 10016

Current printing (last digit):
10 9 8 7 6 5 4 3 2 1

PRINTED IN THE UNITED STATES OF AMERICA

Preface

This book is intended as a text. It is centered on the core material pertaining to functions of a real variable; that is, subsets of the line, the theory of measure, the Lebesgue integral and its relationship to the derivative. This material is contained in the first sections of Chapters 2 through 8. Chapter 1 contains mathematical preliminaries, most of which would normally be found in a course in undergraduate analysis. Chapter 9 deals with integrals which are more general than that of Lebesgue. The second sections of Chapters 2 through 8 each contain supplementary or additional material some of which is usually taught in a graduate course in real variables. These additional subjects include an introduction to the axioms of Zermelo-Fraenkel Set theory, Hausdörff measures, probability measures and additive set functions, the density topology and approximate limits, $\mathcal{L}^p$ spaces, the Radon-Nikodym derivative, the representation of linear functionals and differentiation in Euclidean n-space.

At this time the core material pertaining to set theory and to the theory of functions of a real variable forms a structure which has been thoroughly developed and consists of many interrelated theorems. The theorems presented in this book have been provided with complete proofs; that is, no part of the proof of a theorem is ever left as an exercise. Historical anecdotes and mathematical illustrations are absent from this book. The reader is clearly at liberty to investigate the people involved with the development of this theory and doing so will no doubt add to its interest. There are two reasons, besides simplicity, for the absence of illustrations: for one thing, a picture is never a proof; for another, the pictures required would necessarily be only approximate and would be misleading with regard to the generality, complexity and/or precision of the mathematics involved. Nonetheless, Venn diagrams, rough sketches and outlines are very useful and it is expected that many of these will be made by the reader as aids to understanding. Exercises are provided at the end of each section which while not easy, will challenge the reader's understanding and provide a variety of questions of a secondary nature. There are 183 exercises in the book.

The relationships between the various sections of the book can be represented schematically as follows:

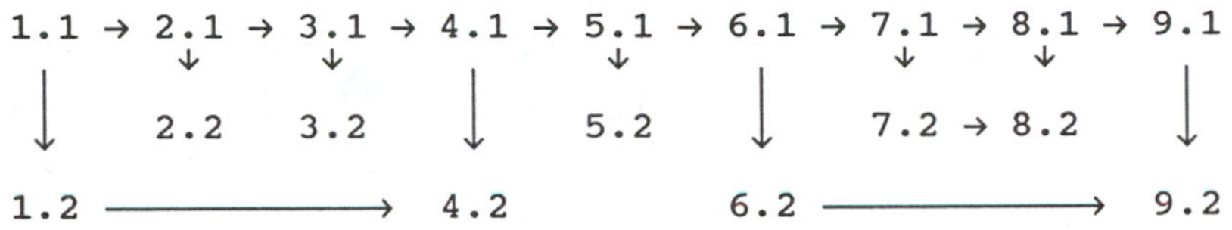

$$
\begin{array}{ccccccccccccccccc}
1.1 & \to & 2.1 & \to & 3.1 & \to & 4.1 & \to & 5.1 & \to & 6.1 & \to & 7.1 & \to & 8.1 & \to & 9.1 \\
\downarrow & & \downarrow & & \downarrow & & \downarrow & & \downarrow & & \downarrow & & \downarrow & & \downarrow & & \downarrow \\
 & & 2.2 & & 3.2 & & & & 5.2 & & & & 7.2 & \to & 8.2 & & \\
\downarrow & & & & & & \downarrow & & & & \downarrow & & & & & & \downarrow \\
1.2 & \multicolumn{5}{c}{\longrightarrow} & 4.2 & & & & 6.2 & \multicolumn{5}{c}{\longrightarrow} & 9.2
\end{array}
$$

A one semester course can be taught from the first sections of Chapters 1 - 8 and a shorter course can be offered by omitting some of this material or presenting it without proof.

Over the past sixteen years the subject matter of this text has been presented by its author in alternate years in the form of a two semester course in Real Analysis. The material is intended as a basic unit preliminary to research in Real Analysis and/or to the study of those subjects which use analysis as a tool.

Background material, summarized in the first section of the book, is available in texts on advanced undergraduate analysis such as Rudin's _Principles of Mathematical Analysis_. The reader is also directed to the following classics in which one may find many of the topics of this book in broader contexts: Hausdörff's _Set Theory_, Kelley's _General Topology_, Natanson's _Theory of Functions of a Real Variable_ and Saks' _Theory of the Integral_.

Acknowledgements: While the responsibility for writing this book lies with myself, a number of mathematicians have been helpful with its preparation and deserve thanks. Leslie Willett studied the material on the way to obtaining a masters' degree, suggested going ahead with the book and typed it. Richard O'Malley made valuable suggestions on presenting the density topology and Lee Hart on the logic of set theory. Richard Delaware and Sandra Meinershagen read the entire book and their suggestions and corrections along with those of the class of '89-'90 are greatly appreciated as are the encouraging results reported by Paul Humke who also taught a course from the manuscript. The staff at the University of Missouri-Kansas City, especially Carol Rust, helped get the book in camera-ready form. Finally, the personnel and consultants of Marcel Dekker have been most patient, helpful and clearly instrumental in its production.

James Foran

Contents

FUNDAMENTALS OF REAL ANALYSIS

Chapter One

1.1 Basic Definitions and Background Material

The definitions occurring in the first section of this chapter delineate mathematical structures which are natural settings for theorems in analysis. The theorems in this section are basic in nature and form the background for latter parts of the book. Some of this material will no doubt be familiar to the reader. The second section of this chapter contains important results which are of interest but which are not developed further in later chapters.

Standard set theoretic notation will be used in this book. We begin by summarizing this notation. Capital letters are used for sets, small letters for their elements. Thus $a \in A$ denotes the fact that the element a belongs to the set A, or briefly, that a is an element of A. That a is not an element of A is denoted by $a \notin A$. Sets are given either by enumeration (for example, $\{1, 2, 3\}$ is the set consisting of the numbers 1, 2 and 3) or by description (for example,

$$\{x: x \text{ is a natural number less than } 4\},$$

the set of all x such that x is a natural number less than 4) and the brackets $\{\ \}$ enclose the enumeration or description. Two <u>sets</u> A and B <u>are equal</u> if they have the same elements, written $A = B$; A <u>is contained in</u> B, written $A \subset B$ or $B \supset A$, if every element of A is an element of B. Thus $A = B$ if and only if $A \subset B$ and $B \subset A$. (The word "iff" is sometimes used as an abbreviation for "if and only if".) The <u>union</u> of two sets A and B is $A \cup B = \{x: x \in A \text{ or } x \in B \text{ (or both)}\}$; the <u>intersection</u> is $A \cap B = \{x: x \in A \text{ and } x \in B\}$. The difference of two sets A and B, written $A \backslash B$, is $\{x: x \in A \text{ and } x \notin B\}$. If all sets under consideration belong to a given set X, then the <u>complement</u> of a set A, written A^c, is $X \backslash A$. A <u>sequence</u> is a set whose elements are indexed with the natural numbers. If for each α belonging to a set A, E_α is a set, then the collection of all sets E_α with $\alpha \in A$ is written $\{E_\alpha\}_{\alpha \in A}$. If the indexing set consists of the natural numbers and for each natural number n, E_n is a set, then the sequence consisting of the sets E_n is written $\{E_n\}_{n=1}^{\infty}$ or sometimes just $\{E_n\}$. The union of an indexed collection of sets $\{E_\alpha\}_{\alpha \in A}$ is written $\bigcup_{\alpha \in A} E_\alpha$; this is the set

$$\{x: \text{there is an } \alpha \in A \text{ with } x \in E_\alpha\}.$$

The intersection of the sets in such a collection is written $\bigcap_{\alpha\in A} E_\alpha$ and this is the set

$$\{x: \text{for each } \alpha \in A,\ x \in E_\alpha\}.$$

If the indexing set consists of the natural numbers and $\{E_n\}$ is a sequence of sets, then $\bigcup_{n=1}^{\infty} E_n$ or $\cup E_n$ denotes the union of the sets in the sequence and $\bigcap_{n=1}^{\infty} E_n$ or $\cap E_n$ denotes the intersection. It is standard to denote <u>the empty set</u>, the set which has no elements, by $\emptyset$. The expression $A = \emptyset$ is used when it has been determined that there are no elements satisfying the description which defines A. Two sets A and B are said to be <u>disjoint</u> if they have no elements in common; that is, if $A \cap B = \emptyset$. A collection of sets $\{E_\alpha\}_{\alpha\in A}$ is said to be a collection of <u>pairwise disjoint</u> sets if for distinct α and β in A, E_α and E_β are disjoint. An <u>ordered pair</u> of elements $a \in A$, and $b \in B$ is written (a,b) indicating a is the first element and b is the second element in the pair; the collection of all ordered pairs (a,b) with $a \in A$ and $b \in B$ is called the <u>cross product</u> of A and B and is written $A\times B$. A <u>function</u> f is a collection of ordered pairs no two of which have the same first element. The set A of all first elements in some pair is called the <u>domain</u> of the function; the set of all second elements is called the <u>image</u> of the function; if f is contained in $A\times B$, B is called the <u>range</u> of f. If $E \subset A$, then

$$f(E) = \{y: \text{there is } (x,y) \in f \text{ with } x \in E\}$$

is called the <u>image of</u> E <u>under</u> f. If $H \subset B$, then

$$f^{-1}(H) = \{x: (x,y) \in f \text{ and } y \in H\}$$

is called the <u>inverse image of</u> H <u>under</u> f. If A is the domain and B is the image of a function f, then f is said to take A <u>onto</u> B; if no two distinct elements of A have a pair in f with the same second element, then f is said to be <u>one to one</u>. When $(a,b) \in f$, one writes $b = f(a)$.

Sets may have other sets as elements and these sets may in turn have sets as elements. A set is sometimes called a collection; this is done when it is desirable to emphasize the fact that its elements are also sets. However, it is basic to the notion of a set that a set is strictly determined by and made up of its elements with no further structure assumed. We thus never allow (or need) a sequence of sets $\{A_n\}_{n=1}^{\infty}$ with $A_{n+1} \in A_n$. (Disallowing this is equivalent to one of the axioms from the form of axiomatic set theory which is in most common use.) This assumption implies that certain combinations of the form $A \in B$ are impossible; for example, it is not possible that there be a set A such that A belongs to A. Also, if $A \in B$, $B \notin A$ and if $A \in B$ and $B \in C$, $C \notin A$, etc. Otherwise, such set inclusions would give rise to an unallowable sequence of sets. This assumption also puts aside certain contradictions which arose historically in the development of the theory of sets. For example, it is impossible to form the set of all sets. If there were such a set, it would have to contain itself as an element. While we will not need to resort to axioms, axiomatic set theory will be discussed briefly later.

Basic familiarity with the arithmetic of the natural numbers, the integers and the rational numbers is presumed in what follows. The discussion of these number systems which follows presents in outline form an approach which is capable of rigorous development. The set of natural numbers (the counting numbers) will refer to the collection of numbers consisting of 1, 2, 3 and so forth. The mathematician, Peano, took some of the mystery out of these numbers by setting down five axioms for a set $\mathbb{N}$ so that $\mathbb{N}$ could be taken as a model of the natural numbers. These axioms are:

i) There is an element $1 \in \mathbb{N}$.

ii) For each element $n \in \mathbb{N}$, there is an element $n^+ \in \mathbb{N}$ (n^+ is called the <u>successor of</u> n).

iii) If $S \subset \mathbb{N}$, $1 \in S$ and $n \in S$ implies $n^+ \in S$, then $S = \mathbb{N}$.

iv) For each $n \in \mathbb{N}$, $n^+ \neq 1$.

v) For each $n, m \in \mathbb{N}$, if $n^+ = m^+$ then $n = m$.

Axiom iii) is the statement of mathematical induction. Induction is frequently used in definitions which involve the natural numbers. For example, addition of natural numbers can be defined for each $n, m \in \mathbb{N}$, by letting $n + 1 = n^+$ and, if $n + m$ has been defined, then $n + (m^+)$ is defined to be $(n + m)^+$. The usual order can be defined for the natural numbers by $n < m$ if there is $k \in \mathbb{N}$ with $n + k = m$. Multiplication can be defined inductively by $n \cdot 1 = n$ and, if $n \cdot m$ has been defined, $n \cdot (m^+) = nm + n$. Induction is, of course, also used to prove theorems as follows: if a statement is true of 1 and whenever the statement is true of n it is true of $n + 1$, then the statement is true of every natural number. An important property of the natural numbers is that every non-empty subset of the natural numbers has a least element. To see this, let $S \subset \mathbb{N}$ and $S \neq \emptyset$. Suppose, if possible, that S does not have a least element. Then $1 \notin S$ and if no number less than or equal to n belongs to S, $n \notin S$ and no number less than or equal to $n + 1$ belongs to S. That is, no natural number belongs to S, a contradiction. It follows that every non-empty subset of the natural numbers has a least element.

A modern approach to modeling the natural numbers consists of defining $0 = \emptyset$, $1 = \{0\} = \{\emptyset\}$, $2 = \{0,1\} = \{\emptyset, \{\emptyset\}\}$, $3 = \{0, 1, 2\} = \{\emptyset, \{\emptyset\}, \{\emptyset, \{\emptyset\}\}\}$, and $n + 1 = \{0, 1, \ldots, n\}$. While this notation, when written out, is quite cumbersome, this approach clearly does not require any specific reference to the world of objects. Also each natural number n is represented by a set which contains n elements and $n < m$ iff $n \in m$. Having the natural numbers, the integers are defined to be numbers of the form $0, +1, -1, \ldots, +n, -n, \ldots,$ and the rational numbers are those of the form p/q where p is an integer, q a natural number and two such numbers p/q and r/s are considered equal if $p \cdot s = q \cdot r$.

Of course, part of the motivation for these numbers is the attempt to represent points on the number line or real line. When 0 and 1 are chosen as two distinct points on a line, the integers can be positioned in the usual fashion using only a compass. With these in place as follows

... -4 -3 -2 -1 0 1 2 3 4 ...

the rational numbers can be positioned at their

respective places on the line. Indeed, if a and b are two distances (positive numbers) then $a \cdot b$ can be constructed with a compass and straightedge by constructing a right triangle with sides 1 and a and a similar triangle with corresponding sides b and $x = a \cdot b$. Also a/b can be constructed with two right triangles, the first having sides b and a, the second having corresponding sides 1 and $x = a/b$.

The real numbers are intended to be a collection of numbers corresponding to the points on a line and the non-negative real numbers are needed to represent all possible distances between points. The set of all non-ending decimal expansions can be taken as a model of the real numbers. Here, a non-ending decimal expansion is one of the form $\pm N.n_1 n_2 \ldots n_k \ldots$ where $+$ and $-$ are two signs indicating whether the point corresponding to the number is to the right or left of 0, N is a natural number or 0, and each n_k is a natural number with $0 \le n_k \le 9$. It is necessary to indicate that some points do not have only one representation in this notation. Specifically $+0.000\ldots$ is the same point as $-0.000\ldots$ and each number with an expansion ending in 9's has another representation ending in 0's; for example, $2.7999\ldots$ and $2.8000\ldots$. Two non-ending decimals of this form will be considered to be equal and to represent the same number. Addition, subtraction, multiplication and division are performed in the usual way by determining the result to more and more decimal places of accuracy.

From an axiomatic standpoint, the real numbers form what is called a complete ordered field. First, a <u>field</u> is a set F along with two operations, $+$

(addition) and $\cdot$ (multiplication), each of which associates with pairs of elements of F a unique element of F. There are five axioms for addition:

a1. (closure) For each $a, b \in F$, there is an element $a + b \in F$.

a2. (associativity) For each $a, b, c \in F$, $a + (b + c) = (a + b) + c$.

a3. (identity) There is an element $0 \in F$ such that for each $a \in F$, $a + 0 = a$.

a4. (inverses) For each $a \in F$ there is an element $-a \in F$ such that $a + (-a) = 0$.

a5. (commutativity) For each $a, b \in F$, $a + b = b + a$.

A set S with one operation $+$ is called a <u>group</u> if the first four of these axioms are satisfied. If all five are satisfied it is called a <u>commutative group</u> or an <u>Abelian group</u>.

There are five parallel axioms for multiplication:

m1. (closure) For each $a,b \in F$ there is $a \cdot b \in F$.

m2. (associativity) For each $a,b,c \in F$, $(a \cdot b) \cdot c = a \cdot (b \cdot c)$.

m3. (identity) There is an element $1 \in F$ with $1 \neq 0$ such that for each $a \in F$, $a \cdot 1 = a$.

m4. (inverses) For each $a \in F$ with $a \neq 0$, there is $a^{-1} \in F$ such that $a \cdot a^{-1} = 1$.

m5. (commutativity) For each $a,b \in F$, $a \cdot b = b \cdot a$.

Thus $F \backslash \{0\}$ is a commutative group under the operation of multiplication.

The eleventh axiom for a field is the distributive law:

(d) For each $a,b,c \in F$, $a \cdot (b + c) = a \cdot b + a \cdot c$.

The axioms for an ordered field can be stated in a most elementary fashion by asserting that there is a subset P of F (the positive elements of F) such that:

p1. For each $a \in F$ exactly one of $a \in P$, $-a \in P$ or $a = 0$ holds.

p2. For each $a,b \in P$, $a + b \in P$.

p3. For each $a,b \in P$, $a \cdot b \in P$.

The order of the elements of F is then defined by $a < b$ iff $b - a \in P$. Note that for fields F, one can use the alternate definition of order which asserts that

o1. For each $a,b \in F$ exactly one of $a < b$, $a = b$ or $b < a$ holds.

o2. For $a,b,c \in F$, $a < b$ and $b < c$ implies $a < c$.

o3. For $a,b,c \in F$ if $a < b$ then $a + c < b + c$.

o4. For $a,b,c \in F$ if $a < b$ and $0 < c$ then $a \cdot c < b \cdot c$; if $a < b$ and $c < 0$ then $b \cdot c < a \cdot c$.

Finally, an ordered field F is a <u>complete ordered field</u> if the completeness axiom (c) holds:

(c) Every nonempty subset A of F which has an upper bound, has a least upper bound.

Here an element b is an <u>upper bound</u> for A if for each $a \in A$, $a \le b$; b is a <u>least upper bound</u> for A if b is an upper bound and, whenever b' is an upper bound, $b \le b'$. The least upper bound of a set A is called the <u>supremum</u> of A, written $\sup A$. The <u>infimum</u> of a non-empty set A which has a lower bound, written $\inf A$, is the greatest lower bound of A; that is, $\inf A = -\sup(-A)$ where $-A = \{x: -x \in A\}$.

One can check that the non-ending decimal expansions satisfy the axioms for a complete ordered field. In particular, the completeness axiom can be shown to hold by starting with a non-empty set A which is bounded above by $b = N.x_1x_2\ldots$ and noting that $N + 1$ is an upper bound for A. There is then a least integer n so that $n + 1$ is an upper bound of A. If $n \geq 0$ (the case where $n < 0$ is similar) one may determine a least number $x_1 = 0, 1, \ldots, 9$ so that $n + (x_1 + 1)/10$ is an upper bound for A. Continuing in this fashion yields a non-ending decimal $n.x_1x_2\ldots x_k\ldots$ which can readily be seen to be the least upper bound of A.

From the axioms for an ordered field, it is not difficult to determine that $(-a)\cdot b = -(a\cdot b)$, $(-1)\cdot(-1) = 1 > 0$, $2 = 1 + 1 > 1$, $3 = 2 + 1 > 2$ and continuing inductively, that any ordered field contains a copy of the natural numbers, with the successor of n equal to $n + 1$. Then, since a field contains additive and multiplicative inverses and products, an ordered field must contain a copy of the integers and of the rational numbers. Then, since the rational numbers form an ordered field, it follows that they are the smallest ordered field in the sense that, given an ordered field F, there is a map Φ from the rational numbers into F so that $\Phi(0) = 0 \in F$, $\Phi(1) = 1 \in F$ and, for rational numbers a and b, $\Phi(a + b) = \Phi(a) + \Phi(b)$, $\Phi(a\cdot b) = \Phi(a)\cdot\Phi(b)$ and, if $a < b$, $\Phi(a) < \Phi(b)$.

In a complete ordered field, the following property, known as the <u>Archimedean property</u> holds:

If $x > 0$, there is a natural number n such that $x < n$. To see this, suppose that there were an x in a complete ordered field F for which this property did not hold. Then the set of natural numbers in F would be bounded above by x and hence would have a least upper bound, say x_o. But then $x_o - 1$ would not be an upper bound for the natural numbers. That is, there would be a natural number N with $x_o - 1 < N$ and x_o would be less than $N + 1$. This contradiction implies the Archimedean property.

A consequence of the Archimedean property is that between any two elements of a complete ordered field F there is a rational number from F. To see this, consider the case where $x, y \in F$ and $0 \le x < y$ (the other cases are similarly shown to hold). Then there is a natural number n such that $1/(y - x) < n$; that is, $y - x > 1/n$. There is a least natural number m such that $nx < m$ and then $x < m/n$ and $m - 1 \le nx$. Hence $(m - 1)/n \le x$ and, since $x + 1/n < y$, $m/n < y$; that is, $x < m/n < y$.

Given a complete ordered field F, let $\mathbb{Q}$ be a copy of the rational numbers contained in F. If $A \ne \emptyset$ is a subset of F, let

$$A' = \{q \in \mathbb{Q}: \text{there is } a \in A \text{ with } q < a\}.$$

It then follows from the above that $\sup A = \sup A'$.

Now, if F_1 and F_2 are complete ordered fields, there are copies of the rational numbers $\mathbb{Q}_1$ and $\mathbb{Q}_2$ with $\mathbb{Q}_1 \subset F_1$ and $\mathbb{Q}_2 \subset F_2$. There is a map Ψ taking $\mathbb{Q}_1$ onto $\mathbb{Q}_2$ so that $\Psi(0) = 0 \in \mathbb{Q}_2$, $\Psi(1) = 1 \in \mathbb{Q}_2$, for $a, b \in \mathbb{Q}_1$, $\Psi(a + b) = \Psi(a) + \Psi(b)$, $\Psi(a \cdot b) = \Psi(a) \cdot \Psi(b)$ and $a < b$ in $\mathbb{Q}_1$ implies $\Psi(a) < \Psi(b)$. We may now define Ψ on all of F_1. If $x \in F_1$, since $x = \sup\{y \in F_1: y \le x\} = \sup\{q \in Q_1: q \le x\}$, let

$$\Psi(x) = \sup\{\Psi(q): q \le x\}.$$

It is now not difficult to show that Ψ takes F_1 onto F_2 and if $a, b \in F_1$ then $\Psi(a + b) = \Psi(a) + \Psi(b)$, $\Psi(a \cdot b) = \Psi(a) \cdot \Psi(b)$ and $a < b$ implies $\Psi(a) < \Psi(b)$. That is, any two complete ordered fields have corresponding elements for which the same arithmetic and the same order relationships hold. Thus, we will refer to any complete ordered field as the real numbers which will be denoted by $\mathbb{R}$. The non-negative real numbers will be denoted by $\mathbb{R}^+$. Sometimes the symbols $+\infty$ and $-\infty$ are adjoined to the real numbers to form the <u>extended real numbers</u> with the understanding that for each $x \in \mathbb{R}$, $-\infty < x < \infty$, $x/\infty = x/-\infty = 0$, $x + \infty = \infty$, $x + (-\infty) = -\infty$ and for $x > 0$, $\infty \cdot x = -\infty \cdot (-x) = \infty$ and $-\infty \cdot x = \infty \cdot (-x) = -\infty$. If A is a non-empty subset of $\mathbb{R}$ which is not bounded above, we will write $\sup A = \infty$ and $\inf(-A) = -\infty$.

We now turn to a brief presentation of vector spaces and then to metric spaces and topological spaces. Brief, but complete proofs of essential theorems about metric spaces and topological spaces are presented.

Euclidean k-dimensional space, symbolized by $\mathbb{R}^k$, has points which can be represented by the collection of ordered k-tuples $(x_1, x_2, \ldots, x_k)$ of real numbers. The space $\mathbb{R}^k$ is an example of a vector space over the field of real numbers when addition of $x = (x_1, x_2, \ldots, x_k)$ and $y = (y_1, y_2, \ldots, y_k)$ is defined by $x + y = (x_1 + y_1, x_2 + y_2, \ldots, x_k + y_k)$ and multiplication by a real number a is defined by

$$ax = (ax_1, ax_2, \ldots, ax_k).$$

In general, a <u>vector space</u> over a field (we will usually be concerned with the case where the field consists of the real numbers) is a set S, whose elements are called vectors, for which addition is defined and satisfies the axioms for a commutative group under addition; namely, for each $x, y, z \in S$

$$x + y \in S$$

$$(x + y) + z = x + (y + z)$$

$$\text{there is } \bar{0} \in S \text{ so that } \bar{0} + x = x$$

$$\text{for } x \in S \text{ there is a } -x \in S \text{ so that } x + (-x) = \bar{0}$$

$$x + y = y + x.$$

There must also be multiplication of vectors by real numbers (scalar multiplication) resulting in other vectors and satisfying for each $a, b \in \mathbb{R}$, $x, y \in S$

$$0 \cdot x = \bar{0}$$

$$a \cdot \bar{0} = \bar{0}$$

$$a(bx) = (ab)x$$

$$a(x + y) = ax + ay$$

$$(a + b)x = ax + bx.$$

Actually, $\mathbb{R}^k$ is also an example of an <u>inner product space</u>; that is, a vector space on which there is defined a product (dot product or inner product) of two vectors yielding a real number and satisfying for all $a, b \in \mathbb{R}$, $x, y \in S$

$$x \cdot y = y \cdot x$$

$$0 \cdot x = 0$$

$$a \cdot \bar{0} = 0$$

$$(ax) \cdot (by) = (ab) x \cdot y$$

$$a \cdot (x + y) = a \cdot x + b \cdot y.$$

The inner product in $\mathbb{R}^k$ is defined for $x = (x_1, \ldots, x_k)$ and $y = (y_1, \ldots, y_k)$ by $x \cdot y =$

$\sum_{i=1}^{k} x_i y_i$.

A <u>norm</u> on a vector space S over the real numbers is a real-valued function $\| \ \|$ defined on S and satisfying for $a \in \mathbb{R}$, $x, y \in S$

$$\|x\| = 0 \text{ iff } x = \overline{0}$$
$$\|ax\| = |a|\|x\|$$
$$\|x + y\| \le \|x\| + \|y\|.$$

The norm on $\mathbb{R}^k$ is defined by $\|x\| = (x \cdot x)^{1/2}$. This norm can also be defined on any inner product space. Clearly $\|x\| = 0$ iff $x = \overline{0}$ and $\|ax\| = |a|\|x\|$. To see that $\|x + y\| \le \|x\| + \|y\|$ we show that for all x, y, $|x \cdot y| \le \|x\|\|y\|$ for then

$$\begin{aligned} \|x + y\|^2 &= (x + y) \cdot (x + y) \\ &= \|x\|^2 + 2x \cdot y + \|y\|^2 \\ &\le \|x\|^2 + 2\|x\|\|y\| + \|y\|^2 \\ &= (\|x\| + \|y\|)^2. \end{aligned}$$

To see that $|x \cdot y| \le \|x\|\|y\|$, fix $y, x \ne \overline{0}$ and consider the function $f(t) = \|tx + y\|^2 = (tx + y) \cdot (tx + y) = t^2\|x\|^2 + 2tx \cdot y + \|y\|^2$. Then $f(t) \ge 0$ and $f(t)$ takes on its minimum when $f'(t) = 0$; that is, when $2t\|x\|^2 + 2(x \cdot y) = 0$ or $t = -x \cdot y / \|x\|^2$. For this value of t, $f(t) = (\|x\|^2\|y\|^2 - x \cdot y)/\|x\|^2$ and from $f(t) \ge 0$ the desired inequality $x \cdot y \le \|x\|\|y\|$ follows.

Finally, every norm $\| \ \|$ on a vector space S gives rise to a distance function d between pairs of points $x, y \in S$ where d is given by $d(x,y) = \|y - x\|$. This function d satisfies for all $x, y, z \in S$

$$d(x,y) = 0 \text{ iff } x = y$$
$$d(x,y) = d(y,x)$$
$$d(x,y) \le d(x,z) + d(y,z).$$

Some of the theorems in the latter chapters of this book can and will be presented in more general settings than the real line or Euclidean k-space. While there is usually no great loss if one thinks in terms of the line or Euclidean spaces, the very basic understanding of metric spaces and topological spaces which will be developed here will allow for this generality of presentation. The theorems on metric spaces and topological spaces provide only a minimal background for these subjects.

A metric space is a set X for which there is given a function d (distance) taking $X \times X$ into $\mathbb{R}^+$ so that d satisfies for each x, y and z in X

i) $d(x,y) = 0$ iff $x = y$

ii) $d(x,y) = d(y,x)$

iii) $d(x,y) \le d(x,z) + d(z,y)$ (triangle inequality)

Elements of a metric space are called its points. A subset X' of a metric space X is also a metric space when the function d is restricted to $X' \times X'$. Thus Euclidean spaces and their subsets provide examples of metric spaces. Given x in a metric space X and $r > 0$, the neighborhood of radius r about x is $\{y \in X: d(x,y) < r\}$; this is denoted by $N_r(x)$. A set $G \subset X$ is said to be an open set or to be open if for every point $x \in G$ there is $r > 0$ so that $N_r(x) \subset G$. Each $N_r(x)$ is an example of an open set. To see this, let $y \in N_r(x)$, $d = d(x,y)$ and $r' = r - d$. Then, if $z \in N_{r'}(y)$, by iii), $d(x,z) \le d(x,y) + d(y,z) < d + r' = r$. Thus $z \in N_r(x)$; that is, $N_{r'}(y) \subset N_r(x)$ and, since y is an arbitrary element of $N_r(x)$, it follows that $N_r(x)$ is open.

The following three properties hold for the collection of open subsets of a metric space X:

i′) The set X and the empty set are open

ii′) if G_1 and G_2 are open, $G_1 \cap G_2$ is open

iii′) if $\{G_\alpha\}_{\alpha \in I}$ is a collection of open sets, $\bigcup_{\alpha \in I} G_\alpha$ is open.

Since i′) and iii′) are easily seen to be true, it remains to show that ii′) holds. But if G_1 and G_2 are open and $x \in G_1 \cap G_2$, then there is r_1 so that $N_{r_1}(x) \subset G_1$ and r_2 so that $N_{r_2}(x) \subset G_2$. If $r = \min(r_1, r_2)$ then $N_r(x) \subset G_1 \cap G_2$. Since x was an arbitrary element of $G_1 \cap G_2$, it follows that $G_1 \cap G_2$ is open. Note that ii′) implies that if $\{G_i\}_{i=1}^n$ is a collection of open sets, then $\bigcap_{i=1}^n G_i$ is an open set.

The conditions i′), ii′) and iii′) allow for a natural generalization of the concept of a metric space. A <u>topological space</u> (X, τ) is a set X and a collection τ of subsets of X, called the open subsets of X, with the provision that i′), ii′) and iii′) must hold for the sets belonging to τ; τ is said to be the <u>topology</u> on X. A subset X' of a topological space X can itself be considered to be a topological space by defining the topology τ' by $G' \in \tau'$ iff there is $G \in \tau$ and $G' = G \cap X'$. Then X' is said to be a <u>subspace</u> of X and to have the <u>subspace topology</u> τ'. In a topological space (or in a metric space) a set E is said to be a neighborhood of a point x if there is an open set G with $x \in G$ and $G \subset E$.

In a metric space X, a point x_o is said to be a _limit point_ of a set A if every neighborhood of x_o contains points of A distinct from x_o. A set $F \subset X$ is said to be _closed_ of every limit point of F belongs to F. The relationship between closed subsets of a metric space X and open subsets of X is given by the following theorem.

(1.1) _Theorem. In a metric space X, a set $F \subset X$ is closed if and only if $G = X \backslash F$ is open._

Proof. Suppose F is closed. Let $x \in G = X\backslash F$. Since F is closed, x is not a limit point of F and thus there is a neighborhood $N_r(x)$ which contains no points of F. That is, $N_r(x) \subset G$ and since x was an arbitrary point of G, each point of G has a neighborhood contained in G and thus G is open. Now suppose $G = X\backslash F$ is open. Let x be a limit point of F and suppose, if possible, that x belongs to G. Then there would be a neighborhood of x contained in G and x would not be a limit point of F. This contradiction implies that every limit point of F must belong to F and that F is closed.

This relationship between the open and the closed subsets of a metric space motivates the definition of closed sets in a topological space; namely, a set F contained in a topological space X is said to be _closed_ if $X\backslash F$ is open. It then follows from the properties of open sets that

i″) both X and the empty set are closed

ii") if F_1 and F_2 are closed sets, $F_1 \cup F_2$ is closed

iii") if $\{F_\alpha\}_{\alpha\in I}$ is a collection of closed sets, $\bigcap_{\alpha\in I} F_\alpha$ is closed.

Note also that a set F' is closed in X', a subspace of X, if and only if there is F closed in X and $F' = F \cap X'$. Furthermore, if F' is a closed subset of X' and X' is closed in X, then $F' = F \cap X'$ is a closed subset of X.

Given a set $E \subset X$, the union of all the open sets contained in E is called the <u>interior</u> of E, written $\mathring{E}$. The intersection of all the closed sets containing E is called the <u>closure</u> of E, written $\overline{E}$. The <u>boundary</u> of a set E is defined to be $\overline{E}\setminus\mathring{E}$. A point is said to be an <u>isolated point</u> of E (or X) if it has a neighborhood which contains no other points of E. If x is an isolated point of a space X, then $\{x\}$ is an open set in X.

A subset of a topological space is said to be a <u>perfect set</u> if it is closed and has no isolated points. The closed interval $[a,b]$ is a perfect subset of the line. An important example on the line of a non-empty perfect set which does not contain an interval is that of the <u>Cantor ternary set</u>

$$C = \{x \in [0,1]: x = \Sigma a_i/3^i \text{ with each } a_i = 0 \text{ or } 2\}.$$

Some frequently used topological properties hold for all metric spaces. A space is said to be a <u>Hausdörff space</u> if any two distinct points of the space have disjoint neighborhoods. It is said to be <u>regular</u> if given any closed set F and point $x \notin F$ there are disjoint open sets G_1 and G_2 with $x \in G_1$ and $F \subset$

G_2. A space is said to be <u>normal</u> if, given any two disjoint closed sets F_1 and F_2, there are disjoint open sets G_1 and G_2 with $F_1 \subset G_1$ and $F_2 \subset G_2$. That metric spaces are Hausdörff follows by considering two points x and y and their disjoint neighborhoods $N_r(x)$ and $N_r(y)$ where $r = (1/2) \cdot d(x,y)$. The Hausdörff property implies that for each x in a Hausdörff space, $\{x\}$ is a closed set; this is because $\{x\}$ is the complement of $\bigcup_{y \neq x} G_y$ where for $y \neq x$ each G_y is an open set containing y but not x. Thus a normal Hausdörff space is necessarily a regular space. It remains then to show that metric spaces are normal. To see this, let F_1 and F_2 be two disjoint closed subsets of a metric space. For each $x \in F_1$, let $d_1(x) = (1/2) \cdot \inf\{d(x,y): y \in F_2\}$ and for each $y \in F_2$ let $d_2(y) = (1/2) \cdot \inf\{d(x,y): x \in F_1\}$. Let $G_1 = \bigcup_{x \in F_1} N_{d_1(x)}(x)$ and $G_2 = \bigcup_{y \in F_2} N_{d_2(y)}(y)$. Then $F_1 \subset G_1$ and $F_2 \subset G_2$; also G_1 and G_2 are open. If a point z belonged to $G_1 \cap G_2$ there would be points $x \in F_1$ and $y \in F_2$ with $z \in N_{d_1(x)}(x) \cap N_{d_2(y)}(y)$ and thus $d(z,x) < d_1(x)$, $d(z,y) < d_2(y)$ and $d(x,y) < d_1(x) + d_2(y)$. This implies that $d(x,y)$ would be less than one of $2d_1(x)$ or $2d_2(y)$ contrary to the definition of $d_1(x)$ and $d_2(y)$. This contradiction implies that the open sets G_1 and G_2 must be disjoint. It follows that metric spaces are normal.

An important and frequently used topological notion is that of compactness. If $\mathcal{U}$ is a collection of subsets of X, $\mathcal{U}$ is said to be a <u>cover</u> of a set K

$\subset X$ if each $x \in K$ belongs to some $U \in \mathcal{U}$. In a topological space X, $\mathcal{U}$ is said to be an <u>open cover</u> of a set $K \subset X$ if each $U \in \mathcal{U}$ is open and $\mathcal{U}$ is a cover of K. A set K contained in a topological space X is said to be <u>compact</u> if every open cover of K contains a finite subcover; that is, K is compact if for each open cover $\mathcal{U}$ of K there is a natural number n and sets $U_1, U_2, \ldots, U_n$ which belong to $\mathcal{U}$ and $K \subset \bigcup_{i=1}^{n} U_i$.

(1.2) <u>Theorem</u>. <u>Compact subsets of a Hausdörff space are closed</u>.

<u>Proof</u>. If not, there is a Hausdörff space X, a compact subset K of X and a point x which is a limit point of K and $x \notin K$. For each point $y \in K$ there are disjoint open sets G_y and G'_y with $y \in G_y$ and $x \in G'_y$. Since $\{G_y\}_{y \in K}$ is an open cover of K, there are $G_{y_1}, G_{y_2}, \ldots, G_{y_n}$ such that $K \subset \bigcup_{i=1}^{n} G_{y_i}$. But then $\bigcap_{i=1}^{n} G'_{y_i}$ is an open set containing x and since x is a limit point of K, there is a point of K in this set. But then this point is not contained in $\bigcup_{i=1}^{n} G_{y_i}$ which is a contradiction. It follows that every compact subset of a Hausdörff space is closed.

(1.3) <u>Theorem</u>. <u>Closed subsets of a compact space are compact</u>.

Proof. Let F be a closed subset of the compact space X and $\{G_\alpha\}_{\alpha\in I}$ be an open cover of F. Let $G = X\setminus F$ be adjoined to this cover forming an open cover of X. Then, if $\{G_{\alpha_i}\}_{i=1}^{n}$ is a finite subcover of X, $\{G_{\alpha_i}\}_{i=1}^{n} \setminus G$ is a finite subcover of F. Since $\{G_\alpha\}_{\alpha\in I'}$ an arbitrary open cover of F has a finite subcover, it follows that F is compact. Thus closed subsets of a compact space are compact.

Note that a set $K \subset X' \subset X$ is compact in X' iff it is compact in X. This is because each open cover $\mathcal{U} = \{G_\alpha\}_{\alpha\in I}$ from X corresponds to an open cover $\mathcal{U}' = \{G'_\alpha\}_{\alpha\in I}$ from X' when $G'_\alpha = G_\alpha \cap X'$. Then K has a finite subcover from $\mathcal{U}$ iff it has a finite subcover from $\mathcal{U}'$. Thus K is a compact subset of X iff the space K is compact and, moreover, closed subsets of a compact set are compact.

Further note that, if $\{K_\alpha\}_{\alpha\in A}$ is a non-empty collection of compact subsets of a Hausdörff space, then $\bigcap_{\alpha\in A} K_\alpha$ is compact; this is because each K_α is closed so that $\cap K_\alpha$ is closed and, being a closed subset of a compact space, $\bigcap_{\alpha\in A} K_\alpha$ is compact.

A point x_o is said to be a <u>limit point of a sequence</u> $\{x_n\}$ if, for each open set G containing x_o and natural number N, there is $n > N$ with $x_n \in G$. We then observe that if $\{x_n\}$ is a sequence of points and $\{x_n\} \subset K$, a compact subset of a Hausdörff space, then $\{x_n\}$ has a limit point in K. For otherwise, it is easily shown that each $x \in K$ has a neighborhood G_x with at most one point of $\{x_n\}$ in it. The collection $\{G_x\}$, being an open cover of K, has a finite subcover $\{G_{x_j}\}_{j=1}^N$. It then follows that one of the points $\{x_n\}$, say x_{n_o}, occurs repeatedly in the sequence $\{x_n\}$ and this point x_{n_o} must itself be a limit point of $\{x_n\}$.

(1.4) Theorem. _Every compact Hausdörff space is a normal space (and hence is also a regular space)._

Proof. Let X be a compact Hausdörff space and let F_1 and F_2 be two disjoint closed subsets of X. Fix $x \in F_1$ and, for each $y \in F_2$, let G_y and G'_y be a pair of disjoint open sets with $y \in G_y$ and $x \in G'_y$. From the open cover $\{G_y\}_{y \in F_2}$ select a finite subcover $\{G_{y_i}\}_{i=1}^{n_x}$ of F_2. Let $G_x = \bigcap_{i=1}^{n_x} G'_{y_i}$ and $H_x = \bigcup_{i=1}^{n_x} G_{y_i}$. Now, if this is done for each $x \in F_1$, the collection $\{G_x\}_{x \in F_1}$ is an open cover of F_1 and contains a finite subcover $\{G_{x_j}\}_{j=1}^m$ of F_1. Let $G_1 = \bigcup_{j=1}^m G_{x_j}$ and $G_2 = \bigcap_{j=1}^m H_{x_j}$. Then $F_1 \subset G_1$, $F_2 \subset G_2$ and $G_1 \cap G_2 = \emptyset$. Since G_1 and G_2 are open sets, it follows that X is a normal space.

A space X is said to be _locally compact_ if every

point of X has a compact neighborhood. If X is a locally compact Hausdörff space, then inside every neighborhood of each point $x \in X$ there is a compact neighborhood. To see this, consider $x \in X$, E a neighborhood of x and K a compact neighborhood of X. Then there is an open set $G \subset K \cap E$ with $x \in G$. Since $K\backslash G = K \cap G^c$ is a closed subset of K, $K\backslash G$ is compact. For each $y \in K\backslash G$ there are disjoint open sets G_y and G'_y with $y \in G_y$ and $x \in G'_y$. Since $\{G_y\}_{y \in K\backslash G}$ is an open cover of $K\backslash G$ there are G_{y_1}, ..., G_{y_n} with $K\backslash G \subset \bigcup_{i=1}^{n} G_{y_i}$. Then $G_o = \bigcap_{i=1}^{n} G'_{y_i}$ is an open set containing x and $\overline{G}_o \subset K\backslash \bigcup_{i=1}^{n} G_{y_i} \subset G$; this makes $\overline{G}_o$ a compact neighborhood of x contained in E.

If X is a metric space and $E \subset X$, the <u>diameter</u> of E, written $\operatorname{diam} E = \sup\{d(x,y): x,y \in E\}$. A set E is said to be <u>bounded</u> if $\operatorname{diam} E < \infty$.

(1.5) <u>Theorem</u>. <u>Compact subsets of a metric space are bounded</u>.

<u>Proof</u>. Let K be a compact subset of a metric space. If $K = \emptyset$, there is nothing to prove. So let $x \in K$. Then $\{N_n(x)\}_{n=1}^{\infty}$ is an open cover of K and hence there is a natural number M so that $K \subset \bigcup_{n=1}^{M} N_n(x)$. Then for any two points y and z in K, $d(y,z) \leq d(y,x) + d(x,z) \leq 2M$. Thus compact subsets of a metric space are bounded.

In Euclidean k-dimensional space, a k-<u>cell</u> is a set of the form

$$\{(x_1, x_2, \ldots, x_k)\colon a_1 \le x_1 \le b_1,\ a_2 \le x_2 \le b_2,\ \ldots,\ a_k \le x_k \le b_k\},$$

written $[a_1, b_1;\ a_2, b_2;\ \ldots;\ a_k, b_k]$.

(1.6) <u>Theorem</u>. <u>The closed interval</u> $[a,b]$ <u>and each k-cell in Euclidean k-dimensional space is compact</u>.

<u>Proof</u>. Given $[a_1, b_1;\ a_2, b_2;\ \ldots;\ a_k, b_k]$, let $\{G_\alpha\}_{\alpha\in I}$ be an open cover of this k-cell. Suppose it is possible that there is no finite subcover from $\{G_\alpha\}_{\alpha\in I}$. Divide each $[a_i, b_i]$ into two equal length intervals $[a_i, (a_i+b_i)/2]$ and $[(a_i+b_i)/2, b_i]$. Then the k-cell is the union of 2^k k-cells

$$[a_1', b_1';\ \ldots;\ a_k', b_k']$$

where each $[a_i', b_i']$ is one of these two equal length intervals. At least one of these can not have a finite subcover from $\{G_\alpha\}_{\alpha\in I}$. Select such a k-cell and denote it by

$$[a_{1,1}, b_{1,1};\ a_{1,2}, b_{1,2};\ \ldots;\ a_{1,k}, b_{1,k}].$$

Repeating this argument for successive k-cells yields a sequence of k-cells none of which has a finite subcover from $\{G_\alpha\}_{\alpha\in I}$; namely,

$$\{[a_{n,1}, b_{n,1};\ a_{n,2}, b_{n,2};\ \ldots;\ a_{n,k}, b_{n,k}]\}_{n=1}^{\infty}.$$

Then there are unique real numbers x_i' in each $\bigcap_{n=1}^{\infty} [a_{n,i}, b_{n,i}]$; namely, $x_i' = \sup\{a_{n,i}\}_{n=1}^{\infty}$. Clearly $x = (x_1', x_2', \ldots, x_n')$ is a point in the intersection of the sequence of k-cells. There is an open set G_{α_0} with $x \in G_{\alpha_0}$ and a number $r > 0$ so that $N_r(x) \subset$

G_{α_o}. Since the diameters of the k-cells tend to 0, there is a k-cell from the sequence contained in $N_r(x)$. But then $\{G_{\alpha_o}\}$ is a finite subcover of this k-cell. It follows that the original collection $\{G_\alpha\}_{\alpha \in I}$ must contain a finite subcover of the original k-cell and that k-cells are compact.

(1.7) Theorem. (Heine-Borel) A subset of Euclidean space is compact if and only if it is closed and bounded.

Proof. Since metric spaces are Hausdörff spaces, compact subsets of a metric space are closed. By Theorem 1.5 compact subsets of metric spaces are also bounded. If K is a closed and bounded subset of Euclidean k-space, since K is bounded, K is contained in a k-cell. Since the k-cell is compact, K is a closed subset of a compact set and, by Theorem 1.3, K is compact.

In a metric space a sequence $\{x_n\}_{n=1}^{\infty}$ is said to be a Cauchy sequence provided that for each $\varepsilon > 0$ there is a natural number N so that $d(x_n, x_m) < \varepsilon$ whenever $n > N$ and $m > N$. A sequence $\{x_n\}$ converges to a point x_o, written $\lim x_n = x_o$, provided that for every $\varepsilon > 0$ there is an N so that $d(x_n, x_o) < \varepsilon$ whenever $n > N$. If a sequence does not converge it is said to diverge or to be divergent. A metric space X is said to be complete if every Cauchy sequence $\{x_n\} \subset X$ converges to some point of X. On the line, the <u>limit</u> <u>supremum</u> of a sequence $\{x_n\}$ is ∞ if the sequence is not bounded above; otherwise, it is the supremum of the set of limit points of $\{x_n\}$ and $-\infty$ if the sequence has no limit points. It is written $\overline{\lim}\, x_n$. The <u>limit</u> <u>infimum</u> of $\{x_n\}$ is $\underline{\lim}\, x_n = -\overline{\lim}(-x_n)$. Then $\lim x_n$ exists if and only if $\overline{\lim}\, x_n = \underline{\lim}\, x_n$ and both are finite; $\lim x_n$ is the common value of the limit supremum and the limit infimum. If $\overline{\lim}\, x_n = \underline{\lim}\, x_n = \infty$, one says $\lim x_n = \infty$; similarly for $\overline{\lim}\, x_n = \underline{\lim}\, x_n = -\infty$. Such a sequence is not considered to be convergent.

In a metric space, if $\{x_n\}$ is a convergent sequence, then $\{x_n\}$ is Cauchy. To see this, given $\varepsilon > 0$ choose N so that for $n > N$ each x_n satisfies $d(x_n, x_o) < \varepsilon/2$ where $x_o = \lim x_n$. Then for $n > N$ and $m > N$, $d(x_n, x_m) \le d(x_n, x_o) + d(x_m, x_o) < \varepsilon$.

The limit of a convergent sequence is unique. For, if $\lim x_n = x_o$ and $\lim x_n = x_o'$, then given $\varepsilon > 0$ there is N so that for all $n \ge N$, $d(x_o, x_n) < \varepsilon/2$ and $d(x_o', x_n') < \varepsilon/2$. Then $d(x_o, x') < d(x_o, x_N) + d(x_o', x_N) < \varepsilon$ and since $\varepsilon > 0$ was arbitrary, $d(x_o, x_o') = 0$ and $x_o = x_o'$.

Furthermore, every Cauchy sequence in a metric space (and thus every convergent sequence) is bounded. For if $\{x_n\}$ is a Cauchy sequence and $\varepsilon = 1$, there is an N so that $n, m \ge N$ implies $d(x_n, x_m) \le 1$. If $M = 1 + \max\{d(x_i, x_j) : i, j \le N\}$ then for all i and j, $d(x_i, x_j) \le d(x_i, x_N) + d(x_N, x_j) \le 2M$.

The following theorems indicate types of metric spaces which are complete. The proof of the first of these theorems requires two observations.

1) For any set E in a metric space $\operatorname{diam} \bar{E} = \operatorname{diam} E$.
2) If $\{K_n\}$ is a sequence of compact subsets of a Hausdörff space such that, for each n, $\emptyset \neq K_{n+1} \subset K_n$, then $\bigcap_{n=1}^{\infty} K_n \neq \emptyset$.

Proof of 1). If $\operatorname{diam} E = \infty$, so is $\operatorname{diam} \bar{E} = \infty$. If $\operatorname{diam} E < \infty$, then given $\varepsilon > 0$, there are points $x_1, y_1 \in E$ such that $d(x_1,y_1) > \operatorname{diam} E - \varepsilon/3$. Since $x_1, y_1 \in \bar{E}$, there are points $x_o, y_o \in E$ so that $d(x_o,x_1) < \varepsilon/3$ and $d(y_o,y_1) < \varepsilon/3$. Then $d(x_1,y_1) \leq d(x_1,x_o) + d(x_o,y_o) + d(y_o,y_1)$ and $\operatorname{diam} \bar{E} - \varepsilon/3 \leq d(x_1,y_1) < d(x_o,y_o) + 2\varepsilon/3$ so that $\operatorname{diam} \bar{E} \leq d(x_o,y_o) + \varepsilon \leq \operatorname{diam} E + \varepsilon$. Since $\varepsilon > 0$ was arbitrary, $\operatorname{diam} \bar{E} \leq \operatorname{diam} E$ and since $E \subset \bar{E}$, $\operatorname{diam} E \leq \operatorname{diam} \bar{E}$ and thus $\operatorname{diam} E = \operatorname{diam} \bar{E}$.

Proof of 2). Suppose, if possible, that $\bigcap_{n=1}^{\infty} K_n = \emptyset$. Since each K_n is closed, $\{K_n^c\}_{n=1}^{\infty}$ is an open cover of K_1 and thus there is a natural number N so that $K_1 \subset \bigcup_{n=1}^{N} K_n^c$. But then $\emptyset \neq K_{N+1} \subset K_1 \setminus \bigcup_{n=1}^{N} K_n^c = \emptyset$, a contradiction. It follows that $\bigcap_{n=1}^{\infty} K_n \neq \emptyset$.

(1.8) <u>Theorem</u>. <u>Compact metric spaces are complete</u>.

<u>Proof</u>. Let X be a compact metric space and let $\{x_n\} \subset X$ be a Cauchy sequence. For each natural number n, let K_n be the closure of $\{x_i\}_{i=n}^{\infty}$. Then each K_n is compact and since $\{x_n\}$ is a Cauchy sequence, by 1),

$\lim \operatorname{diam} K_n = \lim \operatorname{diam}\{x_i\}_{i=n}^{\infty} = 0$. Since $\emptyset \neq K_{n+1} \subset K_n$, by 2), $\bigcap_{n=1}^{\infty} K_n \neq \emptyset$ and since $\lim \operatorname{diam} K_n = 0$, $\bigcap_{n=1}^{\infty} K_n$ consists of a single point, say x_o. Given $\varepsilon > 0$, since $\lim \operatorname{diam} K_n = 0$ there is an N so that $\operatorname{diam} K_n < \varepsilon$ whenever $n > N$. But then $d(x_n, x_o) \leq \operatorname{diam} K_n < \varepsilon$ when $n > N$ and $\lim x_n = x_o$.

(1.9) Theorem. Each Euclidean k-dimensional space is complete. In particular, the real line is complete.

Proof. Let $X = \mathbb{R}^k$ be Euclidean k-dimensional space and $\{x_n\} \subset X$ be a Cauchy sequence. Then $\{x_n\}$ is bounded and hence is contained in a k-cell. Since the k-cell is compact, it follows that x_n converges to a point x_o in the k-cell. Thus each Cauchy sequence in X converges and $X = \mathbb{R}^k$ is complete.

(1.10) Theorem. Closed subsets of a complete metric space are complete.

Proof. Let F be a closed subset of a complete metric space and $\{x_n\} \subset F$ be a Cauchy sequence. Then there is $x_o \in X$ with $\lim x_n = x_o$ and since x_o is clearly a limit point of F, $x_o \in F$. Hence, Cauchy sequences in F converge to a point in F and thus F is complete.

We now turn to the proof of the Baire Category theorem. While this theorem will only be needed in the context of the real line, it holds for complete metric spaces and for locally compact Hausdörff spaces and the proof will be given for such spaces. By the above, $\mathbb{R}^k$ is a complete metric space and since each x in $\mathbb{R}^k$ is contained in a k-cell which is a compact neighborhood of x, $\mathbb{R}^k$ is also a locally compact space.

Several definitions will be needed for the Baire Category theorem. In a topological space X a set E' is said to be _dense_ in a set E if $E \subset \overline{E'}$; that is, provided each neighborhood of each point of E contains a point of E'. A set E is said to be _nowhere dense_ in X if E is not dense in any non-empty open subset of X; that is, provided each non-empty open set G contains a non-empty open set G' such that $E \cap G' = \emptyset$. Note that E is nowhere dense iff $(\overline{E})^\circ = \emptyset$ and this holds iff $\overline{E}$ is nowhere dense. A set $E \subset X$ is said to be of _first category_ (or _meagre_ or of _category_ I) if E is the union of a sequence of nowhere dense sets. If E is not of category I, E is said to be of _category_ II (or of _second category_). If E is of category II and $X \backslash E$ of category I, E is said to be _residual_ (or _comeagre_). A topological space X is said to be a _Baire space_ if each non-empty open set $G \subset X$ is of second category.

(1.11) _Theorem_. (The Baire Category Theorem) _Every complete metric space and every locally compact Hausdörff space is a Baire space._

Proof. Suppose X is not a Baire space. Then there

is a non-empty open set $G \subset X$ which is of first category; that is, $G = \cup F_n$ where each set F_n is nowhere dense and closed. Moreover, no point of G can be an isolated point. For if x were an isolated point and $x \in F_i$ then $\{x\}$ is an open set contained in F_i contrary to the fact that each F_i is nowhere dense. Now suppose, if possible, that X is also a complete metric space. Then there is a point x_1 and $r_1 > 0$ so that $\overline{N}_{r_1}(x)$ is contained in $G \backslash F_1$. In general, if points $x_1, x_2, \ldots, x_n$ and positive numbers $r_1, r_2, \ldots, r_n$ have been chosen with $r_i < r_{i-1}/2$ such that $\overline{N}_{r_i}(x_i)$ is contained in $N_{r_{i-1}}(x_{i-1})$ and $N_{r_i}(x_i) \cap \bigcup_{j=1}^{i} F_j = \emptyset$, since x_n is not an isolated point there is a point x_{n+1} and a positive number $r_{n+1} < r_n/2$ such that $N_{r_{n+1}}(x_{n+1}) \subset N_{r_n}(x_n)$ and $N_{r_{n+1}}(x_{n+1}) \cap \bigcup_{j=1}^{n+1} F_i = \emptyset$. But then the sequence $\{x_n\}$ is a Cauchy sequence because $d(x_i, x_j) \leq \sum_{n=1}^{j-i} d(x_{i+n-1}, x_{i+n}) \leq \sum_{n=i}^{\infty} r_n < r_i < 2^{-N} \cdot r_1$ whenever $i, j > N$. Thus $\{x_n\}$ converges to some point x_o. Moreover, for each natural number n, $x_o \in \overline{N}_{r_n}(x_n)$ and thus $x_o \in G$ but $x_o \notin \bigcup_{i=1}^{\infty} F_i$, a contradiction. It follows that, if X is not a Baire space, X is not a complete metric space; that is, every complete metric space is a Baire space. Now suppose X is a locally compact Hausdörff space but not a Baire space. Then there is an open set G of first category as above and $G \subset \bigcup_{i=1}^{\infty} F_i$ where each F_i is nowhere dense and closed. Since X is locally compact there is a point $x_1 \in G$ and a compact neighborhood $K_1 \subset G$ of x_1 so that $K_1 \cap F_1 = \emptyset$. In general, if $x_1, \ldots, x_n$ and $K_1, \ldots, K_n$ have been chosen with $x_i \in \mathring{K}_i$, $K_{i+1} \subset K_i$, and $K_i \cap \bigcup_{j=1}^{i} F_j = \emptyset$, let $x_{n+1} \in \mathring{K}_n \backslash F_{n+1}$ and K_{n+1} be a compact neighborhood of x_{n+1} contained in $\mathring{K}_n \backslash F_{n+1}$. Then $\bigcap_{n=1}^{\infty} K_n \neq \emptyset$. Let $x_o \in \bigcap_{n=1}^{\infty} K_n$. Then $x_o \notin \bigcup_{i=1}^{\infty} F_i$, a contradiction. This implies that every locally compact Hausdörff space is a Baire space.

We now consider some basic material involving functions. We will usually be concerned with functions whose range is the set of real numbers. However, we will sometimes be interested in functions whose range is a metric space, a topological space or simply a set. If f has domain X and range Y where both X and Y are metric spaces, then the <u>limit as</u> x <u>approaches</u> x_o <u>of</u> $f(x)$ <u>is</u> y_o, written $\lim_{x \to x_o} f(x) = y_o$ provided that for each $\varepsilon > 0$ there is $\delta > 0$ so that $d_Y(f(x), y_o) < \varepsilon$ whenever $d_X(x, x_o) < \delta$ where d_X and d_Y are the distances in X and Y, respectively. The function f is said to be <u>continuous at</u> x_o if

$\lim_{x \to x_o} f(x) = f(x_o)$.

An alternate definition of the limit using sequences is: $\lim_{x \to x_o} f(x) = y_o$ provided that whenever $\{x_n\}$ is a sequence in X with $\lim x_n = x_o$, then $\lim f(x_n) = y_o$. To see that these are equivalent, suppose the former holds. Let $\{x_n\}$ satisfy $\lim x_n = x_o$. Then, given $\varepsilon > 0$, there is $\delta > 0$ so that $d_Y(f(x_n), y_o) < \varepsilon$ whenever $d_X(x_n, x_o) < \delta$. Then there is N so that for $n > N$, $d_X(x_n, x_o) < \delta$. Consequently, for $n > N$, $d_Y(f(x_n), y_o) < \varepsilon$ and it follows that $\lim f(x_n) = y_o$. To see the converse, suppose the latter definition holds and the former fails for some f and x_o. Then there is $\varepsilon > 0$ and no δ can be found for this ε. Thus if $\delta_n = 1/n$ there is a sequence of points $\{x_n\}$ such that $d(x_n, x_o) < 1/n$ but $d(f(x_n), y_o) > \varepsilon$. This contradicts the latter definition of the limit.

A third definition of the limit is given in terms of open sets: $\lim_{x \to x_o} f(x) = y_o$ provided that, for each open set $H \subset Y$ with $y_o \in H$, there is an open set $G \subset X$ with $x_o \in G$ and $f(G) \subset H$. This definition is clearly equivalent to the first definition of the limit. Since it involves only open sets, it is natural that it is the definition of limit for topological spaces. Then a function is said to be a continuous function if it is continuous at each point in its domain. Note that the third definition implies that a function f is continuous if and only if for each open set H contained in the range of f, $f^{-1}(H)$ is an open set in X.

A function f taking X into Y where X and Y are metric spaces is said to be <u>uniformly continuous</u> provided that for each $\varepsilon > 0$ there is $\delta > 0$ so that $d_Y(f(x), f(y)) < \varepsilon$ whenever $d_X(x,y) < \delta$. An important result involving uniform continuity is:

(1.12) <u>Theorem</u>. *If X is a compact metric space and f is a continuous function taking X into a metric space Y, then f is a uniformly continuous function.*

<u>Proof</u>. By the continuity of f it follows that given $\varepsilon > 0$, for each $x \in X$ there is $\delta_x > 0$ so that whenever $d_X(t,x) < \delta_X$ then $d_Y(f(t), f(x)) < \varepsilon/2$. Then $\{N_{1/2\cdot\delta_x}(x)\}$ is an open cover of X and has a finite subcover

$$\{N_{1/2\cdot\delta_{x_1}}(x_1),\ N_{1/2\cdot\delta_{x_2}}(x_2),\ \ldots,\ N_{1/2\cdot\delta_{x_n}}(x_n)\}.$$

Let $\delta = \min\,\{1/2\cdot\delta_{x_i} : i = 1, 2, \ldots, n\}$. Then, if x and t are points of X and $d_X(x,t) < \delta$, there is x_i with $x \in N_{\delta_{x_i}/2}(x_i)$. That is, $x_i \in N_{\delta_{x_i}/2}(x)$ and since $t \in N_{\delta_{x_i}/2}(x)$ this implies $d_X(x_i,t) \le d_X(x_i,x) + d_X(x,t) < \delta_{x_i}$. But then $d_Y(f(t), f(x)) \le d_Y(f(t), f(x_i)) + d_Y(f(x_i), f(x)) < \varepsilon/2 + \varepsilon/2 = \varepsilon$. It follows that f is a uniformly continuous function on X.

A sequence of functions $\{f_n\}$ from a set X to a metric space Y is said to have a <u>pointwise limit</u> f

if for each $x \in X$, $\lim_{n\to\infty} f_n(x) = f(x)$. A sequence $\{f_n\}$ is said to approach f uniformly (or to have f as a uniform limit) provided that for each $\varepsilon > 0$ there is a natural number N so that for each $x \in X$ and $n > N$, $d_Y(f_n(x), f(x)) < \varepsilon$. We will need the following theorem concerning uniform limits.

(1.13) Theorem. If $\{f_n\}$ is a sequence of continuous functions each taking a metric space X into a metric space Y and f_n approaches a function f uniformly, then f is continuous.

Proof. Let x be a point in X and let $\varepsilon > 0$ be given. Then there exists N such that for all $n \ge N$ and $t \in X$, $d_Y(f_n(t), f(t)) < \varepsilon/3$. There is $\delta > 0$ so that if $d_X(t,x) < \delta$, then $d_Y(f_N(t), f_N(x)) < \varepsilon/3$. But then, if $d_X(t,x) < \delta$, by the triangle inequality, $d_Y(f(t), f(x)) \le d_Y(f(t), f_N(t)) + d_Y(f_N(t), f_N(x)) + d_Y(f_N(x), f(x)) < \varepsilon$. It follows that f is continuous at each x in X and thus f is a continuous function.

Rudimentary knowledge of calculus is presumed in the following chapters. In addition, the Riemann and Riemann-Stieltjes integrals on the line are discussed. Their definitions are as follows: If $f(x)$ is a real-valued function defined on $[a,b]$ and $\alpha(x)$ is a non-decreasing function defined on $[a,b]$, then the Riemann-Stieltjes integral of f with respect to α on $[a,b]$, if it exists, is $\int_a^b f(x)\, d\alpha$ which is a number satisfying: for each $\varepsilon > 0$ there is $\delta > 0$

such that if $P = \{x_o, x_1, \ldots, x_n\}$ with $x_o = a$, $x_n = b$, $x_{i-1} < x_i$ and $\min\{ x_i - x_{i-1}: i = 1, 2, \ldots, n\} < \delta$ then

$$\left|\int_a^b f(x)\, d\alpha - \sum_{i=1}^{n} f(w_i)\Delta\alpha_i\right| < \varepsilon$$

whenever $w_i \in [x_{i-1}, x_i]$ and $\Delta\alpha_i = \alpha(x_i) - \alpha(x_{i-1})$. If $\alpha(x) = x$, this is the definition of the <u>Riemann integral</u>.

If P is a partition of $[a,b]$, the <u>upper Riemann sum</u>, $U(P,f)$ is defined to be $\sum_{i=1}^{n} M_i \Delta x_i$ where M_i is the supremum of the values of $f(x)$ with $x \in [x_{i-1}, x_i]$. The <u>lower Riemann sum</u> is $L(P,f) = \sum_{i=1}^{n} m_i \Delta x_i$ where m_i is the infimum of the values of $f(x)$ with $x \in [x_{i-1}, x_i]$. A Cauchy criterion for integrability can be given as follows: f is Riemann integrable on $[a,b]$ if for each $\varepsilon > 0$ there is a partition P of $[a,b]$ such that $U(P,f) - L(P,f) < \varepsilon$. When this criterion holds, the intersection of the intervals $[L(P,f), U(P,f)]$ consists of a single point; the real number corresponding to this point is the value of the integral of f on $[a,b]$.

1.1 Exercises

1. Ordered pairs (a,b) can be represented in a set theoretic fashion by $(a,b) = \{a, \{a,b\}\}$. Show that with this definition $(a,b) = (c,d)$ iff $a = c$ and $b = d$.

2. Using the Peano axiom show that for each $n, m \in \mathbb{N}$, exactly one of $n < m$, $n = m$ or $m < n$ holds.
3. The natural numbers can be represented in set theoretic fashion as: $0 = \emptyset$, $1 = \{0\}$, $2 = \{0,1\}$ and, in general, $n + 1 = \{0, 1, \ldots, n\}$. Show that the collection of sets, so defined as the natural numbers, does not contain a sequence $\{n_m\}$ such that, for each m, $n_{m+1} \in n_m$.
4. For $(x,y) \in \mathbb{R}^2$ show that $\|(x,y)\| = |x| + |y|$ is a norm.
5. Show that $\max(|x|,|y|)$ is a norm on the points $(x,y) \in \mathbb{R}^2$.
6. Show that the norms given in Exercises 4 & 5 give rise to the same topologies as the usual norm $(x^2 + y^2)^{1/2}$; that is, show that $\mathbb{R}^2$ has the same open sets under each norm.
7. Show that the usual definition as given in the text of a metric space is redundant. Show that from $d(x,y) = 0$ iff $x = y$ and the triangle inequality one can prove: $d(x,y) \geq 0$ and $d(x,y) = d(y,x)$.
8. Prove that, if a topological space is normal; then two disjoint closed sets F_1 and F_2 are contained in open sets G_1 and G_2 so that $\overline{G}_1 \cap \overline{G}_2 = \emptyset$.
9. Call a metric space X absolutely closed if whenever $X' \subset Y$ where X' is a metric space which has the same topology as X, then X' is a closed subset of Y. Show that compact metric spaces X are absolutely closed.
10. Show that absolutely closed metric spaces (see Exercise 9) are complete.

11. Show that the real line and $(0,1)$ have the same topologies; that is, there is a one to one function f from one onto the other so that both f and f^{-1} take open sets to open sets. However, the real line is complete and $(0,1)$ is not. What does this signify?

12. Let X be a metric space. Let Y be the collection of all Cauchy sequences $\{x_n\}$ of elements of X where two elements $y = \{x_n\}$ and $y' = \{x'_n\}$ of Y are equivalent (equated) if the sequence $y'' = \{x''_n\}$ with $x''_{2n-1} = x_n$ and $x''_{2n} = x'_n$ is a Cauchy sequence. Show that Y is a complete metric space when $d(y,y')$ is defined by $\overline{\lim}\, d(x_n,x'_n)$.

13. For $x \in X$ with X and Y as in Exercise 12, let $y_x = \{x_n\}$ where each $x_n = x$. Then show $f(x) = y_x$ is a distance preserving one to one map from X into Y. Show also that $f(X)$ is dense in Y.

14. Suppose $d(x,y)$ is a real-valued function defined on all pairs (x,y) with $x,y \in X$ and for each $x \in X$, $d(x,x) = 0$ and d satisfies the triangle inequality. Let $A_x = \{y\colon d(x,y) = 0\}$. Show that $d(A_x,A_y) = d(x,y)$ is well-defined and is a metric on $\{A_x\}_{x \in X}$.

15. If X is a metric space with $x \in X$, show that $f(y) = d(x,y)$ is a continuous function. If $\emptyset \neq A \subset X$ and $d(y,A) = \sup\{d(x,y)\colon y \in A\}$ show that $g(y) = d(y,A)$ is a continuous function.

16. Show that if $f_n(x) = 0$ if $x \in [1/n, 1]$, $f(0) = 0$, $f(2/n) = 1$ and the graph of f is linear on $[0,2/n]$ and $[2/n, 1/n]$, then $f_n(x) \to 0$ at each x but f_n does not approach 0 uniformly.

17. Note that the f_n of the previous exercise satisfy $\int_0^1 f_n(x)dx \to 0$. If $g_n(x) = n \cdot f_n(x)$, show $g_n \to 0$ but not uniformly and $\int_0^1 g_n(x)dx = 1/2$. If $h_n(x) = n^2 \cdot f_n(x)$, show $h_n \to 0$ but not uniformly and $\int_0^1 h_n(x)dx \to \infty$.
18. Prove that if f is uniformly continuous on E, a subset of a metric space, then f can be extended to a continuous function on $\bar{E}$. Show this function is unique and is uniformly continuous on $\bar{E}$.
19. Suppose $\lim_{h \to 0} \dfrac{F(x + h) - F(x)}{h} = F'(x)$ uniformly. Show that this happens iff $F'(x)$ is continuous.

1.2 Additional Basic Theorems

We will mostly be concerned here with real-valued functions defined on a closed interval. Since some of the pertinent results hold in a more general setting, they first will be stated in such a setting and then specialized to the interval.

Thus, consider a continuous function f which takes a topological space X into a topological space Y. It is easily seen that each compact subset K of X has an image which is compact. For let $\{G_\alpha\}_{\alpha \in A}$ be an open cover of $f(K)$. Since f is continuous, each $f^{-1}(G_\alpha)$ is an open set in X. Thus $\{f^{-1}(G_\alpha)\}_{\alpha \in A}$ is an open cover of K and has a finite subcover, say $\{f^{-1}(G_{\alpha_1}), \ldots, f^{-1}(G_{\alpha_n})\}$. But then $\{G_{\alpha_1}, \ldots, G_{\alpha_n}\}$ is a finite subcover of $f(K)$ and it follows that

$f(K)$ is compact and that the continuous image of compact sets is compact. Now, if Y is a metric space and f is continuous on X and takes X into Y, then the image of each compact subset of X is a closed and bounded subset of Y. In the particular case where $X = I$, a closed interval, and f is a continuous real-valued function on I, the above result implies that f is bounded on I and that on each closed interval $I' \subset I$, f takes on its maximum and its minimum on I'. Indeed, if $K \subset I$ is a closed set, then there are points $x_o \in K$, $x'_o \in K$ such that $f(x_o) = \sup_{x \in K} f(x)$ and $f(x'_o) = \inf_{x \in K} f(x)$ because both points $\sup_{x \in K} f(x)$ and $\inf_{x \in K} f(x)$ are clearly limit points of the bounded set $f(K)$ and hence belong to $f(K)$ because $f(K)$ is closed.

Now consider the collection $M(S)$ of all real-valued bounded functions defined on a set S. This collection of functions forms a vector space because for each $f, g \in M(S)$ and $c \in \mathbb{R}$, $f + g$ and cf are bounded and hence belong to $M(S)$. The usual norm which is defined on $M(S)$ is obtained by letting $\|f\| = \sup\{|f(s)|: s \in S\}$; this is called the <u>sup norm</u>. To see that it is a norm, note that $\|f\| = 0$ iff for each $s \in S$, $f(s) = 0$; that is, $f = \overline{0}$ in $M(S)$. Clearly, $\|cf\| = |c|\|f\|$ for each $f \in M(S)$ and finally

$$\begin{aligned}\|f + g\| &= \sup_{s \in S}|f(s) + g(s)| \\ &\le \sup_{s \in S}|f(s)| + \sup_{s \in S}|g(s)| = \|f\| + \|g\|.\end{aligned}$$

The sup norm on $M(S)$ gives rise to a complete metric when $d(f,g)$ is defined by $\|f - g\|$. To see this, let $\{f_n\}$ be a Cauchy sequence of elements of $M(S)$. For each $s \in S$, $|f_n(s) - f_m(s)| \leq \|f_n - f_m\|$ and hence each $\{f_n(s)\}_{n=1}^{\infty}$ is a Cauchy sequence of real numbers and converges to a value $f(s)$. It remains to show that the function $f(s)$ so defined is bounded and that f_n approaches f with respect to the sup norm. Let $\varepsilon < 1$ be a positive real number. Then, since $\{f_n\}$ is Cauchy, there is a natural number N so that for $n,m \geq N$ and each $s \in S$, $|f_n(s) - f_m(s)| \leq \|f_n - f_m\| \leq \varepsilon < 1$. Letting m approach ∞ yields for each $s \in S$, $|f_n(s) - f(s)| \leq 1$ whenever $n \geq N$. Thus $|f(s)| \leq |f(s) - f_N(s)| + |f_N(s)| \leq \varepsilon + \|f_N\| < \infty$ and f is a bounded function. Also, when $n \geq N$

$$\begin{aligned} \|f_n - f\| &= \sup_{s \in S} |f_n(s) - f(s)| \\ &\leq \sup_{s \in S} |f_n(s) - f_N(s)| + \sup_{s \in S} |f_N(s) - f(s)| \\ &\leq 2\varepsilon. \end{aligned}$$

Hence $\|f_n - f\|$ approaches 0 and $\lim f_n = f$ in the metric on $M(S)$.

The sup norm on $M(S)$ is sometimes called the norm of uniform convergence, since if f_n approaches f under this norm, then clearly f_n approaches f uniformly.

Now, if $S = X$, a topological space, the subspace $C(X) \subset M(X)$ of all continuous real-valued bounded functions is obviously a subspace of $M(X)$; moreover, it is closed in $M(X)$. For if $\{f_n\} \subset C(X)$ and f_n approaches $f \in M(X)$, then f_n converges to f uniformly and hence f is continuous; that is, $f \in C(X)$ and $C(X)$ is a closed subspace of $M(X)$. In the case where X is compact, each continuous real-valued function on X is bounded and in this case $C(X)$

consists of all the continuous real-valued functions defined on X.

We will specifically be interested in the case where $X = I$ is a closed interval. Then, applying the Baire category theorem to this space will give us the classic theorem which asserts that there are many (a residual set of) continuous functions which do not have a derivative at a single point. (Such functions are called _nowhere differentiable_.)

(1.14) _Theorem. In the space C of continuous functions defined on $[a,b]$, the set consisting of those functions that have a derivative at a single point of $[a,b]$ is of first category in C._

Proof. Actually a stronger result will be proven. A function f is said to satisfy a _local Lipschitz condition_ at a point x_o if there is a number K and $\delta > 0$ so that $|h| < \delta$ implies $|f(x_o + h) - f(x_o)| < Kh$. At each point x_o of differentiability of a function f, f satisfies such a condition with $K = |f'(x_o)| + 1$. For, if f is differentiable at x_o, by definition, $\lim_{x \to x_o} \dfrac{f(x_o + h) - f(x_o)}{h} = f'(x_o)$ and given $\varepsilon > 0$ with $\varepsilon \le 1$ there is $\delta > 0$ so that $|(f(x_o + h) - f(x_o))/h - f'(x_o)| < \varepsilon$ when $|h| < \delta$. That is, $|(f(x_o + h) - f(x_o))/h| - |f'(x_o)| < \varepsilon$ and, for $\varepsilon = 1$, $|f(x_o + h) - f(x_o)| \le (|f'(x_o)| + 1)|h|$. Now let E_k be the set of $f \in C$ such that there is $x \in [a,b]$ and $|h| < 1/k$ implies $|f(x + h) - f(x)| \le k|h|$. Then $f \in \bigcup_{k=1}^{\infty} E_k$ iff there is a point $x_o \in [a,b]$

at which f satisfies a local Lipschitz condition. To see this, note that if there is a number K, $\delta > 0$ and $x_o \in [a,b]$ such that $|h| < \delta$ implies $|f(x_o + h) - f(x_o)| \le K|h|$, then $f \in E_k$ where $k > \max(K, 1/\delta)$. Conversely, if $f \in E_k$ it satisfies a local Lipschitz condition with constant k and $\varepsilon = 1/k$ at some point of $[a,b]$. It now remains to show that each E_k is a closed subset of C and is also nowhere dense in C. Fix k and suppose $\{f_n\}_{n=1}^{\infty} \subset E_k$ and f_n approaches f uniformly. Then there are points $x_n \in [a,b]$ such that $|h| < 1/k$ implies $|f_n(x_n + h) - f_n(x_n)| \le k|h|$. Now, a subsequence $\{x_{n_i}\}$ of the sequence $\{x_n\}$ converges to some point $x_o \in [a,b]$. Let t satisfy $|t - x_o| < 1/k$ and $h = t - x_o$; then given $\varepsilon > 0$ there is an N so that for $n_i > N$, $\|f_{n_i} - f\| < \varepsilon$. But then, for $t_{n_i} = x_{n_i} + h$,

$$\begin{aligned} |f(t) - f(x_o)| &\le |f(t) - f(t_{n_i})| + |f(t_{n_i}) - f_{n_i}(t_{n_i})| \\ &\quad + |f_{n_i}(t_{n_i}) - f_{n_i}(x_{n_i})| + |f_{n_i}(x_{n_i}) - f(x_{n_i})| \\ &\quad + |f(x_{n_i}) - f(x_o)| \ < \ 2\varepsilon + |t - x_o|\cdot k + 2\varepsilon. \end{aligned}$$

Since ε is an arbitrary positive number, it follows that $|f(t) - f(x_o)| \le |t - x_o|\cdot k$ and $f \in E_k$. Consequently E_k is a closed subset of C. Because E_k is closed, to see that E_k is nowhere dense in C, it is sufficient to show that each $C\backslash E_k$ is dense in C. To do this, fix k, $f \in E_k$ and $\varepsilon > 0$. It will suffice to find $g \in C\backslash E_k$ with $\|g - f\| < \varepsilon$. Since f is uniformly continuous on $[a,b]$, there is a partition $a = x_o < x_1 < \dots < x_n = b$ of $[a,b]$ so that if $x \in [x_{i-1}, x_i]$, $|f(x) - f(x_{i-1})| < \varepsilon/2$ and for each $i = 1, 2, \ldots, n$, $|x_i - x_{i-1}| < \varepsilon/(2k + 2)$. Then $g(x)$ can be defined so that the graph of g consists of connected line segments each with slope having absolute value equal to

$$\max\bigl(|(f(x_i) - f(x_{i-1}))/(x_i - x_{i-1})|, k + 1\bigr)$$

when $x \in [x_{i-1}, x_i]$. Then $g \in C \setminus E_k$ and for each x, if $x \in [x_{i-1}, x_i]$, then

$$\begin{aligned} |g(x) - f(x)| &\le (x - x_{i-1})(k + 1) \\ &\quad + \max\bigl(|f(x) - f(x_{i-1})|, |f(x_i) - f(x)|\bigr) \\ &\le \varepsilon/2 + \varepsilon/2 = \varepsilon. \end{aligned}$$

Thus each E_k is a closed nowhere dense subset of C and the continuous functions which are differentiable at a point of $[a,b]$ are of first category in C.

While Theorem 1.14 shows that continuous nowhere differentiable functions exist and that there are many of them, it is frequently worthwhile to have explicit examples and methods for constructing such examples. One way to achieve this is to use the fact that the uniform limit of continuous functions is continuous. Thus an example of a continuous nowhere differentiable function can be constructed as follows:

For $x \in [0,1]$, let

$$f_n(x) = \min\{d(x, k/10^n) : k = 0, 1, \ldots, 10^n\}.$$

Then each $f_n(x)$ has 10^n "sawteeth" and $\|f_n(x)\| = 1/2 \cdot 10^{-n}$. Let $f(x) = \sum_{n=1}^{\infty} f_n(x)$; that is, $f(x) = \lim_{N\to\infty} g_N(x)$ where $g_N(x) = \sum_{n=1}^{N} f_n(x)$. Each $g_N(x)$ is continuous and $\|g_N - f\| \le \sum_{N+1}^{\infty} 10^{-n} < 2 \cdot 10^{-N}$ and it follows that g_N approaches f uniformly and thus f

is continuous. Clearly, $\limsup_{h\to 0^+} [f(h) - f(0)]/h = \infty$ since $f(0) = 0$ and $f(1/2\cdot 10^n)/(1/2\cdot 10^n) = n$. Similarly, $\liminf_{h\to 0^+} [f(1 - h) - f(1)]/h = -\infty$. For $x \in (0,1)$ and each natural number m with $x > 10^{-m}$ and $x < 1 - 10^{-m}$, let $\delta_m = \pm 10^{-m-1}$ so that either x and $x + \delta_m$ both belong to some interval $[k\cdot 10^{-m}, (k + 1/2)10^{-m}]$ $k = 0, 1, \ldots, 9$ or both belong to $[(k + 1/2)10^{-m}, (k + 1)10^{-m}]$. Let $t_m = x + \delta_m$ and s_m be between $x - \delta_m$ and $x + \delta_m$ so that $|s_m - x| = 10^{-m-1}$ and

$$\frac{f_m(x) - f_m(s_m)}{x - s_m} = 1.$$

Then for $n < m$.

$$\frac{f_n(x) - f_n(t_m)}{x - t_m} = \frac{f_n(x - s_m)}{x - s_m}$$

and for $n > m$, $f_n(s_m) = f_n(t_m) = f_n(x)$; furthermore, $f_m(x) = f_m(t_m)$. Hence

$$\left| \frac{f(x) - f(t_m)}{x - t_m} - \frac{f(x) - f(s_m)}{x - s_m} \right| = 1.$$

Since t_m and s_m approach x as m approaches ∞, it follows that f does not have a derivative at x and thus that f does not have a derivative at any point of $[0,1]$.

An alternative method for constructing continuous functions with given properties (in this case, nowhere differentiability) uses the following theorem:

(1.15) Theorem. *A real-valued function defined on a closed interval I is continuous iff its graph $\{(x,y): x \in I$ and $y = f(x)\}$ is a compact subset of the plane.*

Proof. If f is continuous on $I = [a,b]$ then f is bounded on I and its graph is a bounded subset of the plane. To see that the graph is also closed, let (x_o,y_o) be a limit point of the graph. If $\{(x_n,f(x_n))\}$ is a sequence of points from the graph of f on $[a,b]$ which approaches (x_o,y_o), then $|x_n - x_o| \le d((x_n,f(x_n)), (x_o,y_o))$ and $\lim x_n = x_o$. By the continuity of f, $\lim f(x_n) = f(x_o)$; that is, if

$$\begin{aligned} d_n &= d((x_n,f(x_n)), (x_o,f(x_o))) \\ &\le |x_n - x_o| + |f(x_n) - f(x_o)|, \end{aligned}$$

d_n approaches 0 and it follows that $f(x_o) = y_o$. Hence (x_o,y_o) is a point on the graph of f and the graph is closed and bounded. Conversely, if the graph of a real-valued function f defined on $[a,b]$ is compact, let $x_o \in [a,b]$ and x_n approach x_o. Since f is bounded, $(x_o, \overline{\lim}\, f(x_n))$ is a limit point of $\{(x_n, f(x_n))\}$. Since the graph of f is closed and f is a function, $f(x_o) = \overline{\lim}\, f(x_n)$. Similarly, $f(x_o) = \underline{\lim}\, f(x_n)$. Thus $\lim f(x_n) = f(x_o)$ and, since x_o was an arbitrary point of $[a,b]$, it follows that f is a continuous function on $[a,b]$.

Now, other examples of continuous nowhere differentiable functions (as well as other unusual continuous functions) can be constructed using this last theorem by defining a sequence of compact subsets

$\{K_n\}$ of the plane so that $\cap K_n$ is the graph of a function defined on $[a,b]$; that is, so that for each $x \in [a,b]$ there is only one y with (x,y) in $\cap K_n$. The sets K_n can be taken to be a finite union of closed rectangles which meet each line $x = t$ for each t in an interval $[a,b]$. Indeed, every continuous function on a closed interval can be obtained in this way. For let f be continuous on $[a,b]$ and let $K_n = \bigcup_{k=1}^{2^n} R_{k,n}$ where $R_{k,n} = I_{k,n} \times J_{k,n}$ with $I_{k,n} = [(k - 1)(b - a)/2^n, k(b - a)/2^n]$ and

$$J_{k,n} = [\inf\{f(x): x \in I_{k,n}\}, \sup\{f(x): x \in I_{k,n}\}].$$

It then follows from the uniform continuity of f that $\bigcap_{n=1}^{\infty} K_n$ is the graph of a function; namely, it is the graph of f.

To construct a nowhere differentiable continuous function in this manner, let $I = [0,1]$. For $k = 1, 2, \ldots, 9^n$, let $I_{k,n} = [(k - 1)/9^n, k/9^n]$ and let the rectangle $R_{k,n} = I_{k,n} \times J_{k,n}$ where $J_{k,n} = [a_{k,n}/3^n, (a_{k,n} + 1)/3^n]$ and the numbers $a_{k,n}$ are defined as follows:

Let

$$e_j = \begin{cases} 0 & \text{if } j = 1, 6, \text{ or } 7 \\ 1 & \text{if } j = 2, 5, \text{ or } 8. \\ 2 & \text{if } j = 3, 4, \text{ or } 9 \end{cases}$$

Let $a_{k,1} = e_k$. If each $a_{k,i}$ has been defined for $i = 1, \ldots, n$, let $a_{1,n+1} = 0$ and for $K = 9^{n+1}$ let

$a_{K,n+1} = 2$. If $R_{k-1,n}$ meets $R_{k,n}$ at its lower left hand corner and $R_{k+1,n}$ meets $R_{k,n}$ at its upper right hand corner, let $a_{9k+j,n+1} = 3a_{k,n} + e_j$; otherwise, let $a_{9k+j,n+1} = 3a_{k,n} + 2 - e_j$. This defines $a_{k,n+1}$ for $k = 1, 2, \ldots, 9^{n+1}$.

Then each resulting set K_n is the union of 9^n closed rectangles each of height $1/3^n$ and each line $x = t$ with $t \in [0,1]$ meets K_n in an interval of length at most $2 \cdot 3^{-n}$. Since $K_{n+1} \subset K_n$, $\cap K_n$ is the graph of a continuous function f defined on $[0,1]$. For each fixed n and k with $1 \le k \le 9^n$, if $t' = (k - 1)/3^n$ and $t'' = k/3^n$, the function f satisfies $|f(t'') - f(t')| = 1/3^n$. Hence, if $x \in [t',t'']$, then one of $|f(x) - f(t')|$ or $|f(x) - f(t'')|$ is at least $(1/2)(1/3^n)$. If t is the point t' or t'' at which this occurs, then the absolute value of the difference quotient $[f(x) - f(t)]/(x - t)$ is at least $(1/2)(3^{-n}/9^{-n}) = (1/2)(3^n)$. Since each $x \in [0,1]$ determines a sequence of intervals of the form $[t',t'']$ which contain x, it follows that the function f is not differentiable at any point of $[0,1]$.

We return for the moment to the setting of a topological space X. Two sets A and B in X are said to be <u>separated</u> if $\bar{A} \cap B = A \cap \bar{B} = \emptyset$. A set $E \subset X$ is said to be <u>connected</u> if E is not the union of two nonempty, disjoint separated sets. While connected sets can be complicated, the motivation for this concept is that a connected set E should not be made up of two such "pieces" A and B. If a set is not connected, it is said to be <u>disconnected</u>. The following theorem shows that continuous functions cannot take connected sets to disconnected ones.

(1.16) Theorem. *If* f *is a continuous function taking a topological space* X *into a topological space* Y *and* E *is a connected subset of* X, *then* $f(E)$ *is a connected subset of* Y.

Proof. Given a continuous function f and a connected subset E of X, suppose if possible that $f(E)$ is not connected. Then there are two sets A and B with $f(E) = A \cup B$ where $A \neq \emptyset$, $B \neq \emptyset$ and $\bar{A} \cap B = A \cap \bar{B} = \emptyset$. Let $A_o = f^{-1}(A) \cap E$ and $B_o = f^{-1}(B) \cap E$. Then $A_o \neq \emptyset$, $B_o \neq \emptyset$ and $E = A_o \cup B_o$. By the continuity of f, $f^{-1}(\bar{A})$ and $f^{-1}(\bar{B})$ are closed subsets of X. Hence $A_o \subset \overline{f^{-1}(A)} \subset f^{-1}(\bar{A})$ and $f(\bar{A}_o) \subset \bar{A}$. Likewise, $B_o \subset \overline{f^{-1}(B)} \subset f^{-1}(\bar{B})$ and $f(\bar{B}_o) \subset \bar{B}$. Thus $\bar{A}_o \cap B_o = A_o \cap \bar{B}_o = \emptyset$. But this contradicts the fact that E is a connected set. It follows that the continuous image of a connected set is always connected.

In a topological space X, if $\{E_\alpha\}_{\alpha \in A}$ is a collection of connected sets and $\bigcap_{\alpha \in A} E_\alpha \neq \emptyset$, then $E = \bigcup_{\alpha \in A} E_\alpha$ is a connected set. For, if this were not the case, E would be separated by two sets A and B. There would be a point of $\bigcap_{\alpha \in A} E_\alpha$ in one of the two, say in A, and a point $y \in B$ belonging to some E_{α_o}.

But then it is easily seen that $A \cap E_{\alpha_o}$ and $B \cap E_{\alpha_o}$ would separate E_{α_o} contrary to the fact that E_{α_o} is connected.

Sometimes it is easier to show that a set is connected by showing more. A set E in a space X is said to be arcwise connected if for each pair of points $x, y \in E$ there is a continuous function f taking $[0,1]$ into E with $f(0) = x$ and $f(1) = y$. That arcwise connected sets are connected will follow from the above observation along with Theorem 1.16 once it is shown that $[0,1]$ is a connnected set. In fact, as one would expect, the connected subsets of the line are characterized as follows:

(1.17) Theorem. A set $E \subset \mathbb{R}$ is connected if and only if for each $x, y \in E$ and z between x and y, $z \in E$.

Proof. If the latter property fails for a set $E \subset \mathbb{R}$, then there are $x, y \in E$ and $z \notin E$ with $x < z < y$. But then $A = (-\infty, z) \cap E$ and $B = (z, \infty) \cap E$ are two non-empty separated sets whose union is E. (Clearly, z is the only point in $\overline{A}$ which could be in B and $z \notin B$ and vice versa.) Now suppose that the latter property holds for a set $E \subset \mathbb{R}$. Suppose if possible that E is the union of two non-empty separated sets A and B. Then there is $x \in A$ and $y \in B$ and without loss of generality we may suppose that $x < y$. Let $A_y = \{t \in A: t < y\}$ and let $z = \sup A_y$. By hypothesis, $z \in E$. But if z belongs to A and hence z does not belong to $\overline{B}$, there would be an ε

> 0 so that $(z, z+\varepsilon) \subset A$, contrary to the definition of z as $\sup A_y$. On the other hand, if z belongs to B and not to $\bar{A}$ there would be an $\varepsilon > 0$ so that $(z-\varepsilon, z) \cap A = \emptyset$ and in order that $z = \sup A_y$, z must belong to A, a contradiction. It follows that E cannot be written as the union of two non-empty separated sets and that such a set E is connected.

A real-valued function defined on an interval $[a,b]$ is said to have the _mean value property_ or to be a _Darboux function_ on $[a,b]$ if for each $x,y \in [a,b]$ with $x < y$ and c between $f(x)$ and $f(y)$ there is $z \in (x,y)$ with $f(z) = c$. From Theorems 1.16 and 1.17 it follows that continuous real-valued functions defined on an interval have the Darboux property. For, if $x,y \in [a,b]$ with $x < y$, then the image of $[x,y]$ under f is connected and hence must contain every point between $f(x)$ and $f(y)$.

Several other classes of real-valued functions are also Darboux functions. In particular, as shown in the next theorem, derivatives are Darboux functions. (Derivatives need not be continuous or even bounded; for example, if $F(0) = 0$ and, for $x \neq 0$, $F(x) = x^2\sin(x^{-2})$, then $F(x)$ is differentiable and $F'(0) = 0$ but $F'(x)$ is easily seen to be unbounded in each neighborhood of 0.)

(1.18) _Theorem. If $F(x)$ is differentiable on $[a,b]$ and $x,y \in [a,b]$ with $x < y$, for each c between $F'(x)$ and $F'(y)$ there is $z \in (x,y)$ with $F'(z) = c$._

<u>Proof</u>. Given such a function F and points $x,y \in [a,b]$, assume without loss of generality that $F'(x) > F'(y)$. Let c be any number between $F'(x)$ and $F'(y)$ and let $G(t) = F(t) - ct$. Since G is differentiable, G is continuous and must take on its maximum at some point $z \in [x,y]$. But $G'(x) = F'(x) - c > 0$ and thus $z \neq x$; likewise, $G'(y) = F'(y) - c < 0$ and $z \neq y$. Then $z \in (x,y)$ and the derivative of G at z must be 0; that is, $F'(z) = c$.

We now return to some additional considerations involving the uniform limits of sequences of real functions. Frequently, a class of functions which occurs naturally in analysis forms a vector space; that is, the class is closed under addition and multiplication by constants. Familiar examples of such classes are: polynomials, continuous functions, differentiable functions, Riemann integrable functions and derivatives. While these vector spaces often do not have a natural norm or a norm under which they become complete metric spaces, a number of such classes are nonetheless closed under uniform limits. Indeed, whether a class of functions is closed under uniform limits is useful information and is thus a recurring theme in analysis. Easy examples show that the uniform limit of a sequence of differentiable functions (or a sequence of polynomials) need not be differentiable. This section concludes with proofs that the class of functions which are Riemann-Stieltjes integrable on an interval with respect to a given non-decreasing function and also the class of derivatives are each closed under uniform limits and, finally, that every continuous function defined on a closed interval is the uniform limit of a sequence of polynomials (Weierstrass' approximation theorem).

(1.19) Theorem. *Suppose $\alpha(x)$ is a non-decreasing function on $[a,b]$ and $\{f_n\}$ is a sequence of Riemann-Stieltjes integrable functions on $[a,b]$ with respect to α. If f_n approaches a function f uniformly on $[a,b]$, then f is Riemann-Stieltjes integrable and*

$$\int_a^b f(x)\,d\alpha = \lim_{n\to\infty}\int_a^b f_n(x)\,d\alpha.$$

Proof. Note first that each constant function c satisfies $\int_a^b c\,d\alpha = c(\alpha(b) - \alpha(a))$ since this is the value of every partition sum $\sum_{i=1}^{n} c(\alpha(x_i) - \alpha(x_{i-1}))$. Also, if $|g(x)| \le M$ and g is Riemann-Stieltjes integrable with respect to α, then $\left|\int_a^b g(x)\,d\alpha\right| \le M(\alpha(b) - \alpha(a))$ because every partition sum is bounded by this quantity. By hypothesis, the sequence $\{f_n\}$ approaches f uniformly. So, given $\varepsilon > 0$, there is a natural number N so that for each x in $[a,b]$ and $n \ge N$, $|f_n(x) - f(x)| < \varepsilon$. Then, for $n,m > N$,

$$\left|\int_a^b f_n(x)\,d\alpha - \int_a^b f_m(x)\,d\alpha\right| \le \int_a^b 2\varepsilon\,d\alpha \quad \text{and thus}$$

$$(*) \qquad \left|\int_a^b f_n(x)\,d\alpha - \int_a^b f_m(x)\,d\alpha\right| \le 2\varepsilon(\alpha(b) - \alpha(a)).$$

Thus $\left\{\int_a^b f_n(x)\, d\alpha\right\}$ is a Cauchy sequence of real numbers and this sequence converges to some number L. Indeed, letting m approach ∞ in (*) yields

$$(**) \qquad \left|\int_a^b f_n(x)\, d\alpha - L\right| \le 2\varepsilon\big(\alpha(b) - \alpha(a)\big)$$

whenever $n \ge N$. Choose $\delta > 0$ so that, whenever a partition $x_o < x_1 < \ldots < x_k$ of $[a,b]$ satisfies $|x_i - x_{i-1}| < \delta$, $i = 1, 2, \ldots, k$, then, letting $\Delta\alpha_i = \alpha(x_i) - \alpha(x_{i-1})$,

$$(***) \qquad \left|\sum_{i=1}^{k} f_N(x_i)\Delta\alpha_i - \int_a^b f_N(x)\, d\alpha\right| < \varepsilon.$$

For this δ, one also computes that

$$(****) \qquad |\Sigma f(x_i)\Delta\alpha_i - \Sigma f_N(x_i)\Delta\alpha_i| < \varepsilon\big(\alpha(b) - \alpha(a)\big).$$

Combining (**), (***) and (****) yields

$$|\Sigma f(x_i)\Delta\alpha_i - L| \le 3\varepsilon\big(\alpha(b) - \alpha(a)\big) + \varepsilon$$

for any partition with $x_i - x_{i-1} < \delta$. It follows that the upper and lower integrals of f with respect to α equal L, that $\int_a^b f(x)\, d\alpha = L$ and thus $\int_a^b f(x)\, d\alpha = \lim_{n\to\infty} \int_a^b f_n(x)\, d\alpha$.

(1.20) <u>Theorem</u>. <u>If</u> $\{f_n(x)\}$ <u>is a sequence of derivatives on</u> $[a,b]$ <u>and</u> f_n <u>approaches</u> f <u>uniformly on</u> $[a,b]$, <u>then</u> f <u>is a derivative on</u> $[a,b]$. <u>Moreover, if</u> $F_n(x)$ <u>are chosen so that</u> $f_n(x) = F_n'(x)$ <u>at each</u> $x \in [a,b]$ <u>and</u> $F_n(a) = 0$, <u>then</u>

$F_n(x)$ <u>converges uniformly to a function</u> F <u>on</u> $[a,b]$ <u>and at each</u> $x \in [a,b]$, $F'(x) = f(x)$.

<u>Proof</u>. First note that, if $G(x)$ is differentiable on $[a,b]$ and $|G'(x)| \le M$ on $[a,b]$, then for each $x,y \in [a,b]$, $|G(x) - G(y)| \le M|x - y|$. For otherwise, there would be $x_1, y_1 \in [a,b]$ with $|G(x_1) - G(y_1)| > M|x_1 - y_1|$ and by the standard mean value theorem for differentiable functions there would be z between x_1 and y_1 with $G'(z) = (G(x_1) - G(y_1))/(x_1 - y_1)$ contrary to the fact that $|G'(z)| \le M$. Now suppose $\{f_n\}$ is a sequence of derivatives approaching f uniformly on $[a,b]$. Given $\varepsilon > 0$, choose a natural number N so that for $n,m > N$, $|f_n(x) - f_m(x)| < \varepsilon$. Let $F_n(x)$ satisfy $F_n(a) = 0$ and $F_n'(x) = f_n(x)$. Applying the above note to $G(x) = F_n(x) - F_m(x)$ gives

$$(*) \qquad |(F_n(x) - F_m(x)) - (F_n(y) - F_m(y))| \le |x - y|\cdot\varepsilon \le \varepsilon(b - a).$$

In particular, with $y = a$, $|F_n(x) - F_m(x)| \le \varepsilon(b - a)$ and for each $x \in [a,b]$, $\{F_n(x)\}$ is a Cauchy sequence and thus F_n converges to a function F. By (*)

$$-\varepsilon \le \frac{F_n(x) - F_n(t)}{x - t} - \frac{F_m(x) - F_m(t)}{x - t} < \varepsilon$$

for each $t \in [a,b]$ with $t \ne x$ and $n,m > N$. Letting m approach ∞ yields

$$-\varepsilon \le \frac{F_n(x) - F_n(t)}{x - t} - \frac{F(x) - F(t)}{x - t} \le \varepsilon.$$

By computing the $\overline{\lim}$ and $\underline{\lim}$ as t approaches x in this inequality one has

$$-\varepsilon \le f_n(x) - \overline{\lim_{t\to x}} \frac{F(t) - F(x)}{t - x}$$

$$\le f_n(x) - \underline{\lim_{t\to x}} \frac{F(t) - F(x)}{t - x} \le \varepsilon.$$

Now letting n approach ∞ yields

$$-\varepsilon \le f(x) - \overline{\lim_{t\to x}} \frac{F(t) - F(x)}{t - x}$$

$$\le f(x) - \underline{\lim_{t\to x}} \frac{F(t) - F(x)}{t - x} \le \varepsilon.$$

Finally, since $\varepsilon > 0$ is arbitrary, $F'(x) = f(x)$.

(1.21) Theorem. (Weierstrass) Every continuous function $f(x)$ defined on a closed interval $[a,b]$ is the uniform limit of a sequence $\{P_n(x)\}$ of polynomials.

Proof. Without loss of generality, we may assume that the interval under consideration is $[0,1]$; for otherwise, let $g(x) = f[(x - a)/(b - a)]$ and note that a sequence $\{P_n(x)\}$ approaching $g(x)$ uniformly on $[0,1]$ will have $\{P_n[x \cdot b + (1 - x)a]\}$ approaching $f(x)$ uniformly on $[a,b]$. Also, it is sufficient to consider that $f(0) = f(1) = 0$; for otherwise, if $g(x) = f(x) - [x \cdot f(1) + (1 - x) \cdot f(0)]$ then $g(0) = g(1) = 0$ and, if $P_n(x)$ approaches $g(x)$ uniformly, then $P_n(x) + x \cdot f(1) + (1 - x) \cdot f(0)$ approaches $f(x)$ uniformly. We also assume that $f(x) = 0$ if $x \notin [0,1]$. This makes f continuous on the entire real line. Let $\Phi_n(x) = c_n(1 - x^2)^n$ where c_n is chosen so that $\int_{-1}^{1} \Phi_n(x)\, dx = 1$. Then

$$1/c_n = \int_{-1}^{1}(1 - x^2)^n dx = 2\int_0^1 (1 - x)^n(1 + x)^n dx$$

$$\geq 2\int_0^1 (1 - x)^n dx = 2/(n + 1)$$

and $c_n \leq (n + 1)/2$. Suppose $f(x)$ is continuous on $[0,1]$, $f(0) = f(1) = 0$, and let $f(x) = 0$ if $x \notin [0,1]$. Let $P_n(x) = \int_{-1}^{1} f(x + t)\, \Phi_n(t)\, dt$. Since $f(x + t) = 0$ if $x + t \notin [0,1]$; that is, if $t \notin [-x, 1 - x]$, $P_n(x) = \int_{-x}^{1-x} f(x + t)\, \Phi_n(t)\, dt$. By making a change of variables with $u = x + t$ and $du = dt$, it follows that

$$P_n(x) = \int_0^1 f(u)\, \Phi_n(u - x)\, du$$

which shows that $P_n(x)$ is a polynomial. To see that $P_n(x)$ approaches $f(x)$ uniformly, fix $\varepsilon > 0$. Choose δ so that if $|t| < 2\delta$ then $|f(x + t) - f(x)| < \varepsilon/2$. If $|f(x)| \leq M$ on $[0,1]$, then

$$\begin{aligned}
|P_n(x) - f(x)| &= \left|\int_{-1}^{1}(f(x + t) - f(x))\, \Phi_n(t)\, dt\right| \\
&\leq \int_{-1}^{1}|f(x + t) - f(x)|\, \Phi_n(t)\, dt \\
&\leq \int_{-1}^{-\delta} 2M\, \Phi_n(t)\, dt + \frac{\varepsilon}{2}\int_{-\delta}^{\delta} \Phi_n(t)\, dt \\
&\qquad + \int_{\delta}^{1} 2M\, \Phi_n(t)\, dt \\
&\leq 2M(1 - \delta^2)^n(n + 1)/2 + \varepsilon/2 \\
&\qquad + 2M(1 - \delta^2)^n(n + 1)/2 \\
&< \varepsilon \quad \text{if } n \text{ is sufficiently large.}
\end{aligned}$$

Thus the sequence of polynomials $\{P_n(x)\}$ approaches $f(x)$ uniformly on $[0,1]$.

1.2 Exercises

1. Given the space M of bounded real functions defined on $[0,1]$ with the metric given by the sup norm, show that no sequence $\{f_n\}$ is dense in M.
2. If $f(x)$ is continuous and nowhere differentiable on $[0,1]$, show that the relative maxima of f are dense in $[0,1]$.
3. Show that there is a continuous function taking $[0,1]$ onto $[0,1]$ so that each $f^{-1}(y)$ is a perfect set.
4. Show that there is a continuous function F defined on $[0,1]$ so that for each $t,t' \in [0,x]$, $|F(t) - F(t')| \le |t - t'|^x$ but for each x there are intervals $[x,t'] \subset [x,1]$ for which
$$|F(x) - F(t')| > |x - t'|^x.$$
5. Show that the pointwise limit of a sequence of non-decreasing functions defined on $[0,1]$ is non-decreasing.
6. Show that the limit in Exercise 5 need not be a uniform limit.
7. Show that, if each f_n satisfies $|f_n(x) - f_n(y)| \le M|x - y|$ for all $x,y \in [0,1]$ and if $f_n \to f$ pointwise, then f also satisfies $|f(x) - f(y)| \le M|x - y|$ and the sequence f_n approaches f uniformly.

Chapter Two

2.1 Sets and Cardinal Numbers

A descriptive definition of a set is: a collection of objects such that, given any object whatsoever, that object is either in the collection or is not. As is traditional, objects in a given collection are called its elements and are said to belong to the collection.

A large collection of sets will frequently be called a class. This serves two purposes. It indicates that it is the elements of the class that are of concern and that there is either a logical difficulty with or little interest in forming a set which has the class as an element. If there is a logical difficulty with forming the collection, and it is desirable to indicate this difficulty, the class can be called a proper class. Some of the logical difficulties, that is to say, contradictions, involved with forming very large collections will be pointed out as they occur. It will become clear that these contradictions can be avoided in a simple fashion by referring to such collections as proper classes and not allowing sets to contain them as elements.

Following Cantor, the originator of modern set theory, we say that two sets A and B belonging to a class $\mathcal{S}$ of sets are cardinally equivalent and write $A \simeq_c B$ if there is a one to one function from A onto B. Such a function is sometimes called a one to one correspondence between A and B. When such a function exists between two sets, it shows that their elements can be brought together into a collection of pairs so that every element of A occurs exactly once as a first element of a pair and every element of B exactly once as a second element. In this sense, the two sets have the same number of elements. For a class of sets $\mathcal{S}$, we say that $A \in \mathcal{S}$ is cardinally less than or equal to $B \in \mathcal{S}$ and write $A \leq_c B$ if there is a one to one function with domain A into B. We say that A is cardinally less than B if $A <_c B$ but A is not cardinally equal to B.

More abstractly, a relation R on a set S is a subset of $S \times S = \{(A,B): A,B \in S\}$; that is, R is a collection of ordered pairs of elements occurring in S. The interpretation for a specific relation R is that $(A,B) \in R$ if A is related to B as specified in the definition of R; this is sometimes written $A_R B$. Examples of relations in mathematics are plentiful: $a < b$, $a = b$, $a = b^2$, $a > b^2$ for pairs of real numbers; $A \subset B$, $A = B$, $A \in B$, $A \cap B = \emptyset$ for pairs of subsets of a given set, etc.

A relation $\sim$ on a set S is called an <u>equivalence relation</u> if the following properties hold for each $A, B, C \in S$:

i) $A \sim A$ (reflexive)

ii) $A \sim B$ implies $B \sim A$ (symmetric)

iii) $A \sim B$ and $B \sim C$ implies $A \sim C$ (transitive).

It is easily seen that, for any collection S of sets, $\sim_c$ is an equivalence relation:

i) For $A \in S$ and $x \in A$ let $i(x) = x$. Then i is one to one and onto A and this shows that $\sim_c$ is reflexive.

ii) If $\sim$ takes A one to one onto B, then f^{-1} takes B one to one onto A and thus $\sim_c$ is symmetric.

iii) If f takes A one to one onto B and g takes B one to one onto C, then $g \circ f$ takes A one to one onto C and thus $\sim_c$ is transitive.

Given an equivalence $\sim$ on a set S, a <u>partial order</u> $\leq$ on S is a relationship which satisfies for each $A, B, C \in S$:

i) $A \leq A$ (reflexive)

ii) $A \leq B$ and $B \leq A$ implies $A \sim B$ (antisymmetric)

iii) $A \leq B$ and $B \leq C$ implies $A \leq C$ (transitive)

That $\le_c$ is reflexive and transitive follows from the same argument given for $\sim_c$. That $\le_c$ is antisymmetric is the conclusion of the following theorem.

(2.1) <u>Theorem</u>. (Cantor-Bernstein) <u>If</u> $A \le_c B$ <u>and</u> $B \le_c A$, <u>then</u> $A \simeq_c B$.

<u>Proof</u>. Let f take A into B and g take B into A with f and g one to one. Let $A_o = A$, $A_1 = g(B)$, $A_2 = g\circ f(A)$; then $A_o \supset A_1 \supset A_2$. Continuing in this fashion, let $A_{n+2} = g\circ f(A_n)$. It follows by induction that for each natural number n, $A_n \supset A_{n+1}$. Let $E_n = A_n \backslash A_{n+1}$. Then, the sets E_n are pairwise disjoint and for each natural number n,

$$g\circ f(E_n) = g\circ f(A_n \backslash A_{n+1}) = A_{n+2} \backslash A_{n+3} = E_{n+2}.$$

Let $D = \cap A_n$ and define h as follows:

$$h(x) = \begin{cases} g(f(x)), & \text{if } x \in E_{2n} \quad n = 0, 1, 2, \ldots \\ x, & \text{if } x \in E_{2n+1} \quad n = 0, 1, \ldots \text{ or } x \in D \end{cases}$$

Then h takes each E_{2n} one to one onto E_{2n+2} and takes D and each set E_{2n+1} one to one onto itself. Since $A = \bigcup_{n=0}^{\infty} E_n \cup D$, h takes A one to one onto $A_1 = \bigcup_{n=1}^{\infty} E_n \cup D$. Thus $A \simeq_c A_1$. Since g takes B one to one onto A_1, $B \simeq_c A_1$ and hence $A \simeq_c B$.

With each set occurring in mathematics, we wish to associate a symbol, called the <u>cardinal</u> <u>number</u> of the set, so that, if a set A is shown to be cardinally equivalent to a set B, then A and B will have the same cardinal number. We will then speak of the class of cardinal numbers and, given a set A, the symbol $\|A\|_c$ will denote the cardinal number or cardinality of A.

Familiar sets have the following cardinal numbers: 0 is the cardinal number of $\emptyset$, n is the cardinal number of $S_n = \{0, 1, \ldots, n-1\}$, a is the cardinal number of the set of natural numbers, c is the cardinal number of the real numbers (also called the cardinality of the continuum), f is the cardinal number of the set of all functions from the reals to the reals.

A set is said to be <u>finite</u> if it is empty or its cardinality is some natural number n; otherwise, it is said to be <u>infinite</u>.

A set I will be called an <u>index set</u> if, for each x in I, there is defined a unique A_x which is either a set or an element of a set. If I is a set and for each x in I there is an A_x, it will always be considered legitimate to form $\{A_x: x \in I\}$ and, if each A_x is a set, to form

$$\bigcup_{x \in I} A_x = \{y: \text{there is an } x \in I \text{ and } y \in A_x\}.$$

We will frequently use, without comment, what is called <u>the axiom of choice</u>: If A is a nonempty set each of whose elements is a nonempty set, then there is a function f taking A into $\bigcup_{X \in A} X$ such that $f(X) \in X$. That is, there is a function f which chooses an element from each of the nonempty sets belonging to A.

Cantor defined a set to be finite if it could not be brought into one to one correspondence with a proper subset of itself; otherwise, it was said to be infinite. No set X, which is finite by the definition given above, can be brought into one to one correspondence with a proper subset of itself and, assuming the axiom of choice, every infinite set can. To see this, suppose X is finite in the sense of the original definition. If X is empty, X has no proper subsets and hence can not be brought into one to one correspondence with a proper subset of itself. Suppose, if possible, that n is the least natural number such that there is a set $X \sim_c S_n$ and $X \sim_c Y \subsetneq X$. It follows that $S_n \sim_c S \subsetneq S_n$. Also, $n \neq 1$, since it is not possible to bring $\{0\}$ into one to one correspondence with a proper subset of itself because the only proper subset of $\{0\}$ is $\emptyset$. Suppose that f is a one to one correspondence between S_n and $S \subsetneq S_n$. Then $f(n-1) \neq n - 1$, since n is the least such natural number. On the one hand, if $n - 1 \notin S$, let g take S one to one into S_n as follows: let $g[f(n-1)] = n - 1$ and for $f(x) \neq n - 1$, $g[f(x)] = f(x)$. Then $g \circ f$ restricted to S_{n-1} takes S_{n-1} into a proper subset of itself contained in $S \setminus \{f(n-1)\}$. On the other hand, if $n - 1 \in S$, let g take S one to one into S_n as follows: Select $y \in S_n \setminus S$ and let $g(n-1) = y$, let $g[f(n-1)] = n - 1$ and for $x \in S$ with $x \neq f(n-1)$ let $g(x) = x$. Again $g \circ f$ restricted to S_{n-1} takes S_{n-1} onto a proper subset of itself; namely $S_{n-1} \setminus f(n-1)$. Either case contradicts the fact that n was the least natural number with $S_n \sim_c S \subsetneq S_n$. It follows that no finite set can be brought into a one to one correspondence with a proper subset of itself.

Using the axiom of choice, and the original definition of the term infinite, if X is an infinite set, there is an element $a_o \in X$ and, if distinct elements $a_o, a_1, \ldots, a_{n-1}$ are chosen from X, there is an element a_n in $X \backslash \{a_o, a_1, \ldots, a_{n-1}\}$; for otherwise, X would be finite. A choice function, applied to the non-empty subsets of X, thus results in a sequence $\{a_n\}_{n=0}^{\infty}$ contained in X. The map f which satisfies $f(a_n) = a_{n+1}$ for elements of $\{a_n\}_{n=0}^{\infty}$ and $f(x) = x$ for all other elements of X takes X one to one onto $X \backslash \{a_o\}$.

Note that the above proof also shows that every infinite set contains a sequence of distinct elements and thus that every infinite cardinal number is greater than or equal to a.

A set is said to be _countable_, _enumerable_, or _denumerable_ if it can be brought into one to one correspondence with the natural numbers; that is, if it has cardinality a. If a set is neither finite nor countable it is said to be _uncountable_. If a set A is countable, the set can be written as $\{a_i\}_{i=1}^{\infty}$; that is, the elements of A can be indexed with the natural numbers. To index the set in this fashion implicitly involves using the axiom of choice.

The following sets are countable:

i) the union of a finite set and a countable set,
ii) the union of two countable sets,
iii) the countable union of countable sets,
iv) the product of two countable sets,
v) the integers and the rational numbers,

vi) the set of all finite subsets of a countable set.

To see i), let A be a finite set and B be countable. If $A = \emptyset$ there is nothing to prove because B can be brought into one to one correspondence with the natural numbers and $B = \{b_n\}_{n=1}^{\infty} = A \cup B$ is countable by $f(n) = b_n$. If A is finite and nonempty and B countable, then $A = \{a_i\}_{i=0}^{n-1}$ and $B = \{b_i\}_{i=1}^{\infty}$. If $f(a_i) = i + 1$ for $a_i \in A$ and $f(b_i) = n + i$ for $b_i \in B$, then f is one to one from $A \cup B$ onto the natural numbers.

To see ii), let $A = \{a_i\}_{i=1}^{\infty}$, $B = \{b_i\}_{i=1}^{\infty}$ and let $f(a_i) = 2i$ for $a_i \in A$ and $f(b_i) = 2i - 1$ for $b_i \in B$. Then f takes A one to one onto the even natural numbers and B one to one onto the odd natural numbers. Thus f takes $A \cup B$ one to one onto the natural numbers and $A \cup B$ is countable.

To see iii), let $\{A_i\}_{i=1}^{\infty}$ be a countable collection of sets and, for each natural number i, let $\{a_{i,j}\}_{j=1}^{\infty} = A_i$. A direct way to see that $\bigcup_{i=1}^{\infty} A_i$ is countable is to let $f(a_{1,j}) = 2j - 1$ for each $a_{1,j} \in A_1$; let $f(a_{2,j}) = 2(2j - 1)$ for each $a_{2,j} \in A_2$; in general, let $f(a_{i,j}) = 2^{i-1}(2j - 1)$ for each $a_{i,j} \in A_i$. Each natural number n is the product of a power of two (possibly 2^0) with an odd natural number and the representation $n = 2^{i-1}(2j - 1)$ is unique. It follows that $f(a_{i,j}) = 2^{i-1}(2j - 1)$ is one to one and onto the natural numbers.

To see iv), let $A = \{a_i\}_{i=1}^{\infty}$, $B = \{b_j\}_{j=1}^{\infty}$, $C = \{(a_i,b_j): a_i \in A, b_j \in B\}$. Then, as in iii), $f((a_i,b_j)) = 2^{i-1}(2j - 1)$ is a one to one function from C onto the natural numbers.

For v), the integers

$$\mathbb{Z} = \{n: n \text{ is a natural number}\} \cup \{0\} \cup \{-n: n \text{ is a natural number}\}$$

can be enumerated into a sequence $0, 1, -1, 2, -2, \ldots$ That is, if $f(0) = 1$, $f(n) = 2n$ and $f(-n) = 2n + 1$, then f is one to one from the integers onto the natural numbers. The rational numbers $\mathbb{Q}$ can be brought into one to one correspondence with a subset $\mathbb{Q}'$ of $\mathbb{Z} \times \mathbb{Z}$ by $f(p/q) = (p,q)$ where p/q is in lowest terms and $q > 0$; also $\mathbb{Q}$ contains the natural numbers $\mathbb{N}$. Since $\mathbb{N} \simeq_c \mathbb{N} \subset \mathbb{Q}$ and $\mathbb{Q} \simeq_c \mathbb{Q}' \subset \mathbb{Z} \times \mathbb{Z} \simeq_c \mathbb{N}$, it follows from the Cantor-Bernstein theorem that the rational numbers are a countable set.

Finally, to see vi), note that if A is a countable set, then $A_1 = \{\{a_i\}: a_i \in A\}$ is countable. Also $A_2 = \{\{a_i,a_j\}: a_i,a_j \in A\}$ is cardinally equivalent to a subset of the product of A with itself and hence is countable by iv). In general $A_n = \{\{a_{i,1}, a_{i,2}, \ldots, a_{i,n}\}: a_{i,n} \in A\}$ is cardinally equivalent to a subset of the product of A_{n-1} with A and hence is also countable. But the set of all finite subsets of A is $\bigcup_{n=1}^{\infty} A_n$, a countable union of countable sets which is thus countable.

The <u>power set</u> of a given set X is the set of all subsets of X; it is denoted by $P(X)$ or 2^X. Since $f(x) = \{x\}$ takes X one to one into $P(X)$, it follows that $X \leq_c P(X)$. As might be expected, the cardinality of any set X is less that that of $P(X)$. This is shown by the following:

(2.2) <u>Theorem</u>. <u>Given any set</u> X, $\|X\|_c < \|P(X)\|_c$.

<u>Proof</u>: Let X be any set and let f be any one to one function taking X into $P(X)$. Consider $S = \{x \in X: x \notin f(x)\}$. Suppose, if possible, that there is $y \in X$ with $S = f(y)$. If $y \in S$, then $y \in f(y)$ and, by the definition of S, $y \notin S$. Thus $y \in S$ leads to a contradiction. But if $y \notin S$, then $y \notin f(y)$ and, by the definition of S, y must belong to S which is again a contradiction. It is thus impossible that there be a $y \in X$ with S equal to $f(y)$. Thus f is not onto $P(X)$. Since f was an arbitrary one to one function from X into $P(X)$, it follows that there is no one to one function from X onto $P(X)$. That is, for any set X, X is cardinally less than $P(X)$.

Note that because of Theorem 2.2, it follows that there is no largest cardinal number. In fact, given any collection of sets $S = \{A_\alpha\}_{\alpha \in I}$, if S has a largest element in terms of cardinality, say A_{α_o}, then each element of S is cardinally less than $P(A_{\alpha_o})$. On the other hand, if S does not have a largest element in terms of cardinality, then each A_α is cardinally less than $\bigcup_{\alpha \in I} A_\alpha$.

At this point several logical difficulties or paradoxes arise with respect to the theory of sets. These difficulties and others led to the development of axiomatic set theories. Although most of real analysis proceeds from a simplistic view of set theory, which is sometimes called naive set theory, the developments leading to axiomatic set theory and the consequences of this theory should not be neglected. The logical difficulties in set theory all seem to arise from the fact that certain collections of sets involve many other sets from the collection itself for their construction. For example, the collection of all sets cannot be considered to be a set; nor can all cardinal numbers be brought together into a set. If one assumes that they can, contradictions arise.

Suppose, first of all, that the collection of all sets were a set; denote it by U. Then $\{A \in U: A \notin A\}$ would be a set under ordinary methods of set construction; denote it by U'. (Actually collections which contain themselves as elements are usually not allowed to be sets.) The question arises as to whether $U' \in U'$. If it does, then it is not one of the A such that $A \notin A$ and hence U' does not belong to U'. Conversely, if $U' \notin U'$, then it is one of the A for which $A \notin A$ and U' must belong to itself.

These two contradictions show that there is a difficulty with allowing U' to be a set. Note that the question, "Is the class of all sets a set?" can and must be given a negative answer if the construction of subsets of a set by ordinary methods is allowed. That is, the class of all sets is a proper class.

Now suppose, if possible, that the collection of all cardinal numbers is a set; denote it by C. For each cardinal number m there must be a set $M(m)$ with $\|M(m)\|_c = m$. The collection of pairs where the first element is an $x \in M(m)$ and the second element is m is a set, according to ordinary methods of set construction, and it has the same cardinality as $M(m)$. If M_m denotes this set, then for two distinct cardinal numbers m and n, M_m and M_n are disjoint. Using ordinary methods of set construction $V = \bigcup_{m \in C} M_m$ would be a set. But V cannot have a cardinal number. For if p were the cardinal number of V, then $p \in C$. By the remarks following the previous theorem, there is a cardinal number larger than p, call it q. Now V contains an M_q and hence V must have cardinality larger than p, a contradiction. The class of all cardinal numbers is thus not a set; it is a proper class.

As mentioned earlier the known paradoxes can be resolved in perhaps the simplest way if a collection is called a class when it is not clear whether its elements taken together form a set -- a set being a collection from which, using the standard methods of construction, others sets may be developed without contradictions. This means, among other things, that sets can be elements of other sets and classes cannot be elements of sets. Thus, when a class of sets is referred to, conclusions can be reached about the sets in the class without considering the possibility that the class be a set. The list of axioms for Zermelo-Fraenkel set theory which specify methods for the construction of sets will be presented and discussed in the next section of this chapter. Normally, however, we will be concerned with the natural numbers, rational numbers, real numbers and subsets of the line and the plane and will not be involved with the logical intricacies of axiomatic set theory. We will thus proceed with the development of arithmetic for the cardinal numbers.

Given cardinal numbers m and n of sets M and N, the <u>sum of two cardinal numbers</u> is:

$$m + n = \|M \cup N\|_c$$

assuming M and N are disjoint, and the <u>product of two cardinal numbers</u> is:

$$m \cdot n = \|M \times N\|_c.$$

Furthermore,

$$m^n = \|\{f\colon f \text{ has domain } N \text{ and maps } N \text{ into } M\}\|_c.$$

Note that if M and N are not pairwise disjoint, the sets $\{(x,0)\colon x \in M\}$ and $\{(x,1)\colon x \in N\}$ are pairwise disjoint and these sets have respectively the same cardinality as M and N. Thus $m + n$ can always be defined. To see that $m + n$ is well-defined, suppose $m + n$ is defined using sets M and N and also using sets M′ and N′ where $\|M\|_c = m = \|M'\|_c$, $\|N\|_c = n = \|N'\|_c$, $M \cap N = \emptyset$ and $M' \cap N' = \emptyset$. Then $M \simeq_c M'$ and $N \simeq_c N'$. If f takes M one to one onto M′ and g takes N one to one onto N′ then, if h(x) =

$f(x)$ if $x \in M$ and $h(x) = g(x)$ if $x \in N$, h takes $M \cup N$ one to one onto $M' \cup N'$ and thus $\|M \cup N\|_c = \|M' \cup N'\|_c$. Thus $m + n$ does not depend on the sets used to define it. Similarly, $m \cdot n$ and m^n do not depend on the sets used to define them. Note that $m + n$ and $m \cdot n$ agree with the usual definition of sum and product when m and n are finite and m^n also agrees if $n \neq 0$. Also for any set X, $\|P(X)\|_c = 2^{\|X\|_c}$. To see this, let $Y \subset X$. Then there is a function f_Y taking X into $S_2 = \{0,1\}$ with $f_Y(x) = 0$ if $x \notin Y$, $f_Y(x) = 1$ if $x \in Y$. Since distinct subsets Y of X correspond to distinct f_Y and vice-versa, the map h with $h(Y) = f_Y$ is a one to one correspondence between $P(X)$ and

$$\{f\colon f \text{ takes } X \text{ into } \{0,1\}\}.$$

Thus $\|P(X)\|_c = 2^{\|X\|_c}$.

From the properties of countable sets, it follows that for any natural number n,

$$a = a + a = a \cdot 2 = a \cdot n = a \cdot a.$$

Indeed, if p is an infinite cardinal, then $a + p = p$. To see this, let A and P be sets with $\|P\|_c = p$, $\|A\|_c = a$ and $A \cap P = \emptyset$. Let $\{b_i\}_{i=1}^{\infty} \subset P$ and $A = \{a_i\}_{i=1}^{\infty}$. Then $f(a_i) = b_{2i}$, $f(b_i) = b_{2i-1}$ and, for $x \in P\backslash\{b_i\}_{i=1}^{\infty}$, $f(x) = x$ defines a one to one function from $A \cup P$ onto P.

We will also see that

$$c = 2^a = c + c = c \cdot 2 = c \cdot n = c \cdot a = c \cdot c = c^a$$

and

$$f = 2^c = f + f = f \cdot n = f \cdot a = f \cdot c = f \cdot f = f^a = f^c.$$

This will follow readily from the following theorem:

(2.3) Theorem. For any three cardinal numbers, m, n, and p,

$$(m^n)^p = m^{n\cdot p}.$$

Proof. Let M, N and P be sets of cardinality m, n and p respectively. Let M^N denote the collection of all functions with domain N and range contained in M. Then

$$(m^n)^p = \|\{f\colon f \text{ takes } P \text{ into } M^N\}\|_c$$

$= \|\{f\colon f \text{ takes } P \text{ into } \{g\colon g \text{ takes } N \text{ into } M\}\}\|_c$

and

$$m^{n\cdot p} = \|\{h\colon h \text{ takes } N\times P \text{ into } M\}\|_c.$$

If f takes P into M^N and $f(z) = g_z$ and, for $y \in N$, $g_z(y) = x \in M$, associate with f the function $h = h(y,z)$ where $h(y,z) = x$. This association is a one to one mapping of $(M^N)^P$ onto $M^{N\times P}$. To see that it is one to one, suppose $f_1 \neq f_2$. Then there is a $z \in P$ such that $f_1(z) = g_1 \neq g_2 = f_2(z)$ and since $g_1 \neq g_2$ there is a $y \in N$ such that $g_1(y) \neq g_2(y)$. Then h_1 determined by f_1 is not equal to h_2 determined by f_2 because $h_1(y,z) = g_1(y) \neq g_2(y) = h_2(y,z)$. To see that the mapping is onto, let h be a given function taking $N\times P$ into M. Then for $(y,z) \in N\times P$, $h(y,z)$ is a unique element $x \in M$. Thus for each $z \in P$, $g_z(y) = h(y,z)$ defines a function from N into M. Since $f(z) = g_z$ is then a function from P into M^N, the mapping takes $f \in (M^N)^P$ to $h \in M^{N\times P}$. Therefore the mapping is onto.

We now verify the cardinal arithmetic formulas for c and f. First, $c = \|\mathbb{R}\|_c$, by definition. But $(1/2) + \tan^{-1}(x)/\pi$ is a one to one function from the reals onto $(0,1)$. Also $f(x) = (x - a)/(b - a)$ is a one to one function from (a,b) onto $(0,1)$. Thus $\mathbb{R} \simeq_c (0,1) \simeq_c (a,b)$. Since $[a,b] \subseteq \mathbb{R}$ and $\mathbb{R} \simeq_c (a,b)$, by the Cantor-Bernstein theorem $[a,b] \simeq_c \mathbb{R}$. Now 2^a is the cardinal number of the set of all sequences whose entries are 0's and 1's. Denote the set of all such sequences by S. Then each real number in $(0,1)$ can be written uniquely in the binary system as $\Sigma x_i/2^i$ where each $x_i = 0$ or 1 and infinitely many of the x_i are 1's. This is because numbers in $(0,1)$ for which $x_N = 1$ and for $n > N$, $x_n = 0$ can be written with $x_N = 0$ and for $n > N$, $x_n = 1$. The function f which takes x to $\{x_i\}_{i=1}^{\infty}$ is a one to one map from $(0,1)$ into S. But the function g which takes $\{x_i\}_{i=1}^{\infty}$ in S to $\sum_{i=1}^{\infty} 2x_i/3^i$ is a one to one map from S into $[0,1]$. Applying the Cantor-Bernstein theorem yields that $S \simeq_c [0,1]$; that is, $2^a = c$. It also follows that the Cantor Ternary set $C = \{x: x = \Sigma 2x_i/3^i \text{ with } x_i = 0 \text{ or } 1\}$ has cardinality c. But then, using Theorem 2.3 one obtains

$$c^a = (2^a)^a = 2^{a \cdot a} = 2^a = c.$$

Since

$$c \le c + n \le c + a \le c + c \le c \cdot n \le c \cdot a \le c \cdot c \le c^a = c$$

it follows that all these expressions for cardinal numbers are equal. Similarly, $f = c^c$, by definition. Thus

$$f = c^c = (2^a)^c = 2^{a \cdot c} = 2^c$$

and

$$f^c = (2^c)^c = 2^{c \cdot c} = 2^c.$$

Hence,

$$f \le f + n \le f + a \le f + c \le f + f$$
$$\le f \cdot n \le f \cdot a \le f \cdot c \le f \cdot f \le f^n \le f^a \le f^c = f$$

and all these expressions for cardinal numbers are equal. Note that $c = c^n$ implies that the cardinality of Euclidean n-space is c.

A set which is either finite or countable is said to be <u>at most countable</u>. If X is an uncountable set and $Y \subset X$ is at most countable it follows that $\|X\backslash Y\|_c = \|X\|_c$. To see this let $Y = \{y_n\}$ and choose $Y' = \{y'_n\}_{n=1}^{\infty}$ with $Y' \subset X\backslash Y$. Using the proof that the union of a finite set and a countable set is countable, or that the union of two countable sets is countable, f is defined so that f is one to one and takes $Y \cup Y'$ onto Y'. For $x \in X\backslash(Y \cup Y')$, let $f(x) = x$. Then f takes X one to one and onto $X\backslash Y$. Hence $\|X\backslash Y\|_c = \|X\|_c$.

A number of basic questions in mathematics can be answered by cardinality arguments. Sometimes it is possible to show that certain sets exist by showing that there are many of them; for example, the real numbers have cardinality c and the rational numbers are countable and hence the irrational numbers have cardinality c. In other situations the characterization of mathematical structures can involve cardinality.

With this in mind, we turn our attention to characterizations of open and closed subsets of the line, to some important sets which have cardinality c, and then to a characterization of the set of points of discontinuity of a real-valued function of a real variable.

Suppose $G \subset \mathbb{R}$ is an open set. If $x \in G$, then there is an open interval $(x-r, x+r) \subset G$. If $a_x = \inf\{r: (x-r, x) \subset G\}$ and $b_x = \sup\{r: (x, x+r) \subset G\}$, then it follows that $(a_x, b_x) \subset G$. Thus $G = \bigcup_{x \in G} (a_x, b_x)$. Moreover, if $(a_x, b_x) \cap (a_{x'}, b_{x'}) \neq \emptyset$ and y is in their intersection, then $(a_x, b_x) = (a_y, b_y) = (a_{x'}, b_{x'})$. Thus, each open subset of the reals can be written as a union of pairwise disjoint open intervals. From each such interval a rational number can be selected and, since no two disjoint intervals can contain the same rational number, the collection of disjoint intervals in the union is at most countable. Since the union of open intervals is an open set, we have the following characterization of open subsets of the line: A set $G \subset \mathbb{R}$ is open if and only if G is an at most countable union of pairwise disjoint open intervals.

To obtain a characterization of closed subsets recall that a limit point of a set E is a point y such that each neighborhood of y contains a point of E different from y and that a set is closed if and only if it contains all its limit points. A point y is said to be a _condensation point_ of E provided each neighborhood of y contains uncountably many points of E. The following theorem contains a characterization of closed subsets of the reals.

(2.4) _Theorem. Given any set_ $E \subset \mathbb{R}$, _the set_ Y _of real numbers which are points of condensation of_ E _is either empty or is a perfect subset of_ $\mathbb{R}$; _the set of points of_ E _which are not points of condensation of_

<u>E is at most countable. Thus, if F is a closed subset of $\mathbb{R}$, then $F = P \cup D$ where P is either empty or perfect and D is at most countable.</u>

<u>Proof</u>. Let $E \subset \mathbb{R}$ and let E' be the set of $x \in E$ which are not points of condensation of E. Each $x \in E'$ is contained in an interval (r_x, s_x) where r_x and s_x are rational numbers and $(r_x, s_x) \cap E$ is at most countable. Thus $E' = \bigcup_{x \in E'} (r_x, s_x) \cap E$. Since there are at most countably many pairs of rational numbers, E' is the countable union of the at most countable sets $(r_x, s_x) \cap E$ and hence E' is at most countable. Let Y be the set of real numbers which are points of condensation of E. Since each neighborhood of a point $y \in Y$ contains uncountably many points of E, it also contains uncountably many points of $E \backslash E'$. Since $E \backslash E'$ is contained in Y, it follows that each point of Y is a limit point of Y. Thus, if Y is not empty, Y has no isolated points. To see that Y is closed, let y be a limit point of Y. Then each neighborhood $(y - 1/n, y + 1/n)$ of y contains condensation points of E and

$$(y - 1/n, y + 1/n),$$

being a neighborhood of a condensation point of E, contains uncountably many points of E. Then y is also a point of condensation of E. Thus the set of real numbers which are points of condensation of E is also closed. Thus, if $Y \neq \emptyset$, Y is a perfect set. Now let F be a closed subset of the reals. If D is the set of points of F which are not points of condensation of F, D is at most countable. Since F is closed, F contains its set of points of condensation; that is, $P = F\setminus D$ is the set of points of condensation of F and P is either empty or a perfect set and $F = P \cup D$.

The following sets have cardinality c:

i) the collection of all open subsets of the line
ii) the collection of all closed subsets of the line
iii) the collection of all perfect subsets of the line
iv) the collection of all nowhere dense perfect subsets of the line
v) the collection of all countable subsets of the line
vi) the collection of all continuous real-valued functions of a real variable.

To see i), note first that for each real number a, the set (a,∞) is open and hence there are at least c open subsets of the line. However, each open set G is the union of all the open intervals with rational endpoints contained in G. Since the rational numbers are countable, the collection of all pairs of rational numbers is countable and thus the collection C of all intervals with rational endpoints is countable. Since each open set G is associated in a one to one fashion with the collection of open intervals with rational endpoints contained in it, the cardinality of the collection of open sets is no more than $2^{\|C\|_c} = 2^{a} = c$. (Note that the same argument using open balls of rational radius about a countable collection of points shows that the collection of open subsets of any metric space which has a countable dense subset has cardinality at most c.)

To see ii), simply note that there is a one to one correspondence f between the open subsets and the closed ones; namely, $f(G) = G^c$. Thus, there are always exactly as many closed subsets of a space as there are open subsets.

To see iii), note that there are c perfect subsets; namely, $[x,1]$ for each $x < 1$.

To see iv), let C be the Cantor ternary set. Each set $C \cap [x,1]$ where $x \in C$ and x is not a left hand endpoint of a component interval of the complement of C is perfect. The Cantor set has cardinality c and, since there are only countably many component intervals in its complement, there are c distinct nowhere dense perfect sets of the form $C \cap [x,1]$. It follows that the collection of nowhere dense perfect subsets of the line has cardinality c.

To see v), simply note that the collection of countable subsets of the line has cardinality $c^a = c$.

Finally, to see vi), each continuous function is determined by its values on the rational numbers. Thus there are at most as many continuous functions as there are maps from the rational numbers into the reals; that is, at most $c^a = c$. However, there are c constant functions. Thus there are exactly c continuous real-valued functions of a real variable.

If f is a real-valued function of a real variable, a point x is called a <u>point of discontinuity of the first kind</u> for f if $\lim_{t\to x^+} f(t)$ and $\lim_{t\to x^-} f(t)$

exist but f is discontinuous at x. All other discontinuities of f are called <u>discontinuities of the second kind</u>. If f is non-decreasing or non-increasing, then the discontinuities of f are all of the first kind. To see this, suppose f is non-decreasing. Fix x and let

$$A = \sup\{f(t): t < x\}.$$

Given any $\varepsilon > 0$, there is a point $t' < x$ such that $A \geq f(t')$. Since f is non-decreasing, if $\delta = x - t'$ and $0 < x - t < \delta$, then $t' < t < x$ and $f(t') \leq f(t) \leq A$. Hence $|A - f(t)| < \varepsilon$ whenever $0 < x - t < \delta$. It follows that $A = \lim_{t \to x^-} f(t)$. Similarly, for each x,

$$\lim_{t \to x^+} f(t) = \inf\{f(t): t > x\}.$$

Thus all discontinuities of f are of the first kind. This is likewise true for non-increasing functions.

It is now easy to see that a non-decreasing function can have at most countably many discontinuities. For each discontinuity x, a rational number r_x can be chosen so that $\lim_{t \to x^-} f(t) < r_x < \lim_{t \to x^+} f(t)$. Distinct points of discontinuity clearly are associated with distinct rational numbers. Hence the set of points of discontinuity of a non-decreasing f or of a non-increasing f are at most countable. In fact, if f is an arbitrary real function, the set of points of discontinuity of the first kind for f is at most countable. The fact that the rational numbers are a countable set of points or that the set of intervals with rational endpoints are a countable set of intervals is a frequently used principle for showing that a given set is countable. Another such principle is the fact that the set of isolated points of a linear set or the set of points of a set which are isolated on either the right or the left is at most countable. For, if E is a subset of the reals and the set of points of E which are isolated on the right is $E_r = \{x \in E: \text{there is } \varepsilon > 0 \text{ with } (x, x+\varepsilon) \cap E = \emptyset\}$ then, for each $x \in E_r$, one can choose a rational number $r_x > x$ so that $(x, r_x) \cap E = \emptyset$. Since the map from x to r_x is one to one, E_r is at most countable; similarly, the set of points of E which are isolated on the left is at most countable. Using these principles one can prove a result which implies that the set of points of discontinuity of the first kind for an arbitrary function is at most countable.

(2.5) *Theorem*. *If f is an arbitrary real-valued function of a real variable,*

$$\{x: \overline{\lim_{t\to x}}\, f(t) > \overline{\lim_{t\to x^+}}\, f(t)\}, \quad \{x: f(x) > \overline{\lim_{x\to t^+}}\, f(t)\}$$

as well as

$$\{x: \overline{\lim_{t\to x}}\, f(t) > \overline{\lim_{t\to x^-}}\, f(t)\}, \quad \{x: f(x) > \overline{\lim_{t\to x^-}}\, f(t)\},$$

$$\{x: \underline{\lim}_{t\to x}\, f(t) < \underline{\lim}_{t\to x^+}\, f(t)\}, \quad \{x: f(x) < \underline{\lim}_{t\to x^+}\, f(t)\},$$

$$\{x: \underline{\lim}_{t\to x}\, f(t) < \underline{\lim}_{t\to x^-}\, f(t)\} \quad \text{and} \quad \{x: f(x) < \underline{\lim}_{t\to x^-}\, f(t)\}$$

are at most countable sets.

Proof. Given f, let $E = \{x: \overline{\lim_{t\to x}}\, f(t) > \overline{\lim_{t\to x^+}}\, f(t)\}$. Let $\{r_n\}$ be a enumeration of the rational numbers and let

$$E_n = \{x \in E: \overline{\lim_{t\to x}} f(t) > r_n > \overline{\lim_{t\to x^+}} f(t)\}.$$

Then $E = \bigcup_{n=1}^{\infty} E_n$ and it is easily seen that each set E_n consists of points which are isolated on the right in E_n. For, if $\{x_i\}$ is a decreasing sequence of points of E_n and $x_o = \lim_{i\to\infty} x_i$, then there are points $x_i' \in (x_{i+1}, x_i)$ such that $f(x_i') > r_n$. Hence $\overline{\lim_{t\to x_o^+}} f(t) \geq r_n$ and $x_o \notin E_n$. Thus each point of E_n is isolated on the right in E_n; hence each E_n is an at most countable set and $E = \cup E_n$ is at most countable. Now let

$$A_n = \{x: f(x) > r_n > \overline{\lim_{t\to x^+}} f(t)\}.$$

Then, if $\{x_i\}$ is a decreasing sequence of points from a given A_n and x_i approaches x_o, $\overline{\lim_{t\to x_o^+}} f(t) \geq \overline{\lim_{i\to\infty}} f(x_i) \geq r_n$ and $x_o \notin A_n$. Since each point of each set A_n is isolated on the right in A_n, it follows that $\{x: f(x) > \overline{\lim_{t\to x^+}} f(t)\} = \bigcup_{n=1}^{\infty} A_n$ is an at most countable set. The other sets listed in the statement of the theorem are seen to be at most countable in an analogous fashion.

We come now to the classic characterization of the set of points of discontinuity of a real function:

(2.6) Theorem. *If $f(x)$ is an arbitrary real-valued function of a real variable, the set of points of discontinuity of f is a countable union of closed sets. Conversely, if D is a set which is a countable union of closed sets, then there is a real function g such that D is the set of points of discontinuity of g.*

Proof. Fix f. Recall that x is a point of discontinuity of f if and only if $\omega_f(x) > 0$ where $\omega_f(x) = \lim_{\varepsilon\to 0^+} \sup\{f(t) - f(t'): t, t' \in (x-\varepsilon, x+\varepsilon)\}$. Let D be the set of points of discontinuity of f. For each natural number n, let $D_n = \{x: \omega_f(x) \geq 1/n\}$. Then $D = \bigcup_{n=1}^{\infty} D_n$. To show that the points of discontinuity of f are the countable union of closed sets, it remains to show that each D_n is closed. Fix n and let x_o be a limit point of D_n. Let $\{x_i\}$ be a sequence of points of D_n so that x_i approaches x_o. Then given $\varepsilon > 0$ and $\delta > 0$ there are points x_i in $N_\varepsilon(x_o)$ and thus there are points t_i and t_i' in $N_\varepsilon(x_o)$ such that $f(t_i) - f(t_i') > 1/n - \delta$. It follows that $\omega_f(x_o) > 1/n - \delta$ and, since $\delta > 0$ was arbitrary, $\omega_f(x_o) \geq 1/n$. Thus $x_o \in D_n$ and, since x_o was an arbitrary limit point of D_n, D_n is closed. Now, to see the converse, for each natural number n, let F_n be a closed set. Let $D = \bigcup_{n=1}^{\infty} F_n$, $F_o = \emptyset$ and $E_n = F_n \setminus E_{n-1}$. Fix n and let $\{(r_i, s_i)\}_{i=1}^{\infty}$ be an enumeration of the set of open intervals with rational endpoints. Providing $E_n \cap$

$(r_i,s_i) \neq \emptyset$, let x_i be chosen in $E_n \cap (r_i,s_i)$. Then $A_n = \{x_i\}_{i=1}^{\infty}$ is dense in E_n. Let $B_n = E_n \backslash A_n$. When A_n and E_n are chosen for each natural number n, let $g(x) = 0$ if $x \notin \bigcup_{n=1}^{\infty} F_n$, let $g(x) = 1/2n$ if $x \in A_n$ and let $g(x) = 1/(2n + 1)$ if $x \in B_n$. We now show that D is the set of points of discontinuity of g. For let x belong to D. Then there is a natural number N such that $x \in E_N$. If $x \in B_N$, then $g(x) = 1/(2N + 1)$ and there is a sequence $\{t_i\}$ of points of A_N which approach x. However, $g(t_i) = 1/2N$ and hence g is discontinuous at x. On the other hand, if $x \in A_N$, since x belongs to the complement of the closed set $\bigcup_{n=0}^{N-1} F_n$ and A_N is at most countable, every neighborhood of x contains points $t \notin \bigcup_{n=1}^{N-1} F_n \cup A_N$. For such points t, $g(t) \leq 1/(2N + 1)$ and, since $g(x) = 1/2N$, g is discontinuous at each x in A_N. Thus g is discontinuous at each $x \in D = \bigcup_{n=1}^{\infty} F_n$. Now, let $x \notin D$. Given $\varepsilon > 0$, choose a natural number N so that $1/(2N + 2) < \varepsilon$. Since x belongs to the complement of the closed set $\bigcup_1^N F_n$ and the complement is open, there is $\delta > 0$ so that $(x-\delta,x+\delta)$ is contained in $(\bigcup_1^N F_n)^c$. Then, if $t \in (x-\delta,x+\delta)$, $t \notin \bigcup_{n=1}^{N} F_n$ and hence $g(t) \leq 1/(2N + 2) < \varepsilon$. Since $\varepsilon > 0$ was arbitrary and $g(x) = 0$, g is continuous at x and hence g is continuous at each point x which is not in D. Thus, D is the set of points of discontinuity of g.

Note that, since any set which is a countable union of closed sets can be the set of points of discontinuity for a function, so can any closed set or open set be the set of points of discontinuity for a function. For example, if G is an open set and $G = \bigcup_{i=1}^{\infty} (a_i, b_i)$, then $G = \cup F_n$ where

$$F_n = \bigcup_{i=1}^{n} [a_i + \ell_i/n,\ b_i - \ell_i/n]$$

where $\ell_i = b_i - a_i$.

The following function is also worth noting: let $g(x) = 0$ if x is irrational, $g(0) = 1$ and, if p/q is a rational number expressed in its lowest terms, let $g(p/q) = 1/q$. The rational numbers are the discontinuities of g. It follows from the Baire category theorem that the irrational numbers can not be the set of points of discontinuity of a function.

2.1 Exercises

1. Show that every infinite set can be written as a countable union of pairwise disjoint infinite sets.
2. Define the sum and product of an indexed collection of cardinal numbers $\{m_a\}_{a \in I}$.
3. (Konig's Theorem) Prove that if for each $a \in I$, $m_a > n_a > 0$, then $\prod_{a \in I} m_a > \sum_{a \in I} n_a$.
4. Construct a one to one function from (a,b) onto $[a,b]$.

5. Construct a one to one function from $[0,1]$ onto $[0,1] \times [0,1]$.
6. Construct a one to one function from the set of irrational numbers onto the real numbers.
7. Show that the collection of polynomials with rational coefficients is countable.
8. If F is a closed subset of $\mathbb{R}$ and $\{I_n\}$ is the set of intervals contiguous to F, show that the union of F with the intervals in any subset of $\{I_n\}$ is closed.
9. Show that the complement of a first category subset of $\mathbb{R}$ contains a perfect set and hence has cardinality c.
10. Suppose $\{F_i\}_{i=1}^{\infty}$ is a sequence of pairwise disjoint closed subsets of $[0,1]$ and at least two of the F_i are not empty. Prove $[0,1] \neq \bigcup_{i=1}^{\infty} F_i$.
11. Show that a non-empty complete metric space which has no isolated points has cardinality greater than or equal to c.
12. Consider the collection of all closed sets which consist of only rational numbers. What is the cardinality of this collection?

2.2 Discussion of Axiomatic Set Theory

The development of real analysis does not depend on a strictly axiomatic theory of sets. Indeed, in so far as there does not seem to be any logical problem with the construction of the real numbers, any set theoretic model which does not contain a set representative of the real numbers would be inadequate for analysis and of little use for most mathematics. Furthermore, when one proceeds from a naive (not strictly axiomatic) viewpoint, it is natural to assume that the sets that are carefully constructed from this viewpoint are legitimate and should be consistent with and represented in any normal axiomatic scheme.

However, in the light of modern developments in set theory, it no longer seems advisable to only be aware of a naive approach to the theory of sets. This is especially the case due to the fact that many important questions which were looked upon as problems have now been shown to be independent of the formulations of various axiomatic set theories; that is, the answers to these questions cannot be decided within these theories even though it is unclear as to what additional axioms might be naturally appended to provide answers.

The most frequently used axiomatic set theory is that of Zermelo-Fraenkel (the system ZF) which along with the axiom of choice forms the system ZFC. Its axioms will be listed shortly; one can notice that the concepts behind them are employed in the development of set theory in this text.

The formulas which occur in axiomatic set theory (ZF) are those which can be built up from the two formulas $x \in y$ and $x = y$ where the variables x and y denote sets. Along with these two, the theory allows all formulas Φ which use correctly the logical connectives $\wedge$, $\vee$, $\Rightarrow$ and $\sim$ and the logical quantifiers $\forall$ and $\exists$. That is, if Φ and Ψ are formulas, so are $(\Phi \wedge \Psi)$, $(\Phi \vee \Psi)$, $(\Phi \Rightarrow \Psi)$, $(\sim\Phi)$,

$\forall x\Phi$ and $\exists x\Phi$. While $x \in y$ and $x = y$ are defined only by the axioms, the usual meaning for these expressions is intended to be modeled by the theory. Indeed, it was shown by Gödel that this interpretation of $x \in y$ and $x = y$ is tenable and such models of set theory are called standard models.

In one version of the theory (ZF) all variables $u, v, w, x, y, X, Y, Z, \ldots$ are sets and countably many such variables are needed. This is because any finite number of them can occur in a given formula. All the sets which satisfy a given formula form what is called a class; classes are denoted by $\mathcal{U}$, $\mathcal{V}$, etc. Some classes are not sets and thus it is necessary to prove whether or not a given class is a set. A class which is not a set, a proper class, cannot belong to a set. Only sets can belong to sets. (The sets in a model of ZF set theory can be formed beginning with the empty set $\emptyset$ and utilizing set brackets; another version of ZF allows objects which are not sets to belong to sets.) Each set X is also a class; namely, $X = \{x: x \in X\}$. A class of sets which satisfies all the axioms is called a model of the axioms; within a model, the class of all sets is denoted by $\mathcal{V}$ where $\mathcal{V} = \{x: x = x\}$; $\mathcal{V}$ is necessarily a proper class.

On the list below, each axiom is accompanied by a description of the intention behind its formulation. The axioms $A_o - A_8$ make up the system ZF, $A_o - A_9$ the system ZFC; A_9 is the axiom of choice.

The Axioms of Zermelo-Fraenkel:

A_o The existence axiom: $\exists x\ (x = x)$.
("There is a set" -- here the empty set will do.)

A_1 The axiom of extensionality:

$$\forall u \ (u \in x \Leftrightarrow u \in y) \Rightarrow x = y.$$

("If two sets have the same elements, they are equal.")

A_2 The axiom of pairs:

$$\forall u \ \forall v \ \exists x \ \exists y \ (y \in x \Leftrightarrow (y = u \ \text{ or } \ y = v)).$$

("Given two sets there is a set that has exactly those two sets for its elements.") Due to this axiom, ordered pairs, triples, n-tuples of sets from a model of set theory can be formed and are also sets. For example, $(u,v) = \{u, \{u,v\}\}$ or $(u,v,w) = (u, (v,w))$, etc. Also, if x is a set, $\{x\} = \{x,x\}$ is a set. An ordered n-tuple of specific sets will be represented by the letter p in the remaining axioms.

A_3 The axiom schema of comprehension: (This is actually a countable collection of axioms since for each logical formula Φ it defines an axiom.)

$$\forall p \ \forall x \ \exists y \ \forall u \ (u \in y \Leftrightarrow (u \in x \ \text{ and } \ \Phi(u,p))).$$

Here Φ represents a logically constructed formula which is defined containing u and p. ("Given a formula Φ and a set x, there is a set $y \subset x$ which is exactly the set of $u \in x$ for which $\Phi(u,p)$ holds.") Note that A_3 allows for the definition of the intersection of two sets, the difference of two sets and the empty set.

A_4 The sum axiom:

$$\forall x \ \exists y \ \forall z \ \forall u \ ((u \in z \ \text{ and } \ z \in x) \Rightarrow u \in y).$$

("Given a set of sets, the union of all the sets in the given set is again a set.") The axiom asserts that the union is contained in a set but A_3 implies that it actually is a set.

A_5 The power set axiom:

$$\forall x \ \exists y \ \forall z \ (z \subset x \Rightarrow z \in y).$$

("The class of all subsets of a given set is a set.") Again this axiom asserts that the class of all subsets is contained in a set and A_3 implies that it is a set. Of course, $z \subset x$ is defined by $\forall u \ (u \in z \Rightarrow u \in x)$.

A_6 The axiom schema of replacement: (This again is a countable collection of axioms involving all possible formulas Φ.)

$$\forall x \ \forall y \ \forall z \ \big((\Phi(x,y,p) \text{ and } \Phi(x,z,p)) \Rightarrow y = z\big) \Rightarrow$$
$$\forall X \ \exists Y \ (\forall x) \ x \in X \Rightarrow \big(\forall y \ (\Phi(x,y,p) \Rightarrow y \in Y)\big).$$

("If F is a set valued function defined on a class, the image under F of a set is also a set.")

A_7 The axiom of infinity:

$$\exists y \ \big(\emptyset \in y \text{ and } \forall x \ (x \in y \Rightarrow \{x\} \in y)\big).$$

("There is a set which contains $\emptyset$, $\{\emptyset\}$, $\{\{\emptyset\}\}$, $\{\{\{\emptyset\}\}\}$, ... among its elements.")

A_8 The axiom of regularity:

$$(\forall y \neq \emptyset) \ \exists x \in y \ (x \cap y = \emptyset).$$

(Assuming the axiom of choice, A_8 is equivalent to "there does not exist a sequence of sets x_o, x_1, x_2, ... such that for every n, $x_{n+1} \in x_n$.").

A_9 The axiom of choice:

$$\forall x \neq \emptyset \ \ \forall y \ (y \in x \Rightarrow y \neq \emptyset)$$
$$\exists F \in \text{Fnc} \ \ \forall y \in x \ \big(F(y) \in y\big).$$

Here $F \in \text{Fnc}$ is the logical description that asserts that F is a function.

It is worthwhile to note several things at this point. First, a model of these axioms, a class of sets for which the axioms hold, can be obtained by starting with the empty set and the set guaranteed by the axiom of infinity. Other sets in the model can be made up of empty sets and set brackets enclosing them providing these satisfy A_8. Second, the introduction of proper classes avoids the paradoxes which arose by allowing any collection whatsoever to be considered to be a set. Third, it is possible to start with the empty set and the set guaranteed by A_7 and by a sequence of operations to form new sets as guaranteed by the axioms and the countable number of formulas of logic. (This can result in a countable model. It is clear that there are only countably many numbers and other mathematical objects which can be precisely described by formulae. Indeed, such formulae must consist of finite sentences formed from an alphabet of symbols which is at most countable.) Finally, while a model can be countable, the real numbers, for example, which are represented by a set in the model cannot be counted within the model; that is, they cannot be brought into one to one correspondence with the natural numbers in the model by means of any function which is in the model.

Consistency of the ZF axioms (a collection of axioms is said to be <u>consistent</u> if no contradiction can be reached as a theorem from the axioms) would, of course, be of considerable interest. But it is known that a proof of consistency of such an infinite schema cannot be obtained. It is also known that no finite axiom scheme can be given for ZF and that finite systems of axioms of finite length can be proven consistent within ZF. There is also a proof that consistency of ZF implies consistency of ZFC. This was obtained by showing that any standard model of ZF contains a submodel in which the axiom of choice holds. Briefly, this is achieved by defining recursively the collection of constructible sets in the model. While we will not go into the details at all, these sets form a model of ZFC in which the generalized continuum hypothesis holds; that is, for each cardinal number m there are no cardinal numbers between m and 2^m. (Later developments, originating with that of Cohen, showed that models could be expanded to larger ones in which the continuum hypothesis was forced to not hold; that is, there are models that have cardinal numbers between a and c.)

The brief presentation here will be concluded with an illustration of how the real numbers are represented in any model of ZFC. The key is to first obtain the natural numbers from the axiom A_7. Indeed if $0 = \emptyset$, $1 = \{\emptyset\}$, $2 = \{0,1\}$, $n+1 = \{0,1, \ldots, n\}$ then each of these is a set in any model of ZF and it is necessary to show that $\mathbb{N} = \{0, 1, 2, \ldots, n, \ldots\}$ is a set in the model. An outline of how this can be shown follows. Note that all of the terms given here and below are expressible in the language of logic by the logical formulas which define them. A set X is called <u>transitive</u> if $\forall x\ (x \in X \Rightarrow x \subset X)$. In a model of ZFC an <u>ordinal number</u> is a transitive set which is ordered by $\in$ such that every non-empty subset of it has a least element. Each natural number n is an ordinal number in the model. Let A be the infinite set guaranteed by A_7 and let B be the set of subsets of A. Then B contains sets of any given finite cardinality. To collect the natural numbers into a set, let $\Phi(x,y) =$ (y is an ordinal number and y is finite and $|x|_c = |y|_c$) or (x is infinite and $y = \emptyset$). Upon noting that $\Phi(x,y) = \Phi(x,z) \Rightarrow y = z$, it follows from A_6 that there is a set C satisfying $\forall x \in B, \forall y\ \Phi(x,y) \Rightarrow y \in C$. Now

$$D = \{y \in C: y \text{ is a finite ordinal}\}$$

can be shown to be the set of natural numbers; that is, $D = \mathbb{N}$. This is because $\emptyset \in B$ implies $\emptyset \in C$ and thus $\emptyset \in D$. Suppose $n \in D$. Then there is $y \in B$ with $|n|_c = |y|_c = n$. But $z \in y$ implies $z \in A$ and $\{z\} \in A$. Thus $E = \{\{z\}: z \in y\}$ belongs to B. Also $|E|_c = n$ and since $\emptyset \notin E$ if $F = E \cup \{\emptyset\}$, then $|F|_c = n + 1$ and it follows that $n + 1 \in D$.

Once the natural numbers are a set in the model, the integers $\mathbb{Z}$ are obtainable, for example by letting $k = \{(m,n): m,n \in \mathbb{N}$ and $m-n = k\}$ along with $-k = \{(m,n): m,n \in \mathbb{N}$ and $n-m = k\}$. Addition and multiplication can be defined on these equivalence classes of natural numbers and they form a set representative of the integers in the model. The rational numbers $\mathbb{Q}$ can be defined by

$$p/q = \{(p,q): p \in \mathbb{N},\ q \in \mathbb{Z} \text{ and } r \cdot q = s \cdot p\}$$

and these equivalence classes represent the rational numbers in the model. Now the method of Dedekind, for example, gives rise to the real numbers in the model. A <u>Dedekind cut</u> is an ordered pair of sets (A,B) with $A \subset \mathbb{Q}$, $B \subset \mathbb{Q}$, $A \cup B = \mathbb{Q}$ and $a \in A$, $b \in B$ implies $a < b$ and the set B has no least element with respect to the order on $\mathbb{Q}$. It can be shown that the collection of Dedekind cuts form a complete ordered field. That is, the real numbers are represented in ZFC.

2.2 Exercises

1. Show that A6 implies A3. (The axiom A3 is included in ZF for historical reasons and because it is a simple statement of a special case of A6.)
2. Prove that, under the assumption of the axiom of choice, A8 is equivalent to "there does not exist a sequence $\{x_n\}$ of sets with $x_{n+1} \in x_n$".
3. Given Dedekind cuts $\alpha = (A,B)$ and $\beta = (A',B')$ define $\alpha > 0$ by $0 \in A$ and for $\alpha,\beta > 0$ define $\alpha + \beta$ and $\alpha \cdot \beta$. For $\alpha > 0$ define $-\alpha$. Then define $\alpha + \beta$ and $\alpha \cdot \beta$ in general. Prove that under the definitions the Dedekind cuts form a field.
4. For Dedekind cuts $\alpha = (A,B)$ and $\beta = (A',B')$ let $\alpha \le \beta$ if $A \subset A'$. Show that the field of cuts from Exercise 3 is an ordered field.
5. For Dedekind cuts $x = (A_x, B_x)$ show that $\sup_{x \in X}\{(A_x, B_x)\} = (A,B)$ where $A = \bigcup_{x \in X} A_x$ provided X is bounded above and non-empty; that is, show the Dedekind cuts are a complete order.

Chapter Three

3.1 Well-Ordered Sets and Ordinal Numbers

A relationship $\leq$ between the elements of a class $\mathscr{A}$ is a total order or linear order or simply an order provided that any two elements a and b in $\mathscr{A}$ are comparable; that is, either $a \leq b$ or $b \leq a$. A set along with such a relationship is called an ordered set. Two ordered sets $(A, \leq_1)$ and $(B, \leq_2)$ are said to be order equivalent and we write $A \simeq_o B$, if there is a one to one function f from A onto B such that, if $x, y \in A$ and $x <_1 y$, then $f(x) <_2 f(y)$. Such a function is called a similarity map.

Two ordered sets which are order equivalent are clearly also cardinally equivalent. If two sets are order equivalent they will be said to have the same <u>order type</u>. Just as with cardinal numbers, we wish to associate with each of the ordered sets that occur in mathematics a symbol representing the set's order type. A specific set with a given order can serve as the symbol for a given order type. We will then speak of the class of order types and, given an ordered set A, the symbol $\|A\|_o$ will denote the order type of A.

Familiar sets have the following order types: The symbol 0 stands for the order type of $\emptyset$, n is the order type of $S_n = \{0, 1, \ldots, n-1\}$, ω is the order type of the natural numbers, ω^* is the order type of the negative integers, η is the order type of the rational numbers, λ is the order type of the real numbers.

As with cardinal numbers, the ordered sum and ordered product of order types can be defined. If $(A, \le_1)$ and $(B, \le_2)$ are ordered sets and $A \cap B = \emptyset$ and $\|A\|_o = \alpha$, $\|B\|_o = \beta$,

$\alpha + \beta = \|C\|_o$

where $C = A \cup B$ and for $x, y \in C$

$$
\begin{aligned}
x < y \quad & \text{if} \quad x, y \in A \text{ and } x <_1 y \\
& \text{or} \quad x, y \in B \text{ and } x <_2 y \\
& \text{or} \quad x \in A \text{ and } y \in B,
\end{aligned}
$$

$\alpha \cdot \beta = \|D\|_o$

where $D = A \times B$ and for $(u,v), (x,y) \in A \times B$

$$
\begin{aligned}
(u,v) < (x,y) \quad & \text{if} \quad v <_2 y \\
& \text{or} \quad v = y \text{ and } u <_1 v.
\end{aligned}
$$

In the definition of the product of order types, the sets A and B need not be assumed disjoint.

It is an easy exercise to show that the sum and product are well defined. Note also that the associative law holds for both addition and multiplication; that is, $(\alpha + \beta) + \gamma = \alpha + (\beta + \gamma)$ and $(\alpha\cdot\beta)\cdot\gamma = \alpha\cdot(\beta\cdot\gamma)$. The commutative law does not hold; for example, $n + \omega = \omega \neq \omega + n$ and $2\cdot\omega = \omega \neq \omega\cdot 2 = \omega + \omega$. A one-sided distributive law does hold; namely, $\alpha\cdot(\beta + \gamma) = \alpha\beta + \alpha\gamma$. The other in general does not; for example,

$$(\omega + 1)\cdot 2 = \omega + 1 + \omega + 1 = \omega + \omega + 1$$
$$\neq \omega\cdot 2 + 1\cdot 2 = \omega + \omega + 2.$$

One use which is made of addition and multiplication of order types is the production of new symbols for the order types of ordered sets; thus, $\omega^* + \omega$ is the order type of the integers, $1 + \eta + 1$ is the order type of the rational numbers in $[0,1]$, $1 + \lambda + 1$ is the order type of a closed interval.

Countable order types are of particular importance. We begin with a theorem asserting that each countable order type can be represented by a subset of the rational numbers with its natural order.

(3.1) <u>Theorem</u>. <u>If</u> $(A,\leq)$ <u>is an ordered set and</u> $\|A\|_c = a$, <u>then there is a subset</u> D <u>of the rational numbers such that</u> $D \simeq_o A$ <u>where</u> $(D,\leq)$ <u>is the order within the rational numbers of the elements of</u> D.

<u>Proof</u>. Let $\{a_n\}_{n=1}^{\infty}$ be an enumeration of A. Let $f(a_1)$ be any rational number r_1. Suppose f has been defined for $a_1, a_2, \ldots, a_k$ so that for $i, j \leq k$, $a_i < a_j$ implies $f(a_i) < f(a_j)$. Let

$$B_k = \{a_i\colon a_i < a_{k+1},\quad i = 1, 2, \ldots, k\}$$

and let $C_k = \{a_i\colon a_i > a_{k+1},\ i = 1, 2, \ldots, k\}$. Let $b_k = \max\{f(a_i)\colon a_i \in B_k\}$ and $c_k = \min\{f(a_i)\colon a_i \in C_k\}$.

Since $b_k < c_k$, a rational number r_{k+1} can be chosen strictly between b_k and c_k. Let $f(a_{k+1}) = r_{k+1}$. Then, proceeding inductively, f is defined on A and is a one to one function from A into the rational numbers. To see that f preserves order, let $D = \{f(a_i): a_i \in A\}$ and let a_i and a_j be two elements of A with $i < j$. Then, if $a_i < a_j$, it follows that $a_i \in B_j$ and $f(a_i) < f(a_j)$; conversely, if $a_i > a_j$, then $a_i \in C_j$ and $f(a_i) > f(a_j)$. Thus f is order preserving and $\{a_n\}_{n=1}^{\infty} \simeq_o D$.

An ordered set is said to be _unbordered_ if it has no first or last element. It is said to be _order dense-in-itself_, if given any two elements of the set, there is a third element between the two. The rational numbers are an example of a countable, unbordered, dense-in-itself ordered set. That this characterizes sets which have the order type of the rational numbers is the substance of the next theorem.

(3.2) _Theorem_. _If_ $(A,<_1)$ _and_ $(B,<_2)$ _are two countable, unbordered, dense-in-itself ordered sets, then_ $A \simeq_o B$.

Proof. Let $\{a_i\}_{i=1}^{\infty}$ and $\{b_j\}_{j=1}^{\infty}$ be enumerations of A and B. Let $a_{k_1} = a_1$ and $f(a_1) = b_1$. If $b_2 < b_1$, let a_{k_2} be such that $a_{k_2} < a_1$ and let $f(a_{k_2}) = b_2$. If $b_1 < b_2$, let a_{k_2} be such that $a_1 < a_{k_2}$ and let $f(a_{k_2}) = b_2$. Suppose that for $i = 1, 2, \ldots, j$, $f(a_i)$ has been defined and that $a_{k_1}, \ldots, a_{k_j}$

have been determined so that $f(a_{k_i}) = b_i$. Suppose further that, if a_m and a_n are among the a's for which f has been defined, then $a_m < a_n$ implies $f(a_m) < f(a_n)$. Consider a_{j+1}. If f has not already been defined for a_{j+1}, let A_j be the set of a_i for which f has been defined. If a_{j+1} is less than every element in A_j, let $b_{k_{j+1}}$ be less than every element of $\{f(a): a \in A_j\}$ and let $f(a_{j+1}) = b_{k_{j+1}}$. Similarly if a_{j+1} is larger than every element in A_j, let $b_{k_{j+1}}$ be larger than every element of

$$\{f(a): a \in A_j\}$$

and let $f(a_{j+1}) = b_{k_{j+1}}$. If neither of these cases apply, let $a' = \max\{a \in A_j: a < a_{j+1}\}$ and let $a'' = \min\{a \in A_j: a > a_{j+1}\}$. Choose $b_{k_{j+1}} \in B$ so that $f(a') < b_{k_{j+1}} < f(a'')$ and let $f(a_{j+1}) = b_{k_{j+1}}$. Now let B_j be the set of b, including possibly $f(a_{j+1})$, such that b is the image under f of one of the a_i for which f has been defined. If $b_{j+1} \notin B_j$, no $a \in A$ has been defined to map into b_{j+1}. If b_{j+1} is less than every element of B_j, since A is unbordered, there is an $a \in A$ such that f has not yet been defined at a and a is less than each element for which f has been defined. Let $f(a) = b_{j+1}$. Similarly, if b_{j+1} is larger than every element of B_j, let a be larger than every element in A for which f has been defined and let $f(a) = b_{j+1}$. If neither of these cases apply to b_{j+1}, choose $a \in A$ so that f has not been defined at a and so that, if

$$a' = \max\{a \in A: f \text{ has been defined at } a \text{ and } f(a) < b_{j+1}\}$$

and

$$a''_j = \min\{a \in A: f \text{ has been defined at } a \text{ and } f(a) > b_{j+1}\},$$

then $a'_j < a < a''_j$. Let $f(a) = b_{j+1}$. Continuing this process inductively results in a function f which is one to one from A onto B. Furthermore, if $a_i, a_j \in A$ and $f(a_j)$ was defined later than $f(a_i)$ was defined, then $a_i < a_j$ implies $f(a_i) < f(a_j)$ and vice-versa. Thus f is order preserving and $A \simeq_o B$.

While Theorem 3.2 is of value in its own right, it will be be used to show that bounded nowhere dense perfect subsets of the real line all have the same order type.

(3.3) <u>Theorem</u>. <u>If P_1 and P_2 are two bounded nowhere dense perfect subsets of the reals, there is an increasing continuous function f such that $f(P_1) = P_2$; consequently, $\|P_1\|_o = \|P_2\|_o$.</u>

<u>Proof</u>. Let $[a,b]$ and $[c,d]$ be closed intervals which contain P_1 and P_2 with $a,b \in P_1$ and $c,d \in P_2$. Let (a_i,b_i) be an enumeration of the intervals contiguous to P_1; that is, $a_i,b_i \in P_1$ and $(a_i,b_i) \cap P_1 = \emptyset$. Let (c_j,d_j) be the intervals contiguous to P_2. Then $\{a_i\}_{i=1}^{\infty}$ and $\{c_j\}_{j=1}^{\infty}$ are sets which have order type η when given their natural order on the real line. This is true because P_1 and P_2 are

closed sets which do not have any isolated points. For, if (a_i, b_i) and (a'_i, b'_i) are two intervals contiguous to P_1 with $b_i \le a'_i$, then $b_i < a'_i$ since $b_i \in P_1$ and b_i is not an isolated point of P_1. Thus there is an interval contiguous to P_1, (a''_i, b''_i) with $a_i < a''_i < a'_i$. Thus $\{a_i\}$ is a dense in itself order. There is also an interval contiguous to P_1 between a and a_i and one between b'_i and b. Thus $\{a_i\}$ and also $\{c_j\}$ are countable, unbordered dense in themselves orders and by Theorem 3.2, are of order type η. Let g be a one to one order preserving map between $\{a_i\}$ and $\{c_j\}$. If $x \in [a_i, b_i]$ and $g(a_i) = c_j$, let

$$f(x) = \frac{x - a_i}{b_i - a_i}(d_j - c_j) + c_j.$$

Then f is one to one and increasing on $\bigcup_{i=1}^{\infty} [a_i, b_i]$ and takes this set onto $\bigcup_{j=1}^{\infty} [c_j, d_j]$. If $x \le a$, let $f(x) = c - (a - x)$; if $x \ge b$, let $f(x) = d + (x - b)$. For all $x \in P_1$ for which f has not been defined, let

$$f(x) = \inf\{f(t): t \in \bigcup_{i=1}^{\infty} (a_i, b_i), \quad t > x\}.$$

Now, let $x_1 \in P_1 \setminus \bigcup_{i=1}^{\infty} \{a_i, b_i\}$ and x_2 be any other point. Then there is (a_o, b_o) contiguous to P_1 with a_o and b_o between x_1 and x_2. If $x_2 < x_1$, then it is easily seen that $f(x_2) < f(x_1)$. Otherwise, if $x_1 < x_2$, since $f(a_o) < f(b_o)$, it follows that

$$f(x_1) = \inf\{f(t): t \in \bigcup_{i=1}^{\infty} (a_i, b_i), \quad t > x_1\} < f(x_2).$$

It then follows for all x_1 and x_2 that $x_1 < x_2$ implies $f(x_1) < f(x_2)$; thus f is increasing. Suppose, if possible, that f is not continuous at a point x_o. Since f is increasing, f can only have discontinuities of the first kind. Since f is continuous on each interval (a_i,b_i) and on $(-\infty,a)$ and (b,∞), x_o must belong to P_1. Let

$$y'_o = \lim_{t\to x_o^-} f(t) \quad \text{and} \quad y''_o = \lim_{t\to x^+} f(t).$$

Then $y'_o < y''_o$ and, since P_2 is nowhere dense, there is an interval (c_o,d_o) contiguous to P_2 which meets (y'_o,y''_o). If $f(a_o) = c_o$, it is not possible that $a_o > x_o$, because then

$$f(x_o) = \inf\{f(t): t \in \bigcup_{i=1}^{\infty} (a_i,b_i), \quad t > x_o\}$$

would be less than c_o, $y''_o = f(x_o^+)$ would be less than c_o and this would imply $(y'_o,y''_o) \cap (c_o,d_o) = \emptyset$. Since $x_o \in P_1$, we must have $x_o \geq b_o$. If $x_o = b_o$, then $\lim_{x\to x_o^+} f(x) = f(x_o) = \lim_{x\to x_o^-} f(x)$. Finally, if $x_o > b_o$ then $f(x_o^-) = y'_o > d_o = f(b_o)$ and $(c_o,d_o) \cap (y'_o,y''_o) = \emptyset$. Since it is not possible that $y'_o < y''_o$, it follows that $y'_o = y''_o$ and f is continuous at x_o. This implies that f is a one to one, continuous, increasing map from the reals to the reals and that f takes P_1 onto P_2. Thus $\|P_1\|_o = \|P_2\|_o$.

Note that, if P is an unbounded nowhere dense perfect subset of the reals, provided $a < b$, a is not a left endpoint of a contiguous interval of P and b is not a right endpoint of a contiguous interval, then $P \cap [a,b]$ is either empty or is a bounded nowhere dense perfect set. This fact, along with the above theorem, provides a complete picture of the order structure of nowhere dense perfect sets. Moreover, every nowhere dense perfect set has cardinality c. This follows readily from the fact that the Cantor ternary set C has been shown to have cardinality c. Since every bounded nowhere dense perfect set is order equivalent to C, each such set is cardinally equivalent to C and hence has cardinality c.

We now turn to examine the structure of a specific class of ordered sets, well-ordered sets. An ordered set is said to be <u>well-ordered</u> if every non-empty subset of the set has a least element; that is, an element which is less than every other element. An <u>ordinal number</u> is the order type of a well-ordered set. Some examples of well-ordered sets are the empty set, any finite ordered set and the natural numbers. Thus, 0, n and ω are ordinal numbers. Likewise the sum and product of two ordinal numbers is an ordinal number. To see this, let α and β represent the order types of $(A, \leq_1)$ and $(B, \leq_2)$, where A and B are two well-ordered sets which are pairwise disjoint. Then if $C \subset A \cup B$ and $C \neq \emptyset$ either C has an element in A, in which case it has a least element in A, or $C \cap A = \emptyset$ and C has a least element in B. Thus $\alpha + \beta$ is an ordinal number. Let $D \subset A \times B$ with $D \neq \emptyset$. Then $\{b\text{: there is } a \in A \text{ with } (a,b) \in D\}$ has a least element in B, say b_o. Also $\{a \in A\text{: } (a, b_o) \in D\}$ has a least element in A, say a_o. Then (a_o, b_o) is the least element in D. Thus, additional examples of ordinal numbers are

$$\omega + n = \|\{1, 2, 3, \ldots, 1', 2', \ldots, n'\}\|_o,$$

$\omega + \omega = \|\{1, 2, 3, \ldots, 1', 2', 3', \ldots,\}\|_o = \omega\cdot 2$, $\omega\cdot 3$, $\omega\cdot n$, $\omega\cdot\omega$, $(\omega\cdot\omega) + n$, $(\omega\cdot\omega) + \omega$, $\omega\cdot\omega\cdot\omega$, etc.

Under the assumption of the axiom of choice, it will be shown below that every set can be well-ordered; that is, an order can be defined on the set for which the set is well-ordered. One device, which will be used in proving this, holds true for any ordered set; namely, if A is an ordered set, for each $a \in A$, let $A_a = \{b \in A: b < a\}$, let $\mathcal{A} = \{A_a: a \in A\}$ and let $\mathcal{A}$ be ordered by $\subset$, then $A \simeq_o \mathcal{A}$. Indeed, the map $f(a) = A_a$ is one to one and onto $\mathcal{A}$ and if $a \le b$, $A_a \subset A_b$ and thus f is order preserving and $A \simeq_o \mathcal{A}$.

The following definitions are needed:

In an ordered set A, a is an <u>immediate predecessor</u> of b if $a < b$ and whenever $c < b$, $c \le a$. Likewise, a is an <u>immediate successor</u> of b if $b < a$ and whenever $b < c$, $a \le c$. Given $a \in A$, the set $A_a = \{b \in A: b < a\}$ is called the <u>initial segment</u> of A determined by a.

We begin by considering a single well-ordered set.

(3.4) <u>Theorem. If A is a well-ordered set, then</u>

i) <u>every subset of A, and hence every initial segment of A, is well-ordered,</u>

ii) <u>every element of A except the largest, if there is a largest, has an immediate successor,</u>

iii) <u>there does not exist a sequence $\{a_n\}_{n=1}^{\infty} \subset A$, satisfying for each natural number n, $a_{n+1} < a_n$.</u>

iv) <u>if f is a similarity map from A onto $B \subset A$, $f(a) \ge a$,</u>

v) there is only one similarity map from A onto A; namely, the identity map,

vi) A is not order equivalent to any of its initial segments,

vii) no two distinct initial segments of A are order equivalent.

Proof.

i) Clearly, if $B \subset A$, then every non-empty subset C of B has a least element in A. This is also a least element of C in B and hence each subset B of A is well-ordered.

ii) If $a \in A$ and a is not the largest element of A, then $\{b \in A: b > a\}$ is not empty and hence this set has a least element. This element is clearly an immediate successor to a.

iii) If $\{a_n\}_{n=1}^{\infty} \subset A$ and for each n, $a_{n+1} < a_n$, then $\{a_n\}_{n=1}^{\infty}$ would be a non-empty subset of A which does not have a least element. Since this is a contradiction, it follows that there is no such sequence contained in A.

iv) Let f be a similarity map from A onto a subset of A. Suppose, if possible, that there is $a \in A$ with $f(a) < a$. Let $a_1 = a$, $a_2 = f(a)$ and define inductively $a_{n+1} = f(a_n)$. Then $a_2 < a_1$ and, assuming $a_{k+1} < a_k$ it follows from the fact that f is a similarity that $a_{k+2} = f(a_{k+1}) < f(a_k) = a_{k+1}$. Hence, for each natural number n, $a_{n+1} < a_n$. But, by iii), this is impossible. Hence, if f is a similarity map it follows that for each $a \in A$, $f(a) \geq a$.

v) Let f and g be similarity maps from A onto A. Suppose that $f \neq g$; that is, there is $a \in A$ so that $f(a) \neq g(a)$. Without loss of generality, suppose $f(a) < g(a)$. Let $h = f \circ g^{-1}$ and note that h is a similarity from A onto A. However, $h(g(a)) = f(a) < g(a)$. That is, h takes $g(a)$ to an element less than $g(a)$. By iv), this cannot happen. Thus any two similarity maps from A onto A must be equal and, since the identity is such a map, the identity is the only similarity map from a well-ordered set onto itself.

vi) Suppose, if possible, that there is a well-ordered set A and $a \in A$ such that $A \simeq_o A_a$. Then there is a similarity map f from A onto A_a. But $f(a) \in A_a$ and hence $f(a) < a$. By iv), this cannot happen. Thus no similarity map from A onto an initial segment of A is possible.

vii) Let A_a and $A_{a'}$ be distinct initial segments of a well-ordered set A. Without loss of generality, let $a > a'$. Suppose, if possible that f is a similarity map from A_a onto $A_{a'}$. Then $a' \in A$, $A_{a'} \subset A_a$ and $f(a') \in A_{a'}$ which implies $f(a') < a'$. By iv), no such similarity map exists from A_a into its subset $A_{a'}$. Thus no two distinct initial segments of a well-ordered set are order equivalent.

We now make some comparisons between two well-ordered sets.

(3.5) <u>Theorem</u>. <u>If</u> A <u>and</u> B <u>are two well-ordered sets then exactly one of the following holds</u>:

i′) $A \simeq_o B$,

ii′) *there is* $b \in B$ *and* $A \simeq_o B_b$,

iii′) *there is* $a \in A$ *and* $B \simeq_o A_a$.

Moreover, if $A \simeq_o B$ *the similarity map taking* A *onto* B *is unique.*

Proof. We first show that a similarity map from a well-ordered set A onto a well-ordered set B is unique. For if f and g are two similarity maps from A onto B, then $h = f^{-1} \circ g$ is a similarity map from A onto A. By v) above, h is the identity map. But then $f = g$ because, for each $x \in A$, $x = h(x) = f^{-1}(g(x))$ and since f is one to one, $f(x) = g(x)$. Now i′) and ii′) cannot both hold because $A \simeq_o B$ and $A \simeq_o B_b$ implies $B \simeq_o B_b$ which, by vi) above, cannot happen. Also, i′) and iii′) cannot both hold by the same reasoning. If ii′) and iii′) both hold, and the similarity map f takes A onto B_b and the similarity map g takes B onto A_a, then $g \circ f$ is a similarity map from A onto A_a and this cannot happen by vi) above. Thus at most one of i′), ii′) and iii′) can hold. It remains to show that one of i′), ii′) or iii′) holds. To see this, we define a similarity map whose domain is either A or some A_a and whose range is either B or some B_b as follows: Let

$$f = \{(a,b): \text{there is } a \in A,\ b \in B \text{ and } f_{a,b} \text{ a similarity map from } A_a \text{ onto } B_b\}.$$

The remainder of the proof that f is a similarity map satisfying i′), ii′), or iii′) is given in six parts.

1. The set of pairs, f, is a function. For, if $(a,b) \in f$ and $(a,b') \in f$, there are similarity maps

$f_{a,b}$ and $f_{a,b'}$ from A_a onto B_b and $B_{b'}$ respectively. Suppose, if possible, that $b < b'$. Then $f_{a,b} \circ f^{-1}_{a,b'}$ is a similarity map from $B_{b'}$ onto its initial segment B_b and this cannot happen. Likewise, it is impossible that $b' < b$; b' must equal b. Hence f is a function.

2. The function f is one to one. For, if there are $a, a' \in A$ with $a < a'$ and $f(a) = f(a') = b \in B$, then there are $f_{a,b}$ and $f_{a',b}$ which are similarity maps from A_a onto B_b and from $A_{a'}$ onto B_b. Then $f^{-1}_{a,b} \circ f_{a',b}$ is a similarity map from $A_{a'}$ onto its initial segment, which cannot happen. Hence f is one to one.

3. The function f is order preserving. For suppose $a, a' \in \text{dom } f$ with $a < a'$. By 2., it is not possible that $f(a) = f(a')$. Suppose $b = f(a) > f(a') = b'$. Then $f^{-1}_{a,b} \circ f_{a',b'}$ is a similarity map from $A_{a'}$ onto A_a which cannot happen. It follows that f preserves order on its domain.

4. The domain of f is either A or an initial segment of A. For suppose $\text{dom } f \neq A$. Let a_o be the least element of $A \backslash \text{dom } f$. Then if $a < a_o$, $a \in \text{dom } f$. If $a \geq a_o$, it is not possible that $a \in \text{dom } f$. For, if $a \in \text{dom } f$, there is $b \in B$ and some $f_{a,b}$ which is a similarity from A_a onto B_b. Then $f_{a,b}$ restricted to A_{a_o} is a similarity from A_{a_o} to an initial segment of B and $a_o \in \text{dom } f$ contrary to the fact that a_o is the least element of $A \backslash \text{dom } f$. Thus $\text{dom } f = A_{a_o}$ or $\text{dom } f = A$.

5. The range of f is either B or an initial segment of B. For suppose $\text{rng } f \neq B$. Let b_o be

the least element of $B \setminus \mathrm{rng}\, f$. Then if $b < b_o$, $b \in \mathrm{rng}\, f$. If $b \geq b_o$, it is not possible that $b \in \mathrm{rng}\, f$. For, if $b \in \mathrm{rng}\, f$, there is $a \in A$ and some $f_{a,b}$ which is a similarity from A_a onto B_b. If $b = b_o$, $b_o \in \mathrm{rng}\, f$ contrary to the fact that b_o is the least element of $B \setminus \mathrm{rng}\, f$. If $b > b_o$ then there is a_o such that $f_{a,b}(a_o) = b_o$ and $f_{a,b}$ restricted to A_{a_o} is a similarity from A_{a_o} to B_{b_o}. This implies $b_o \in \mathrm{rng}\, f$, but $b_o \notin \mathrm{rng}\, f$. Thus $\mathrm{rng}\, f$ is either B or an initial segment of B.

6. It remains to show that one of i′), ii′) or iii′) holds. But, if $\mathrm{dom}\, f = A$, either i′) or ii′) holds. If $\mathrm{dom}\, f = A_a$ and $\mathrm{rng}\, f = B$, iii′) holds. The only other possibility is that $\mathrm{dom}\, f = A_a$ and $\mathrm{rng}\, f = B_b$; this cannot happen. For if $\mathrm{dom}\, f = A_a$ and $\mathrm{rng}\, f = B_b$, then f is a similarity between A_a and B_b. By the definition of f, the element a must belong to $\mathrm{dom}\, f$ and because $a \notin A_a$ the set A_a is not the domain of f. It now follows that one of i′), ii′) or iii′) holds.

We now consider sets of ordinal numbers and show that every ordered set has the same order type as a set of ordinal numbers and that every set of ordinal numbers is well-ordered. The <u>natural order on the class of ordinal numbers</u> is defined by: $\alpha < \beta$ if there are sets A and B such that A has order type α and B has order type β and there is $b \in B$ with $A \simeq_o B_b$.

(3.6) <u>Theorem. If</u> W <u>is a set of ordinal numbers, then</u> W <u>is well-ordered. If</u> α <u>is an ordinal number</u>

and

$$W_\alpha = \{\beta: \beta \text{ is an ordinal number and } \beta < \alpha\},$$

then W_α _is a set and_ $\|W_\alpha\|_o = \alpha$.

Proof. Let W be a set of ordinal numbers. Suppose $\alpha, \beta \in W$. Then there are sets A and B such that $\|A\|_o = \alpha$, $\|B\|_o = \beta$. Now, exactly one of i′), ii′) or iii′) of the previous theorem holds for A and B. That is, either i′) $A \simeq_o B$ and $\alpha = \beta$ or ii′) $A \simeq_o B_b$ and $\alpha < \beta$ or iii′) $A_a \simeq_o B$ and $\alpha > \beta$. Thus W is an ordered set. If W were not well-ordered, there would be a subset W' of W which does not have a least element. Then it would be possible to choose, for each natural number n, $\alpha_1 > \alpha_2 > \ldots > \alpha_n > \ldots$ with $\alpha_n \in W'$. Suppose $\{A_n\}$ is a sequence of sets with $\|A_n\|_o = \alpha_n$. Let $C = A_1$. Then since $\alpha_1 > \alpha_2$ there must be $c_1 \in C$ such that $A_2 \simeq_o C_{c_1}$. Continuing by induction one finds $c_k \in C$ with $C_{c_k} \simeq_o A_{k+1}$ and C_{c_k} is an initial segment of $C_{c_{k-1}}$; that is, $c_k < c_{k-1}$. But then the sequence $\{c_k\}$ is contained in C but does not have a least element; this is a contradiction to the fact that $C = A_1$ is a well-ordered set. Thus W is a well-ordered set. Now let α be an ordinal number and

$$W_\alpha = \{\beta: \beta \text{ is an ordinal number and } \beta < \alpha\}.$$

Let A be a set of order type α and for each $\beta < \alpha$ let A_β be a set of order type β. Each A_β is order equivalent to some initial segment of A. Thus the ordinal numbers less than α are all order types of sets from the collection $\{A_a: a \in A\}$. Since, for each element of $\{A_a: a \in A\}$ there is an ordinal number β

$< \alpha$, by the usual methods of set construction,

$$\{\beta\colon \text{there is } a \in A,\ \beta = \|A_a\|_o\}$$

is a set. This set is W_α. But the map from A to W_α which takes $a \in A$ to $\beta \in W_\alpha$ when $\|A_a\|_o = \beta$ is easily seen to be one to one, onto and order preserving. Thus, $\|W_\alpha\|_o = \alpha$ and this completes the proof of the theorem.

Since the ordinal numbers are well-ordered and there is no largest ordinal number, each ordinal number has an immediate successor. Clearly an ordinal number, for example ω, need not have an immediate predecessor. An ordinal number which has an immediate predecessor is called a <u>successor</u> <u>ordinal</u>. Ordinals which do not have an immediate predecessor are called <u>limit</u> <u>ordinals</u>. Every non-zero ordinal α which is not a limit ordinal is of the form $\lambda + n$ where λ is a limit ordinal and $n < \omega$. If λ is a limit ordinal and $\alpha = \lambda + n$ with n even, α is called an <u>even</u> <u>ordinal</u>; if $\alpha = \lambda + n$ with n odd, α is called an <u>odd</u> <u>ordinal</u>.

Consider two ordinal numbers α and β with $\beta \neq 0$ and their sum $\alpha + \beta$. Since $n + \omega = \omega$, one can only conclude in general that $\beta \leq \alpha + \beta$. However, one can conclude that $\alpha < \alpha + \beta$. To see this, let A have order type α and B have order type β with $A \cap B = \emptyset$. Then let the order be defined on $C = A \cup B$ so that $\|C\|_o = \alpha + \beta$. If $\alpha + \beta$ were equal to α, then $C \simeq_o C_{b_o}$ where b_o is the least element of B; if $\alpha + \beta$ were less than α, then $C \simeq_o C_a$ for some $a \in A$; both of these situations cannot occur and hence $\alpha + \beta > \alpha$.

We note at this point that the class of ordinal numbers is a proper class. For, if the class of ordinal numbers were a set, it would be a well-ordered set and would have an ordinal number, say α. Then the class of ordinal numbers would be equivalent to $W_\alpha = \{b: \beta < \alpha\}$. However, $\alpha \notin W_\alpha$; that is, if α were the ordinal number of the class of ordinal numbers, α would not be an ordinal number. Nonetheless, Theorem 3.5 shows that $\leq_o$ is an order on the class of ordinal numbers. Moreover, every class of ordinal numbers which is not empty has a least element. For, if β belongs to a given class $\mathscr{C}$ of ordinal numbers and β is not the least element in $\mathscr{C}$, then $C = \{\alpha \in W_\beta: \alpha \in \mathscr{C}\}$ is a subset of W_β and hence is a set. Since $C \neq \emptyset$ and C is well-ordered, C has a least element, say β_o. Given any $\gamma \in \mathscr{C}$, either $\gamma < \beta$ or $\gamma \geq \beta$ and, in either case, $\beta_o \leq \gamma$; that is, β_o is the least element in $\mathscr{C}$. Thus, the ordinal numbers form a well-ordered class.

Recall that the axiom of choice asserts that if A is a non-empty set each of whose elements is a non-empty set, then there is a function f taking A into $\bigcup_{X \in A} X$ such that $f(X) \in X$. Now, if $\bigcup_{X \in A} X$ can be well-ordered, then such a choice function exists; indeed, let $f(X)$ equal the least element of X in the well-ordering of $\bigcup_{X \in A} X$. Historically, it was surprising that a converse also holds; that is, assuming the axiom of choice, every set can be well-ordered. This is the content of the next theorem, the well-ordering theorem.

(3.7) <u>Theorem</u>. (Zermelo) <u>Assuming the axiom of choice, any set X can be well-ordered. Specifically, if f is a function which selects an element from each non-empty subset of X, then f determines a well-ordering of X.</u>

<u>Proof</u>. The idea behind the proof is to define x_o to be $f(X)$ and X_o to be $X\setminus\{x_o\}$. In general, if X_α is $X\setminus \bigcup_{\beta<\alpha} \{x_\beta\}$ and $X_\alpha = \emptyset$ the set X has been well-ordered; otherwise, if $X_\alpha \neq \emptyset$, x_α is defined to be $f(X_\alpha)$. That such a process well-orders the set X involves a somewhat complicated proof. To begin, let A be any subset of X which can be well-ordered by an order $\leq_A$ satisfying:

(*) if A_a is an initial segment in the well-ordering $(A, \leq_A)$, then $a = f(X\setminus A_a)$.

(Examples of such well-orderings of subsets of X are: $\emptyset$, $\{a_1 = f(X)\}$,

$\{a_1, a_2: a_1 < a_2,\ a_1 = f(X),\ a_2 = f(X\setminus\{a_1\})\}$,

etc.) Clearly, for each A satisfying (*) and each $a \in A$, A_a also satisfies (*) with the order from A; this is because every initial segment of A_a is an initial segment of A. Given two sets, $A, B \subseteq X$ well-ordered by some $\leq_A$ and $\leq_B$ each satisfying (*), we know that exactly one of $\|A\|_o < \|B\|_o$, $\|B\|_o < \|A\|_o$ or $\|A\|_o = \|B\|_o$. Suppose $\|A\|_o < \|B\|_o$ and thus A is similar to B_b, an initial segment of B. Let g be the similarity map from A to B_b. We claim that g is the identity map. (The exact same proof that g is the identity holds when B is similar to A_a and when A is similar to B.) If g is not the identity map, there is a least element $x_o \in A$ such that $g(x_o) \neq x_o$. Then g is the identity map on A_{x_o} and since g cannot take an $x \in A$ with $x_o <_A x$ to $y \in B$ with $y = g(x) <_B g(x_o)$, g clearly takes A_{x_o} onto $B_{g(x_o)}$. But then by (*) and the fact that initial segments of sets satisfying (*) also satisfy (*), $x_o = g(x_o)$. Thus, either $A \subset B$ or $B \subset A$ and moreover the order $\leq_A$ agrees with $\leq_B$ on $A \cap B$. Also in the case $A = B$, it follows that a set A can have at most one such order $<_A$ on it. Let $Y = \cup A$ where the union is over all subsets $A \subseteq X$ for which there is a well-ordering satisfying (*). For $x,y \in Y$, let $x \leq_Y y$ if there is A in the union with $x,y \in A$, $x \leq_A y$. If $a,b,c \in Y$, $a \in A$, $b \in B$, $c \in C$, then $a,b,c \in A \cup B \cup C$ and the order on $A \cup B \cup C$ is reflexive, antisymmetric and transitive which implies $\leq_Y$ is an order on Y. We now show that $\leq_Y$ well-orders Y. Let $D \subset Y$, $D \neq \emptyset$. Let $d \in D$. If d is not the least element of D, d belongs to some A with A well-ordered and satisfying (*); hence $D \cap A$ has a least element d_o in A. If d_o is not the least element in D, there is $d_1 \in D\backslash A$ with $d_1 <_Y d_o$. Then d_1 belongs to some B which is well-ordered and satisfies (*) and $d_1 <_B d_o$. Since either $A \subset B$ or $B \subset A$, $d_1 \in A$. But $d_1 \in D\backslash A$. This contradiction shows that d_o is the least element of D. Since D, an arbitrary non-empty subset of Y, has a least element, it follows that Y is well-ordered.

We now show that $(Y, \leq_Y)$ satisfies (*). Let y

$\in Y$. There is then an A contained in X which is well-ordered and satisfies (*) such that $y \in A$. If $x \in Y$ and $x <_Y y$ then there is B with $x \in B$ and $x <_B y$; and thus $x <_A y$ and, since this is true for all $x \in Y$ with $x <_Y y$, $Y_y = A_y$. Since $y = f(X\setminus A_y) = f(X\setminus Y_y)$ and y is an arbitrary element of Y, it follows that Y satisfies (*). Finally, we show that $Y = X$; that is, f determines a well-order on X. If $X\setminus Y \neq \emptyset$, let $x = f(X\setminus Y)$. Then, it is apparent that $Y \cup \{x\}$ is a well-ordered set which satisfies (*) providing the order on $Y \cup \{x\}$ is defined to agree with the order on Y and satisfy for each $y \in Y$, $y < x$. But this is contrary to the fact that Y is the union of all such subsets of X. Thus $Y = X$ and f determines a well-ordering of the entire set X.

The cardinal number of an infinite well-ordered set is called an aleph and the first letter of the Hebrew alphabet $\aleph$ (aleph) along with a subscript, which is an ordinal number, is used to designate the alephs. Note that the well-ordering theorem and the axiom of choice are equivalent to the fact that every cardinal number is the cardinal number of a well-ordered set; that is, every cardinal number is an aleph. Thus, if A and B are two sets, A and B can be well-ordered and, if α and β are the order types of well-orderings of A and B respectively, then $A \simeq_o W_\alpha$, $B \simeq_o W_\beta$ and since either $W_\alpha \subset W_\beta$ or $W_\beta \subset W_\alpha$, either $A \leq_c B$ or $B \leq_c A$. That is, any two cardinal numbers are comparable and the class of cardinal numbers is linearly ordered. Moreover, the class of cardinal numbers is well-ordered and any set of cardinal numbers is a well-ordered set, ordered by $\leq_c$. To see this, let $\{p_\gamma\}_{\gamma\in I}$ be a non-empty collection of cardinal numbers, let $\{A_\gamma\}_{\gamma\in I}$ be a collection of well-ordered sets with $\|A_\gamma\|_c = p_\gamma$. Then $\{\alpha$: α is an order type of some A_γ, $\gamma \in I\}$ has a least element, say $\alpha_o = \|A_{\gamma_o}\|_o$. Clearly, for each $\gamma \in I$, $\|A_{\gamma_o}\|_c \leq \|A_\gamma\|_c$. Thus $\{p_\gamma\}_{\gamma\in I}$ has a least element; it follows that any collection of cardinal numbers is well-ordered.

Ordinal subscripts for the alephs are provided as follows:

Each infinite set A can be ordered in several ways. By the Cantor-Bernstein theorem, if A can be well-ordered with order type α and also with order type β and $\alpha < \gamma < \beta$, then $A \sim_c W_\alpha \subset W_\gamma \subset W_\beta \sim_c A$ and hence A can be well-ordered with order type γ. Given ordinal numbers α and β with $\alpha < \beta$, a <u>segment</u> $[\alpha,\beta)$ of the ordinal numbers is the set consisting of all ordinal numbers γ with $\alpha \leq \gamma < \beta$. It is now apparent that the order types of well-orderings of an infinite set A form the segment of the ordinal numbers $[\alpha,\beta)$ where β is the least ordinal number such that $\|A\|_c < \|W_\beta\|_c$. The least ordinal number in this segment is called the <u>initial ordinal</u> of the cardinal number of A. These initial ordinals are denoted by the symbol "ω" along with a subscript which is an ordinal number. The cardinal number a which is the smallest infinite cardinal number clearly has ω as its initial ordinal; thus ω_o is used to denote ω and $\aleph_o$ denotes a. The least uncountable cardinal number is denoted by $\aleph_1$ and its initial ordinal is denoted by ω_1. In order to continue assigning ordinal subscripts to each of the alephs and to their initial ordinals, the following principle of transfinite induction is needed. This principle has many uses involving the definition of classes, the construction of sets and the proofs of theorems. This principle is analogous to one form of the principle of induction for the natural numbers.

(3.8) Theorem. (Principle of Transfinite Induction) *If S is a statement about ordinal numbers and if S is true of 0 and whenever S is true for every ordinal number $\beta < \alpha$, S is true of α, then S is true for every ordinal number.*

Proof. Suppose the hypotheses hold for a statement S but that S is not true of some ordinal number. Let α_0 be the least ordinal number for which S is not true. By hypothesis, $\alpha_0 \neq 0$. Also S is true for each ordinal number $\beta < \alpha_0$. Thus, by hypothesis, S is true of α_0. This contradiction shows that S is true of each ordinal number.

Ordinal number subscripts are now defined for each of the alephs and for their initial ordinals as follows: The cardinal number a and its initial ordinal have been assigned the subscript 0; that is, $a = \aleph_0$ and $\omega = \omega_0$. Suppose the subscript β has been assigned for each ordinal number $\beta < \alpha$. Since $\{\aleph_\beta\}_{\beta<\alpha}$ is a set of cardinal numbers, there is a cardinal number larger than each element of this set. Then $\aleph_\alpha$ is defined to be the least such cardinal number and the initial ordinal of $\aleph_\alpha$ is denoted by ω_α. It is now an easy matter to prove that each aleph has been assigned a subscript among the ordinal numbers and the initial ordinal of the aleph has been denoted by the symbol "ω" followed by the subscript of the aleph. One may also note that each ω_α is a limit ordinal; for otherwise ω_α would have an immediate predecessor β and W_β would have smaller cardinality than $W_\alpha = W_{\beta+1}$, but for any infinite cardinal p, p + 1 = p + a = p.

The assignment of subscripts to the alephs and to their initial ordinals illustrates the use of transfinite induction to produce definitions. It is worth noting that the principle of transfinite induction can be specialized to any segment $[\alpha,\beta)$ of the ordinal numbers as follows: If S is a statement which is true of an ordinal number α and if, for each γ with $\alpha \le \gamma < \beta$, whenever S is true of each ordinal number δ with $\alpha \le \delta < \gamma$, S is also true of γ, it follows that S is true of each ordinal in $[\alpha,\beta)$. The situation where $\alpha = 0$ and β is some initial ordinal occurs frequently in proofs.

Transfinite induction occurs in the proof of the following theorem. This theorem shows that the alephs behave with respect to addition and multiplication in the same way that the infinite cardinals which we have so far encountered behave.

(3.9) Theorem. For each ordinal number α, $\aleph_\alpha + \aleph_\alpha = \aleph_\alpha$ and $\aleph_\alpha \cdot \aleph_\alpha = \aleph_\alpha$.

Proof. Since $a + a = a \cdot a = a$ and $\aleph_0 = a$, the theorem is clearly true of the ordinal number 0. Suppose that for each $\beta < \alpha$,

$$\aleph_\beta = \aleph_\beta + \aleph_\beta = \aleph_\beta \cdot \aleph_\beta .$$

Since $\aleph_\alpha \leq \aleph_\alpha + \aleph_\alpha \leq \aleph_\alpha \cdot \aleph_\alpha$ clearly holds, it suffices to show that $\aleph_\alpha \cdot \aleph_\alpha = \aleph_\alpha$. By the definition of ω_α, the set of ordinal numbers less than ω_α has cardinality $\aleph_\alpha$ and ω_α is the least ordinal number with this property. Define an order on the ordered pairs of ordinal numbers less than ω_α as follows: if γ', γ'', δ', δ'' are less than ω_α and $\gamma = \max(\gamma', \gamma'')$ and $\delta = \max(\delta', \delta'')$, let

$$(\gamma', \gamma'') < (\delta', \delta'') \quad \text{if } \gamma < \delta$$
$$\text{or } \gamma = \delta \text{ and } \gamma' < \delta'$$
$$\text{or } \gamma = \delta \text{ and } \gamma' = \delta' \text{ and } \gamma'' < \delta''.$$

It is easy to check that this is not only a linear order but a well-ordering. If $\aleph_\alpha < \aleph_\alpha \cdot \aleph_\alpha$, then the set of ordinals less than ω_α is similar to an initial segment of the collection of ordered pairs, say to the segment determined by (γ_0', γ_0''). Let $\delta < \omega_\alpha$ so that $\gamma_0' < \delta$ and $\gamma_0'' < \delta$. Then the initial segment of the collection of pairs determined by (δ, δ) contains that determined by (γ_0', γ_0''). Since $\delta < \omega_\alpha$, $\|W_\delta\|_c < \aleph_\alpha$. Hence if $\|W_\delta\|_c = \aleph_\beta$, then $\aleph_\alpha$ is less than or equal to the cardinality of the segment determined by (δ, δ). But this segment has cardinality $\aleph_\beta \cdot \aleph_\beta = \aleph_\beta < \aleph_\alpha$ by the induction hypothesis. This contradiction implies that $\aleph_\alpha \cdot \aleph_\alpha = \aleph_\alpha$ and by the principle of induction, $\aleph_\alpha \cdot \aleph_\alpha = \aleph_\alpha$ for each ordinal number α.

Note that for two infinite cardinal numbers $\aleph_\alpha$ and $\aleph_\beta$ with $\aleph_\alpha \leq \aleph_\beta$, we have $\aleph_\beta \leq \aleph_\alpha + \aleph_\beta \leq \aleph_\alpha \cdot \aleph_\beta$

$\leq \aleph_\beta \cdot \aleph_\beta = \aleph_\beta$. Thus the sum or product of two infinite cardinal numbers is the maximum of the two cardinal numbers.

Cantor originally posed the question as to whether there were any uncountable subsets of the real numbers with cardinality less than c. This became known as the continuum problem. This problem, along with several other natural problems have been shown to be independent of the Zermelo-Fraenkel axioms along with the axiom of choice (ZFC). Currently, there is no known principle or axiom which would decide this question in a natural way. The statement, "every uncountable subset of the real numbers has cardinality c", is known as the <u>continuum</u> <u>hypothesis</u> (CH). In the context of the axiom of choice, this hypothesis consists of the assertion that $2^{\aleph_o} = \aleph_1$. The continuum hypothesis is occasionally used in analysis to prove theorems or provide examples. Of course, proving a theorem or providing an example without the hypothesis is preferable and it is always the case that the use of the continuum hypothesis is done explicitly.

Because of the continuum problem, the position (subscript) of c among the alephs is not determined. In this regard, the following convention is usually adopted: the symbol $\aleph$ without a subscript or $\aleph_c$ is used to denote c and the initial ordinal of $\aleph_c$ is written ω_c. The position of other cardinals of the form $2^{\aleph_\alpha}$ is also not determined. Note that if $\aleph_\beta \leq \aleph_\alpha$ then

$$2^{\aleph_\alpha} \leq \aleph_\beta^{\aleph_\alpha} \leq (2^{\aleph_\beta})^{\aleph_\alpha} = 2^{\aleph_\beta \cdot \aleph_\alpha} = 2^{\aleph_\alpha}$$

and thus equality holds for each of the above. The generalized continuum hypothesis (GCH) consists of the statement, "for every ordinal number α, $2^{\aleph_\alpha} = \aleph_{\alpha+1}$". One other question stands out in the development of analysis which has also proven to be undecidable using the axioms of ZFC; namely, Souslin's problem. Souslin stated this problem, in 1920 in terms of the real numbers as follows:

> If L is a totally ordered set which is unbordered, dense in itself, complete, and satisfies the condition that every pairwise disjoint collection of open intervals in L is at most countable, is L order equivalent to the real line?

Here an order is said to be complete provided every non-empty subset which is bounded above has a least upper bound; an open interval in L is a set of the form $(a,b) = \{x \in L: a < x < b\}$. The assertion that Souslin's problem has a positive answer is called the Souslin Hypothesis (SH).

The least uncountable cardinal and its initial ordinal ω_1 play a role in many proofs. An ordinal number α is less than ω_1 if and only if α is the order type of an at most countable well-ordered set. An α which is the order type of a countable well-ordered set is called a countable ordinal.

Note that ω_1 has the following property: if $\{\alpha_n\}_{n=1}^{\infty}$ is a sequence of ordinal numbers each less than ω_1 and if α_o is the least ordinal which is larger than or equal to every α_n, then $\alpha_o < \omega_1$. To see this, note that

$$W_{\alpha_o} = \{\alpha\colon \text{there is an } n \text{ with } \alpha < \alpha_n\}.$$

Thus W_{α_o} is the countable union of the at most countable sets W_{α_n} and hence is at most countable.

We note in passing that, while for each $\alpha < \omega_1$, there is a subset of the rational numbers which has order type α, there is no subset of the reals with order type ω_1. For if there were, say $\{x_\alpha\}_{\alpha<\omega_1}$, then between each x_α and $x_{\alpha+1}$ there would be a rational number r_α. Each of these rational numbers would be distinct, but this is impossible since the rational numbers are countable and $\{\alpha\colon \alpha < \omega_1\}$ is uncountable.

It is also not possible to have a transfinite sequence of closed subsets of the reals $\{F_\alpha\}_{\alpha<\omega_1}$ with each $F_\alpha \subset F_\beta$ whenever $\alpha < \beta$. This is a consequence of the following theorem.

(3.10) __Theorem__. __If, for each__ $\alpha < \omega_1$, F_α __is a closed subset of the line and if, for each__ $\alpha < \beta < \omega_1$, $F_\beta \subset F_\alpha$, __then there is__ $\alpha_o < \omega_1$ __such that for__ each $\gamma \geq \alpha_o$ the set F_γ is identically F_{α_o}.

__Proof__. Suppose that, for each $\alpha < \omega_1$, $\{F_\alpha\}$ is a transfinite sequence of closed sets satisfying the hypothesis of the theorem. Let $F'_o = F_o$ and for each $\alpha < \omega_1$, if α is a successor ordinal let $F'_\alpha = F_\alpha$, if α is a limit ordinal, let $F'_\alpha = \bigcap_{\beta<\alpha} F_\beta$. For each α for which $F'_{\alpha+1} \subsetneq F'_\alpha$, there is a rational interval (r_α, r'_α) so that $(r_\alpha, r'_\alpha) \cap F'_\alpha \setminus F'_{\alpha+1} \neq \emptyset$. Then if α

$< \beta$ and $F'_{\alpha+1} \subsetneq F'_\alpha$ and $F'_{\beta+1} \subsetneq F'_\beta$ there is $x_\beta \in (r_\beta, r'_\beta) \cap F'_\beta \backslash F'_{\beta+1}$ and $x_\beta \notin (r_\alpha, r'_\alpha)$. Thus, distinct α are associated with distinct rational intervals. Since there are at most countably many rational intervals so associated, there are at most countably many α with $F'_{\alpha+1} \subsetneq F'_\alpha$. Thus there is $\alpha_o < \omega_1$ with α_o larger than all of these α. For this α_o and each $\gamma > \alpha_o$, $F'_{\gamma+1} = F'_\gamma$. It follows that for each $\gamma > \alpha_o$, $F_\gamma = F_{\alpha_o}$.

We are now in a position to give another characterization of closed subsets of the line. We begin with a description of the various limit points of a closed set. Given a set E, let E' denote the set of limit points of E. This set E' is a closed set and is called the <u>derived set of</u> E. Given a closed set F, let $F_o = F'$. If, for each $\beta < \alpha < \omega_1$, F_β has been defined, if $\alpha = \beta + 1$ is a successor ordinal, let $F_\alpha = F'_\beta$, if α is a limit ordinal, let $F_\alpha = \bigcap_{\beta<\alpha} F_\beta$. The sets F_α are called the α-<u>derived sets of</u> F. Since the derived set of a closed set is closed and the intersection of closed sets is closed, each F_α is a closed set. Moreover, for each $\beta < \alpha < \omega_1$, $F_\alpha \subset F_\beta$. It follows from Theorem 3.10 that there is $\alpha_o < \omega_1$ such that for $\gamma > \alpha_o$, $F_\gamma = F_{\alpha_o}$. Consequently F_{α_o} is a perfect set or is empty. However, each point of $F_\alpha \backslash F_{\alpha+1}$ is an isolated point of F_α. Thus each set $F_\alpha \backslash F_{\alpha+1}$ is at most countable. But it is easily seen that $F \setminus F_{\alpha_o} = \bigcup_{\alpha<\alpha_o} F_\alpha \setminus F_{\alpha+1}$ which is a countable union of at most countable sets. Thus each closed set $F = P \cup D$ where $P = F_{\alpha_o}$ is perfect or empty and $D = \bigcup_{\alpha<\alpha_o} F_\alpha \setminus F_{\alpha+1}$ is at most countable. This discussion as well as Theorem 3.10 also hold in the more general setting where the line is replaced with a separable metric space.

The remainder of this section consists of the construction of three examples of sets. While these sets are applicable within the theory of measures, they are of interest in their own right and illustrate the method of transfinite induction. Considerable effort, which would otherwise be lost in attempts to prove false statements, can be saved by the ability to construct examples; frequently such examples involve transfinite induction.

(3.11) Example. There is a subset S of the real numbers such that neither S nor S^c contains a perfect set. (Such a set is sometimes called totally imperfect.)

Construction. The collection of perfect subsets of the real numbers has cardinality c. Let $\{P_\alpha\}_{\alpha<\omega_c}$ be a well-ordering of these sets. Select $x_o \in P_o$, $y_o \in P_o$ with $x_o \neq y_o$. Suppose that for every $\beta < \alpha < \omega_c$, x_β and y_β have been chosen from P_β with $x_\beta \neq y_\beta$ and so that neither x_β nor y_β belong to $\bigcup_{\gamma<\beta}\{x_\gamma, y_\gamma\}$. Since P_α is a perfect set, it has cardinality c. Since $\bigcup_{\beta<\alpha}\{x_\beta, y_\beta\}$ is the union of fewer than c sets each of which has two elements, $\bigcup_{\beta<\alpha}\{x_\beta, y_\beta\}$ has cardinality $2\cdot\|\omega_\alpha\|_c = \|\omega_\alpha\|_c < c$. Hence x_α and y_α can be chosen in $P_\alpha \setminus \bigcup_{\beta<\alpha}\{x_\beta, y_\beta\}$. Now let $S = \{x_\alpha: \alpha < \omega_c\}$ and let P be any perfect set. Then P is some P_α and $x_\alpha \in S$ but $y_\alpha \in S^c$. Since P is an arbitrary perfect set, neither S nor S^c contains a perfect set.

(3.12) <u>Example</u>. <u>There is a real-valued function f defined on the line such that for each perfect set P, the image of P under f is the entire real line.</u>

<u>Construction</u>. The collection of pairs (P,y), where P is a perfect set and y is a real number, has cardinality $c \cdot c = c$. Let $\{A_\alpha\}_{\alpha<\omega_c}$ be a well-ordering of this set and note that each perfect set P occurs c times as a first element of a pair (P,y). Suppose $A_0 = (P_0, y_0)$. Choose $x_0 \in P_0$ and define $f(x_0)$ to be y_0. Suppose for each $\beta < \alpha$ that $A_\beta = (P_\beta, y_\beta)$ and that a point x_β has been chosen in P_β so that for each $\gamma < \beta$, $x_\beta \neq x_\gamma$ and that $f(x_\beta)$ has been defined to be y_β. Let $A_\alpha = (P_\alpha, y_\alpha)$. Since there are fewer than c points in the set $\{x_\beta: \beta<\alpha\}$, there is a point x_α in

$$P_\alpha \setminus \{x_\beta: \beta < \alpha\}.$$

Define $f(x_\alpha)$ to be y_α. Then, for each $\alpha < \omega_c$, a point x_α has be chosen so that for $\beta < \alpha$, $x_\alpha \neq x_\beta$ and $f(x_\alpha) = y_\alpha$. If $x \notin \{x_\alpha : \alpha < \omega_c\}$, let $f(x) = 0$. Then f is defined on the entire real line. If P is a perfect set and y is a real number, then there is $\alpha < \omega_c$ so that $(P, y) = A_\alpha$. Thus there is $x = x_\alpha$ with $x \in P$ and $f(x) = y$. Hence f takes each perfect set P onto the entire real line.

(3.13) Example. Assuming the continuum hypothesis, there is an uncountable subset E of the real numbers such that each dense open subset of the reals contains all but at most countably many points of E.

Construction. The collection of dense open sets has cardinality c. Let $\{G_\alpha\}_{\alpha<\omega_1}$ be a well-ordering of these sets. Note that the continuum hypothesis is used here; that is, $\omega_c = \omega_1$. Let $x_0 \in G_0$. Given α with $0 < \alpha < \omega_1$, suppose that for each $\beta < \alpha$ a distinct point x_β has been chosen so that each $x_\beta \in \bigcap_{\gamma \le \beta} G_\gamma$. Since $\alpha < \omega_1$, the set of ordinal numbers $\beta < \alpha$ is at most countable. However, the set $\bigcap_{\beta \le \alpha} G_\beta$ is residual and thus uncountable. Hence x_α can be chosen with x_α in $\bigcap_{\beta \le \alpha} G_\beta$ so that for $\beta < \alpha$, $x_\alpha \neq x_\beta$. Let $E = \{x_\alpha\}_{\alpha<\omega_1}$. Then E is uncountable. Now consider any dense open set G. There is $\alpha < \omega_1$ so that $G = G_\alpha$. For each $x \in E$ if $x = x_\gamma$ with $\gamma > \alpha$ then x has been chosen so that it belongs to G_α. Thus $\{x \in E : x \notin G\}$ is contained in $\{x_\beta\}_{\beta<\alpha}$ and this is an at most countable set. Since G was an arbitrary dense open set, each dense open set contains all but at most countably many points of E.

3.1 Exercises

1. Show that there are exactly c distinct countable order types.
2. If f is a function defined on the ordinal numbers less than a limit ordinal λ with $f(\alpha) \geq f(\beta)$ when $\alpha > \beta$, then $\lim_{\alpha<\lambda} f(\alpha) = \gamma$ provided γ is the least ordinal greater than or equal to each $f(\alpha)$. Show that for any α and any limit ordinal λ, $\alpha + \lambda = \lim_{\beta<\lambda}(\alpha + \beta)$.
3. Using the definition in Exercise 2, show that for each ordinal α and limit ordinal λ,
$$\alpha\cdot\lambda = \lim_{\beta<\lambda} \alpha\cdot\beta.$$
4. If λ is a limit ordinal, $\alpha > 0$ and $n < \omega$, show $\alpha\lambda = (\alpha + n)\lambda$.
5. If for ordinal numbers α, γ, $0 \leq \alpha \leq \gamma$, show there is a unique ordinal β so $\alpha + \beta = \gamma$.
6. If γ is any ordinal number and $\beta > 0$, show there is a unique ordinal α such that $\beta\cdot\alpha \leq \gamma < \beta(\alpha + 1)$.
7. For infinite limit ordinals α, let $\mathrm{cof}(\alpha)$ be the least limit ordinal λ such that there is a set $\{\alpha_\xi\}_{\xi<\lambda}$ with each $\alpha_\xi < \alpha$ and $\alpha = \lim_{\xi<\lambda} \alpha_\xi$. Show $\mathrm{cof}(\alpha) = \mathrm{cof}[\mathrm{cof}(\alpha)]$ and $\mathrm{cof}(\omega_\alpha) = \mathrm{cof}(\alpha)$.
8. Using the definition in Exercise 7, show $\mathrm{cof}(\alpha)$ is always the initial ordinal of a cardinal.
9. Show there are ordinal numbers α, β, γ, δ with

$0 < \alpha < \beta$, $0 < \gamma < \delta$ and $\alpha + \gamma > \delta + \beta$, $\alpha\cdot\gamma > \delta\cdot\beta$.

10. Show that $c \neq \aleph_\omega$. In fact, show if m_i are cardinals for $i < \omega$, then $c \neq \Sigma m_i$.
11. If A is any subset of the plane with $\|A\|_c \geq 2$, then $A = B \cup C$ such that $B \cap C = \emptyset$ and if h is any horizontal line and v any vertical line $\|B \cap h\|_c < \|A\|$ and $\|C \cap v\|_c < \|A\|$.
12. Show that the continuous hypothesis implies that there exists a function f from $\mathbb{R}$ into the collection of all sequences of real numbers such that for each $x, y \in \mathbb{R}$, either $x \in f(y)$ or $y \in f(x)$.
13. Prove that the continuum hypothesis implies that each perfect subset of the real line can be written as the union of $\aleph_1$ pairwise disjoint non-empty perfect sets.
14. Let $\{F_\alpha\}_{\alpha<\omega_1}$ be closed subsets of $\mathbb{R}$ with $F_\alpha \subset F_\beta$ whenever $\beta > \alpha$. Prove there is $\alpha_o < \omega_1$ such that for $\alpha > \alpha_o$, $F_\alpha = F_{\alpha_o}$.

3.2 Applications of the Axiom of Choice

The axiom of choice and statements equivalent to the axiom of choice are frequently used to show the existence of mathematical structures which one would not have otherwise thought possible. In simple situations the axiom of choice can be applied directly. Transfinite induction, when applicable, allows for the construction of a mathematical structure through repeated selections from a well-ordered set. Some constructions, however, can be performed using principles equivalent to the axiom of choice which are of intermediate complexity. While there are quite a few statements equivalent to the axiom of choice, in the study of analysis two of these appear with some frequency; namely, Zorn's lemma and the Hausdörff maximality principle. It is for this reason that they are presented here.

These principles are applicable within the setting of partially ordered sets. Recall that $(S,\leq)$ is a partially ordered set if $\leq$ is a relationship which is reflexive, antisymmetric and transitive. Two elements $a,b \in S$ do not need to be comparable; that is, it is not necessary that either $a \leq b$ or $b \leq a$ hold. For any collection S which consists of subsets of a set X, $(S,\subset)$ is an example of a partial order. A subset C of $(S,<)$ is called a chain if C is linearly ordered; an element $s \in S$ is a maximal element of S if no element of S is larger than s. The two principles are as follows:

Zorn's lemma. If $(S,\leq)$ is a partially ordered set and if every chain $C \subset S$ has an upper bound in S, then S has a maximal element.

Hausdörff maximality principle. In a partially ordered set $(S,\leq)$ every chain $C \subset S$ is contained in a maximal chain; that is, each C is contained in a chain C' so that C' is not a proper subset of any chain.

To see that these are equivalent to the axiom of choice, we first show that Zorn's lemma is implied by the well-ordering theorem. Suppose $(S,\leq)$ is a partially ordered set and suppose each chain in S has an upper bound in S. Let $s_o, s_1, \ldots, s_\alpha, \ldots$ be a well-ordering of S and define a chain $C \subset S$ by $s_o \in C$ and $s_\alpha \in C$ if for each $\beta < \alpha$, $s_\beta \in C$ implies $s_\beta < s_\alpha$. Then, since C has an upper bound in S, there is $s_{\alpha_o} \in S$ so that s_{α_o} is greater than or equal to each $s_\alpha \in C$. This s_{α_o} is thus the last element in C and is necessarily a maximal element of S.

To see that Zorn's lemma implies the maximality principle, let $(S,\leq)$ be a partially ordered set and C_o be a chain in S. Let P be the collection of all chains in S which contain C_o. Then $(P,\subset)$ is partially ordered. If D is a chain in P, then the union of all the chains $C \in D$ is a chain in S and an upper bound for D in P. Since each chain in P has an upper bound, by Zorn's lemma, P has a maximal element; that is, there is a maximal chain in S which contains C_o.

Finally, to see that the Hausdörff maximality principle implies the axiom of choice, let $S = \{X_\alpha\}_{\alpha\in A}$ be a non-empty collection of non-empty sets. Let P be the collection of all functions f which satisfy $f(X_\alpha) \in X_\alpha$ for each X_α in the domain of f where the domain of f is contained in S. By the maximality principle, there is a maximal chain C in $(S,\subset)$. Then g can be defined by letting $(X_\alpha,x_\alpha) \in g$ iff there is $f \in C$ with $(X_\alpha,x_\alpha) \in f$. Since C is a chain, g is well-defined. Also each $f \in C$ is a subset of g. Since C is a maximal chain, there cannot be an element X_{α_o} of S which is not in the domain of g; for otherwise, if $x_{\alpha_o} \in X_{\alpha_o}$ and h is defined by $g \cup (X_{\alpha_o}, x_{\alpha_o})$ then the chain $C \cup \{h\}$ would be a larger chain than C. Consequently, g is a choice function on S. It follows that the Hausdörff maximality principle implies the axiom of choice.

The results which will be considered next have been selected for their applicability within analysis. In each case, the proof relies heavily on the axiom of choice. Though the constructions involved do not always have practical uses, they usually indicate that attempts to improve certain theorems without further restrictions on the hypotheses are doomed to failure. Often the constructions also indicate that extensions of a given structure to a larger one is not only possible but also somewhat arbitrary; that is to say that the number of ways in which the extensions can be achieved is infinite.

We begin with some applications which involve linear spaces. A <u>basis</u> for a linear space X over a field F is a subset B of X such that each non-zero $x \in X$ has a unique representation as a finite sum of multiples of non-zero elements of F with elements of B; in other words each $x \in X$, $x \neq \overline{0}$, can be written uniquely as $\sum_{i=1}^{n} c_i b_i$ where the c_i are non-zero elements of F and the b_i are distinct elements of B. Each Euclidean space $\mathbb{R}^k$ has an explicit basis; for example, any set of k mutually perpendicular vectors such as the standard unit vectors forms a basis for $\mathbb{R}^k$.

(3.14) <u>Theorem</u>. <u>Every linear space has a basis</u>.

<u>Proof</u>. Let X be a linear space over F, where F is a field of scalars. The trivial space $\{\overline{0}\}$ obviously has $\emptyset$ as a basis. If $X \neq \{0\}$, let x_o, $x_1, \ldots, x_\alpha, \ldots$ be a well-ordering of $X\backslash\{\overline{0}\}$. Let B be the set of all x_α such that $x_\alpha \neq \Sigma c_i x_{\beta_i}$ for any finite sum for which each β_i is less than α. Clearly, every $x \in X\backslash\{\overline{0}\}$ can then be written as a finite sum of the form $\Sigma c_i x_{\beta_i}$ with each $x_{\beta_i} \in B$ occurring only once in the sum. Moreover, this can be done with each $\beta_i \leq \alpha$ when $x = x_\alpha$. To see that the representation of each x is unique, suppose $x = \Sigma c_i x_{\beta_i} = \Sigma d_j x_{\gamma_j}$ where each x_{β_i} occurs only once in the first sum and each x_{γ_j} occurs only once in the second sum. Then $\overline{0} = \Sigma c_i x_{\beta_i} - \Sigma d_j x_{\gamma_j} = \sum_{k=1}^{n} e_k x_{\delta_k}$ where the latter sum combines the terms of the former two so that each x_{δ_k} occurs only once in this sum and no e_k equals 0. Presumably one of the e_k is not 0. Otherwise, each c_i equals respectively a d_j with the corresponding x_{β_i} equal to an x_{γ_j} and then the two representations were identical. But then there is a largest ordinal index δ_k such that $e_k \neq 0$. With no loss of generality, we may assume this is δ_1. Then $\sum_{k=1}^{n} e_k x_{\delta_k} = \overline{0}$ implies $x_{\delta_1} = \sum_{k=2}^{n} (e_k/e_1) x_{\delta_k}$. This is a contradiction because $x_{\delta_1} \in B$ and thus cannot be written as a sum of constant multiples of previous elements from the well-ordering of $X\backslash\{\overline{0}\}$. It follows

that all such representations of an $x \in X\backslash\{\overline{0}\}$ are the same.

<u>Corollary</u>. <u>If</u> $B' \subset X$ <u>is a set of independent vectors, then there is a basis</u> B <u>of</u> X <u>with</u> $B' \subset B$.

<u>Proof</u>. By definition, B' is a set of <u>independent vectors</u> if whenever $\sum_{i=1}^{n} c_i b_i = 0$, where the b_i in the sum are distinct elements of B', then each $c_i = 0$. Now, if the elements of B' are well-ordered and the well-ordering of B' is followed by a well-ordering of $X\backslash(B' \cup \{0\})$, then the construction of B in the proof of the theorem results in a basis with $B' \subset B$.

When the real numbers are considered to be the vector space X over the field $\mathbb{Q}$ of rational numbers, the above theorem gives rise to a basis which is called a <u>Hamel basis</u>. A Hamel basis H necessarily has the cardinality of the continuum. This is because the set of all finite sums $\Sigma q_i h_i$ with $h_i \in H$, $q_i \in \mathbb{Q}$ has cardinality $\aleph_o \cdot \aleph_o \cdot \|H\|_c$ and this would be less than c if $\|H\|_c$ were less than c. Let h_o belong to a Hamel basis H and, for each $a \in \mathbb{Q}$, let $\mathbb{Q}_a$ be the set of finite sums $ah_o + \sum_{i=1}^{n} q_i h_i$ where each $q_i \in \mathbb{Q}$ and each $h_i \in H\backslash\{h_o\}$. Then the sets $\mathbb{Q}_a$ are pairwise disjoint and each set $\mathbb{Q}_a$ is a translation of $\mathbb{Q}_o$. Suppose F is the real-valued function defined by $F(x) = a$ iff $x \in \mathbb{Q}_a$. Then for each pair x, y of real numbers, F satisfies

$$F(x + y) = F(x) + F(y)$$

because $F(x + y) = a + a' = F(x) + F(y)$ when $x = ah_o + \Sigma q_i h_i$ and $y = a'h_o + \Sigma q_i' h_i'$. A real-valued function defined on a linear space satisfying this last equation is said to be linear. The example just given of a linear function F defined on the line shows that such functions need not be of the form $f(x) = kx$. In fact, for each $a \in \mathbb{Q}$ and $h_1 \in H$ the numbers of the form $ah_o + bh_1$ with $b \in \mathbb{Q}$ are dense, thus for each $a \in \mathbb{Q}$, $F^{-1}(a)$ is dense and hence F is not continuous at a single point.

One restriction on a linear function F which guarantees that F is of the form $f(x) = kx$ is that F be continuous at a single point or, weaker still, that F be bounded in a neighborhood of a single point. To see this, we first note some facts about linear functions. If F is linear, $F(0 + 0) = F(0) + F(0)$ so that $F(0) = 0$. Then, for each x and natural number n, $F(nx) = F(x) + \ldots + F(x)$ (n times) $= n \cdot F(x)$; also, $F(-x) + F(x) = F(0) = 0$ so that $F(-x) = -F(x)$ and thus $F(-nx) = -n \cdot F(x)$; furthermore $F(x) = F[n \cdot (x/n)] = n \cdot F(x/n)$ so that $F(x/n) = F(x)/n$. Putting these facts together yields that, for each x and rational number q, $F(qx) = q \cdot F(x)$.

Now suppose F is linear and is bounded in a neighborhood of x_o, say $F(x) \le M$ when $x \in (x_o - \varepsilon, x_o + \varepsilon)$. Let $q \in (\varepsilon/2, \varepsilon)$ be a rational number. Then, for $t \in (-1,1)$, $x_o + qt = x \in (x_o - \varepsilon, x_o + \varepsilon)$ and thus

$$|F(t)| = |[F(x) - F(x_o)]/q| \le 2M \cdot 2/\varepsilon = M'$$

and thus F is bounded by M' on $(-1,1)$. Consequently, if $x \in (-1/n, 1/n)$, $|F(nx)| = n|F(x)| \le$

M' and $|F(x)| \le M'/n$. This implies that when $\{x_n\}$ is a sequence of numbers which approaches 0, $F(x_n)$ also approaches 0. Likewise, given any number x, if $t_n \to x$, then $F(t_n) - F(x) = F(t_n - x)$ approaches 0 and thus F is continuous at each point x. Again fix x and let $\{q_n\}$ be a sequence of rational numbers with $x = \lim q_n$. Let $k = F(1)$. Then $F(x) = \lim F(q_n) = \lim q_n F(1) = xF(1) = kx$ and thus the function F is of the form $F(x) = kx$.

Discontinuous linear functions such as the above function provide counterexamples to seemingly natural conjectures. For example, a function f is said to be symmetric provided that at each x, $\lim_{h\to 0} f(x+h) + f(x-h) = 2f(x)$. One might guess that such functions must be continuous at least at a single point. However, each linear function F satisfies for all h, $F(x+h) + F(x-h) = 2F(x)$. Thus, without further restrictions on symmetric functions, such functions need not be continuous at any point.

Now suppose X is a linear space over the field of real numbers. A function L taking X into $\mathbb{R}$ is called a linear functional if, for each $x,y \in X$ and $a,b \in \mathbb{R}$,

$$L(ax + by) = a\cdot L(x) + b\cdot L(y).$$

If X' is a subspace of X and L is a linear functional defined on X', using the axiom of choice, it is possible to extend L to all of X. Simply define L on the first element x_o of a well-ordering of $X\setminus X'$ arbitrarily. Then, if $y = ax + bx_o$ with $x \in X'$, let $L(y) = a\cdot L(x) + b\cdot L(x_o)$. Continuing in this fashion with the least element of the well-ordering for which L is not defined leads by transfinite induction to an extension of L to a linear functional defined on all of X.

A linear functional L defined on X', a subspace of a normed linear space X, is said to be a _bounded linear functional_ on X' if there is $M > 0$ so that for $x \in X'$, $|L(x)| \le M\|x\|$. It is frequently desirable to extend a bounded linear functional on X' to one which is bounded on all of X. That this can be done is one of the consequences of the Hahn-Banach theorem. This theorem is usually stated in a slightly more general context. If p is a real-valued function defined on X, p is said to be _homogeneous_ if for each $a \ge 0$, $p(ax) = a \cdot p(x)$; p is said to be _subadditive_ if $p(x + y) \le p(x) + p(y)$ for each $x, y \in X$. A norm is an example of a homogeneous subadditive functional, and such functionals are called _semi-norms_. The extensions obtained are not generally unique.

(3.15) _Theorem_. (The Hahn-Banach Theorem) _Suppose_ p _is a homogeneous, subadditive functional defined on_ X _and that_ L _is a linear functional defined on a subspace_ $X' \subset X$ _and_ L _satisfies_ $L(x) \le p(x)$ _for each_ $x \in X'$. _Then there is an extension of_ L _to_ X _so that_ L _is a linear functional on_ X _and satisfies_ $L(x) \le p(x)$ _for each_ $x \in X$.

Proof. Well order $X \backslash X'$. Let x_o be the least element of the well-ordering and $x, y \in X'$. Then

$$\begin{aligned} L(x) - L(y) &= L(x - y) \\ &\le p(x - y) \le p(x + x_o) + p(-x_o - y); \end{aligned}$$

that is, $-p(-x_o - y) - L(y) \le p(x + x_o) - L(x)$.

Taking the supremum over all $y \in X'$ and then the infimum over all $x \in X'$ yields

$$\sup_{y \in X'} -p(-x_o - y) - L(y) \le \inf_{x \in X'} p(x - x_o) - L(x).$$

Thus there is a real number t such that for all $x, y \in X'$,

$$-p(-x_o - y) - L(x) \le t \le p(x + x_o) - L(x).$$

Select such a t and define $L(x_o) = t$ and, for $x \in X'$, let $L(ax_o + x) = at + L(x)$. By the choice of t, if $a > 0$, then

$$t \le p(x/a + x_o) - L(x/a) = \frac{1}{a}\bigl(p(ax_o + x) - L(x)\bigr)$$

and $L(ax_o + x) = at + L(x) \le p(ax_o + x)$; on the other hand if $a < 0$, then $-p(-x/a - x_o) - L(x/a) \le t$ and $\frac{1}{a}\bigl(p(ax_o + x)\bigr) \le t + \frac{1}{a} \cdot L(x)$ so that again $L(ax_o + x) = at + L(x) \le p(ax_o + x)$. Now, if for each $\beta < \alpha$, L has been defined on each subspace X_β consisting of all finite linear sums of $x \in X$ and x_{γ_i} with each $\gamma_i \le \beta$, and for $z \in X_\beta$, $L(z) \le p(z)$, then the above extension allows one to extend L to X_α so that for $z \in X_\alpha$, $L(z) \le p(z)$. By transfinite induction, L can be extended to all of X so that for each $z \in X$, $L(z) \le p(z)$.

While the Hahn-Banach theorem has numerous applications in functional analysis, we will not pursue any of them here. Instead we turn to a third result of interest involving the axiom of choice.

Using methods which are similar to the above proof, it is possible to define a generalized length $L(E)$ for each subset E of the line so that $L(I)$ agrees with the length of each interval I, L is non-negative, L takes on the same value on congruent sets and, if $\{A_i\}_{i=1}^{n}$ is a finite collection of pairwise disjoint sets, $L(\bigcup_{i=1}^{n} A_i) = \sum_{i=1}^{n} L(A_i)$. Here A is <u>congruent</u> to A', written $A \simeq A'$, if there is a one to one distance preserving function ϕ from A onto A'; moreover, ϕ is distance preserving on A if for $x,y \in A$, $d(x,y) = d(\phi(x),\phi(y))$. The function L satisfies for each set $A \subset \mathbb{R}$, $L(A) \leq \inf \sum_{n=1}^{\infty} L(I_n)$ where $L(I) = b - a$ when $I = [a,b]$ and the infimum is taken over all countable sequences $\{I_n\}$ of intervals whose union contains A. This result along with a similar extension of area in $\mathbb{R}^2$ were shown to be possible by Tarski. The construction of such extensions involves the axiom of choice and is necessarily somewhat arbitrary in nature; that is, such extensions are not unique. Crucial to their existence is the fact that no subset A of the line and no bounded subset of the plane can be divided into two parts, B and C, with $B \cap C = \emptyset$, $A = B \cup C$ and A congruent to both B and C. Such a decomposition of a set A is called a <u>paradoxical decomposition</u>. Surprising, if not paradoxical, is the fact that bounded subsets of $\mathbb{R}^3$ can have such decompositions. Roughly, this is due to the existence of additional congruency maps, specifically rotations, from $\mathbb{R}^3$ to $\mathbb{R}^3$. That such a decomposition exists requires the axiom of choice and is the conclusion of the next theorem. The specific decomposition will then be shown to make it impossible to extend volume in $\mathbb{R}^3$ to all subsets of $\mathbb{R}^3$.

(3.16) Theorem. (Hausdörff Paradoxical Theorem) *If S is the surface of the unit sphere in $\mathbb{R}^3$, then $S = A \cup B \cup C \cup D$ where $A \simeq B \simeq C \simeq B \cup C$, D is countable and A, B, C and D are pairwise disjoint.*

Proof. The construction of these sets will be outlined below; the details which are missing are fairly apparent but require considerable calculation in order to verify them completely. To begin, two points p_1 and p_2 are to be chosen on the surface of the unit sphere. The relationship of p_2 to p_1 will be specified shortly. Let ϕ be the map which takes each point x on the surface to the point $\phi(x)$ by rotating the sphere 180° with p_1 on the axis of rotation. Let ψ be the map which takes each point x on the surface to the point $\psi(x)$ obtained by rotating the sphere 120°, say in a counterclockwise fashion, with p_2 on the axis of rotation. Then, for each x, $\phi^2(x) = x$ and $\psi^3(x) = x$. Let G be the group which consists of the identity map e and all elements g where g is a map which can be obtained by a finite number of compositions of ϕ and ψ. Clearly, each $g \in G$ has an inverse and each $g \in G$ other than the identity can be represented uniquely by a finite length word made up of the compositions which can be written with the letters ϕ, ψ and ψ^2 where in each word ϕ is never preceded by ϕ, and ψ and ψ^2 are never preceded by another ψ or ψ^2. Now p_1 and p_2 are supposed to have been specified in such a fashion that each distinct word represents a unique map of the sphere to itself. (That this is possible requires some proof.) Then each such map, being a composition of the rotations ϕ and ψ, is itself a rotation. Let D be the countable set of points on the surface of the unit sphere which lie on an axis of one of these rotations. The group G is now partitioned into three sets, G_A, G_B and G_C. The identity e belongs to G_A. If $g \in G_A$, then ϕg and ψg belong to G_B; if $g \in G_B$, $\phi g \in G_A$ and $\psi g \in G_C$; if $g \in G_C$, $\phi g \in G_A$ and $\psi g \in G_A$. These specifications allow inductively for the partitioning of G so that each element of G belongs to exactly one of G_A, G_B or G_C. Proof of this requires careful examination of the cases. Note also that $\phi G_A = G_B \cup G_C$, $\psi G_A = G_B$ and $\psi^2 G_A = G_C$. For each $x \notin D$, let G_x be the countable collection of points which can be reached by applying elements of G to x. Given G_x and G_y, it is clear that either $G_x \cap G_y = \emptyset$ or $G_x = G_y$. Thus $\{G_x: x \notin D\}$ is a collection of pairwise disjoint sets. By the axiom of choice, it is possible to form a set X which contains exactly one element from each set G_x. Having done this, let $A = \{g(x): x \in X, \ g \in G_A\}$,

$$B = \{g(x): x \in X, \ g \in G_B\}$$

and $C = \{g(x): x \in X, \ g \in G_C\}$. Then A, B and C are pairwise disjoint and $A \cup B \cup C \cup D$ is the set of points consisting of the entire surface of the unit sphere; indeed, $A \cup B \cup C = \bigcup_{x \notin D} G_x$ consists of all the points on the surface but those in D. But now from the definitions of G_A, G_B and G_C, it follows that ϕ takes A onto $B \cup C$, ψ takes A onto B and ψ^2 takes A onto C making $A \simeq B \simeq C \simeq B \cup C$.

Now consider the closed unit sphere. If X is any set contained in the surface of the sphere, let X' be the set of points of the sphere other than its center which lie on some line segment connecting the center to some point of X. To show that volume cannot be extended to all subsets of $\mathbb{R}^3$ so that congruent sets have the same value for their volume, it remains to show that under such an extension the volume of D' would have to be 0 so that it would not be possible to define volume for the sets A', B' or C'. To see that the volume of D' would necessarily be 0, we show that S' and $S' \backslash D'$ can each be decomposed into pairs of sets so that the first sets in each pair are congruent to each other and the second sets are also congruent to each other. From this it follows that S' and $S' \backslash D'$ would have the same volume and hence the volume of D' would have to be 0. Since D is a countable subset of S, there is a rotation θ of S so that for each n, $\theta^n(D) \cap D = \emptyset$. Select such a rotation, and note that the sets in the sequence $\{\theta^n(D)\}$ are pairwise disjoint. Let $E = D \cup \bigcup_{n=1}^{\infty} \theta^n(D)$. Then $X = (S \backslash E) \cup E$ and $S \backslash D = (S \backslash E) \cup \theta(E)$. The sets S' and $S' \backslash D'$ can thus be decomposed into $(S' \backslash E') \cup E'$ and $(S' \backslash E') \cup \theta(E')$ as required.

It is not difficult to show from the above that a sphere can be decomposed into a finite number of sets which, when rearranged, make up two spheres of the same volume as that of the original one. We state without proof the more startling result that given any two bounded subsets X and X' of $\mathbb{R}^3$ each having non-empty interior, X can be decomposed into a finite number of sets which are pairwise congruent respectively to the sets in a decomposition of X'.

3.2 Exercises

1. Give a direct proof that Zorn's lemma implies the well-ordering theorem.
2. Give a direct proof that the Hausdörff maximality principle implies the axiom of choice.
3. Show that the set $\mathbb{Q}_o$ is of second category.
4. Show that $\mathbb{Q}_o$ is a subgroup of $\mathbb{R}$.
5. Construct a second category subfield of the reals.
6. Produce a proof of the Hahn-Banach theorem using Zorn's Lemma.
7. Let M be the space of all bounded sequences of real numbers. Show that there is a linear functional L defined on this space so that $\underline{\lim}\, x_n \le L(x) \le \overline{\lim}\, x_n$ when $x = \{x_n\}$.
8. If A is a countable subset of the plane lying on the circumference of the unit circle, show there is an angle θ so that the sets

$$A_n = \{z\colon z = ae^{in\theta} \text{ for some } a \in A\}$$

are pairwise disjoint.

Chapter Four

4.1 Borel Sets and Baire Functions

The open subsets of the line or of a general topological space form a collection of sets which is closed under arbitrary unions and finite intersections, but is not generally closed under countable intersections; the closed subsets accordingly are a collection of sets which is closed under arbitrary intersections and finite unions. For the purposes of measure theory we will be interested in a collection $\mathscr{A}$ of subsets of a set X, with $\emptyset \in \mathscr{A}$ and $X \in \mathscr{A}$ such that $\mathscr{A}$ is closed under countable unions, countable intersections and complements. Such a collection $\mathscr{A}$ is called a σ-<u>algebra</u> of subsets of X. Given any set

X, the set of all subsets of X is a σ-algebra. The intersection of any collection of σ-algebras is readily seen to be a σ-algebra. Thus, if $\mathcal{S}$ is a collection of subsets of X, the intersection of all σ-algebras of subsets of X which contain $\mathcal{S}$ is the smallest σ-algebra containing $\mathcal{S}$. When X is a topological space, the collection $\mathcal{B}$ of Borel sets of X is the smallest σ-algebra containing all open subsets of X. Clearly, since each σ-algebra $\mathcal{A}$ contains the complement of each set in $\mathcal{A}$, the Borel sets are also the smallest σ-algebra containing all the closed subsets of X. The Borel sets are a large collection of subsets determined by the topology of X.

An analogous situation holds for collections of functions. The continuous real-valued functions defined on a space are closed under uniform limits but not, in general, under pointwise limits. For example, $f_n(x) = x^n$ for each $x \in [0,1]$ approaches f where $f(x) = 0$ if $x \in [0,1)$ and $f(1) = 1$. While the limit of the above sequence $\{f_n\}$ exists at every point in $[0,1]$, it is not a continuous function on $[0,1]$.

For purposes of integration theory, we will be interested in a collection of functions which is closed under pointwise limits. Such a collection of functions is sometimes called complete. The collection of all real-valued functions defined on a space X is trivially closed under pointwise limits. Also, the intersection of any number of complete collections of functions is readily seen to be complete. Thus, if τ is a collection of functions, the intersection of all complete collections which contain τ is the smallest complete collection containing τ. The <u>Baire functions</u> on a topological space X are those real-valued functions which belong to the smallest complete collection of real-valued functions containing the continuous functions. In some circumstances the term Baire function is allowed to include the extended real-valued functions in the smallest complete collection containing the continuous functions.

The Borel sets and the Baire functions can also be defined using transfinite induction. The Borel class $\mathcal{G}_0$ is the class $\mathcal{G}$ of all open subsets of X; the Borel class $\mathcal{F}_0$ is the class $\mathcal{F}$ of all closed sets of X. If $\mathcal{S}$ is a collection of sets, $\mathcal{S}_\sigma$ denotes the collection of all sets which are the countable union of a sequence of sets from $\mathcal{S}$; $\mathcal{S}_\delta$ denotes the collection of all sets which are the countable intersection of a sequence of sets from $\mathcal{S}$. The lower level Borel classes are as follows: $\mathcal{G}_1 = \mathcal{G}_\delta$, $\mathcal{F}_1 = \mathcal{F}_\sigma$, $\mathcal{G}_2 = \mathcal{G}_{\delta\sigma} = (\mathcal{G}_\delta)_\sigma$, $\mathcal{F}_2 = \mathcal{F}_{\sigma\delta} = (\mathcal{F}_\sigma)_\delta$, $\mathcal{G}_3 = \mathcal{G}_{\delta\sigma\delta}$, $\mathcal{F}_3 = \mathcal{F}_{\sigma\delta\sigma}$, etc. In general, if α is an even ordinal, $\mathcal{G}_\alpha = (\bigcup_{\beta<\alpha} \mathcal{G}_\beta)_\sigma$ and $\mathcal{F}_\alpha = (\bigcup_{\beta<\alpha} \mathcal{F}_\beta)_\delta$; if α is an odd ordinal $\mathcal{G}_\alpha = (\bigcup_{\beta<\alpha} \mathcal{G}_\beta)_\delta$ and $\mathcal{F}_\alpha = (\bigcup_{\beta<\alpha} \mathcal{F}_\beta)_\sigma$.

This defines the Borel classes for each ordinal number α. Note that if β is a successor ordinal, say $\beta = \alpha + 1$, then, if β is even, $\mathcal{G}_\beta = \mathcal{G}_{\alpha\sigma}$ and $\mathcal{F}_\beta = \mathcal{F}_{\alpha\delta}$; if β is odd, $\mathcal{G}_\beta = \mathcal{G}_{\alpha\delta}$ and $\mathcal{F}_\beta = \mathcal{F}_{\alpha\sigma}$. The following theorems indicate the relationship between the Borel classes and the Borel sets.

(4.1) <u>Theorem</u>. <u>If</u> X <u>is a metric space, for each closed set</u> $F \subset X$ <u>there is a sequence of open sets</u> G_n <u>such that</u> $F = \cap G_n$.

<u>Proof</u>. Given a closed set F, let

$$G_n = \{x \in X: d(x,F) < 1/n\}$$

where $d(x,F) = \inf\{d(x,y): y \in F\}$. To see that each G_n is open, let $y_o \in G_n$. Then there is a point $x \in F$ so that $d(x,y_o) = d < 1/n$. Let $r = 1/n - d$. Then, if $d(y,y_o) < r$ it follows from the triangle inequality that

$$d(y,x) \leq d(y,y_o) + d(y_o,x) < r + d = 1/n.$$

Thus $N_r(y_o) = \{y: d(y,y_o) < r\} \subset \{y: d(y,x) < 1/n\} \subset G_n$. Hence, for each $y_o \in G_n$ there is an open neighborhood of y_o contained in G_n and therefore G_n is an open set. Now, for each $y \in F$, $d(y,F) = 0$ and thus F is contained in each G_n and $F \subset \underset{n}{\cap} G_n$. On the other hand, if $y \in \cap G_n$, for each natural number n there exists $x_n \in F$ such that $d(x_n,y) < 1/n$. Then x_n approaches y and, since each x_n belongs to the closed set F, $y \in F$. Thus $\underset{n}{\cap} G_n \subset F$ and, since $F \subset \underset{n}{\cap} G_n$, $F = \cap G_n$.

(4.2) <u>Corollary</u>. <u>If</u> X <u>is a metric space, for each open set</u> $G \subset X$ <u>there is a sequence of closed sets</u> F_n <u>such that</u> $G = \underset{n}{\cup} F_n$.

<u>Proof</u>. This follows from considering the union of the sequence of complements of those open sets from Theorem 4.1 whose intersection is G^c.

(4.3) Theorem. Given a topological space X and ordinal numbers α and β with $\alpha < \beta$, the class $\mathcal{G}_\alpha$ is contained in $\mathcal{G}_\beta$ and $\mathcal{F}_\alpha$ is contained in $\mathcal{F}_\beta$; moreover, a subset E of X belongs to $\mathcal{G}_\alpha$ if and only if E^c belongs to $\mathcal{F}_\alpha$.

Proof. Suppose $E \in \mathcal{G}_\alpha$ and let $E_n = E$, $n = 1, 2, \ldots$. If β is even, $\mathcal{G}_\beta$ consists of all countable unions of sequences of sets from previous classes. If β is odd, $\mathcal{G}_\beta$ consists of all countable intersections of sequences of sets from previous classes. Trivially, $E = \cup_n E_n = \cap_n E_n$ belongs to $\mathcal{G}_\beta$ and since E is an arbitrary element of $\mathcal{G}_\alpha$, $\mathcal{G}_\alpha \subset \mathcal{G}_\beta$. Similarly, $\mathcal{F}_\alpha \subset \mathcal{F}_\beta$. Since a set is open if and only if its complement is closed, $E \in \mathcal{G}_o = \mathcal{G}$ if and only if $E^c \in \mathcal{F}_o = \mathcal{F}$. Thus the second part of the theorem is true if $\alpha = 0$. Suppose it is true for every $\beta < \alpha$. Suppose α is an even ordinal. Then $E \in \mathcal{G}_\alpha$ iff there are $\beta_n < \alpha$ and $E_n \in \mathcal{G}_{\beta_n}$ such that $E = \bigcup_{n=1}^{\infty} E_n$. By the induction hypothesis, $E_n \in \mathcal{G}_{\beta_n}$ iff $E_n^c \in \mathcal{F}_{\beta_n}$. Since $E^c = \bigcap_{n=1}^{\infty} E_n^c$ and since $E^c \in \mathcal{F}_\alpha$ iff E^c is a countable intersection of sets from previous classes, $E \in \mathcal{G}_\alpha$ iff $E^c \in \mathcal{F}_\alpha$. Suppose α is an odd ordinal. In that case $E \in \mathcal{G}_\alpha$ iff there are $\beta_n < \alpha$ and $E_n \in \mathcal{G}_{\beta_n}$ with $E = \bigcap_{n=1}^{\infty} E_n$. Again by the induction hypothesis $E_n \in \mathcal{G}_{\beta_n}$ iff $E_n^c \in \mathcal{F}_{\beta_n}$. Likewise $E^c = \bigcup_{n=1}^{\infty} E_n^c$ and $E^c \in \mathcal{F}_\alpha$ iff E is a countable union of sets from previous classes; that is $E \in \mathcal{G}_\alpha$ iff $E^c \in \mathcal{F}_\alpha$. By induction, $E \in \mathcal{G}_\alpha$ iff $E^c \in \mathcal{F}_\alpha$ holds for each ordinal number α.

(4.4) Theorem. Given a metric space X, for each ordinal number α, the class $\mathcal{G}_\alpha$ is contained in $\mathcal{F}_{\alpha+1}$ and the class $\mathcal{F}_\alpha$ is contained in $\mathcal{G}_{\alpha+1}$; Moreover, the Borel sets $\mathcal{B}$ are identical with $\bigcup_{\alpha<\omega_1} \mathcal{G}_\alpha$ which is also $\bigcup_{\alpha<\omega_1} \mathcal{F}_\alpha$.

Proof. By Theorem 4.1 and its corollary, the class $\mathcal{G}_0 = \mathcal{G} \subset \mathcal{F}_\sigma = \mathcal{F}_1$ and $\mathcal{F}_0 = \mathcal{F} \subset \mathcal{G}_\delta = \mathcal{G}_1$. Thus the first part of this theorem is true for $\alpha = 0$. Suppose it is true for each $\beta < \alpha$. First, suppose that α is odd. Then, if $E \in \mathcal{G}_\alpha$, there are $\beta_n < \alpha$ and $E_n \in \mathcal{G}_{\beta_n}$ such that $E = \bigcap_{n=1}^{\infty} E_n$. By the induction hypothesis $E_{\beta_n} \in \mathcal{F}_{\beta_n+1}$. Each $\beta_n + 1 < \alpha + 1$ and $\alpha + 1$ is even and thus $\mathcal{F}_{\alpha+1}$ consists of intersections of sequences of sets from classes previous to $\mathcal{F}_{\alpha+1}$; that is, $E = \bigcap_{n=1}^{\infty} E_n$ belongs to $\mathcal{F}_{\alpha+1}$. Similarly, if $H \in \mathcal{F}_\alpha$, with α odd, there are $\gamma_n < \alpha$ and sets $H_n \in \mathcal{F}_{\gamma_n}$ such that $H = \bigcup_{n=1}^{\infty} H_n$. But $\alpha + 1$ is even and $\mathcal{G}_{\alpha+1}$ consists of all unions of sequences of sets from classes previous to $\mathcal{G}_{\alpha+1}$. By the induction hypothesis each H_n belongs to $\mathcal{G}_{\gamma_n+1}$. Since $\gamma_n + 1 < \alpha + 1$, it follows that $H = \cup H_n$ belongs to $\mathcal{G}_{\alpha+1}$. On the other hand, suppose α is even. Then, if $E \in \mathcal{G}_\alpha$, there are $\beta_n < \alpha$ and $E_n \in \mathcal{G}_{\beta_n}$ such that $E = \cup E_n$. Again by the induction hypothesis each $E_{\beta_n} \in \mathcal{F}_{\beta_n+1}$, each $\beta_n + 1 < \alpha + 1$ and $\alpha + 1$ is odd so that $\bigcup_{n=1}^{\infty} E_n \in \mathcal{F}_{\alpha+1}$. Likewise, if $H \in \mathcal{F}_\alpha$ with α even, there are $\gamma_n < \alpha$ and sets $H_n \in \mathcal{F}_{\gamma_n}$ such that $H = \bigcap_{n=1}^{\infty} H_n$. But $\alpha + 1$ is odd and by the induction hypothesis $H_n \in \mathcal{G}_{\gamma_n+1}$ so that $\bigcap_{n=1}^{\infty} H_n \in \mathcal{G}_{\alpha+1}$ because each $\gamma_n + 1$ is less than $\alpha + 1$. It follows that $\mathcal{G}_\alpha \subset \mathcal{F}_{\alpha+1}$ and $\mathcal{F}_\alpha \subset \mathcal{G}_{\alpha+1}$ and by transfinite induction this is true for each ordinal number α. To prove the second part of the theorem, first note that $\mathcal{G}_0 = \mathcal{G} \subset \mathcal{B}$ and $\mathcal{F}_0 = \mathcal{F} \subset \mathcal{B}$ and, if for every $\beta < \alpha < \omega_1$, $\mathcal{G}_\beta \subset \mathcal{B}$ and $\mathcal{F}_\beta \subset \mathcal{B}$, then each set in $\mathcal{G}_\alpha$ or $\mathcal{F}_\alpha$ is formed by countable unions or countable intersections of sets from previous classes and each set in $\mathcal{G}_\alpha$ or in $\mathcal{F}_\alpha$ belongs to $\mathcal{B}$. Since this is true for each $\alpha < \omega_1$, it follows that $\bigcup_{\alpha<\omega_1} \mathcal{G}_\alpha$ and $\bigcup_{\alpha<\omega_1} \mathcal{F}_\alpha$ are contained in $\mathcal{B}$. To show that $\mathcal{B} = \bigcup_{\alpha<\omega_1} \mathcal{G}_\alpha = \bigcup_{\alpha<\omega_1} \mathcal{F}_\alpha$, it suffices to show the union of the Borel classes forms a σ-algebra. Clearly, $\emptyset$ and X are in $\mathcal{G}$ and thus belong to $\bigcup_{\alpha<\omega_1} \mathcal{G}_\alpha$. If $E \in \mathcal{G}_\alpha$, by the previous theorem, $E^C \in \mathcal{F}_\alpha$ and by the above $E^C \in \mathcal{G}_{\alpha+1}$. Thus the union of the Borel classes is closed with respect to complements. To show $\bigcup_{\alpha<\omega_1} \mathcal{G}_\alpha$ is a σ-algebra let $E_n \in \bigcup_{\alpha<\omega_1} \mathcal{G}_\alpha$, then there are α_n such that for each n, $E_n \in \mathcal{G}_{\alpha_n}$. Recall that there is an ordinal number $\alpha_o < \omega_1$ so that α_o is larger than each α_n. Then each E_n belongs to $\mathcal{G}_{\alpha_o}$. Consequently, both $\bigcup_{n=1}^{\infty} E_n$ and $\bigcap_{n=1}^{\infty} E_n$ belong to $\mathcal{G}_{\alpha_o+2}$. Thus $\bigcup_{\alpha<\omega_1} \mathcal{G}_\alpha$ is a σ-algebra and, since B is the smallest σ-algebra containing $\mathcal{G}$, $\bigcup_{\alpha<\omega_1} \mathcal{G}_\alpha = \mathcal{B}$. The identical argument with "$\mathcal{F}$" and "$\mathcal{G}$" interchanged shows that $\bigcup_{\alpha<\omega_1} \mathcal{F}_\alpha$ is a σ-algebra.

Note that the construction of the Borel sets by transfinite induction can be accomplished in a general topological space by letting $\mathcal{B}_o = \mathcal{F} \cup \mathcal{G}$ and $\mathcal{B}_\alpha = (\bigcup_{\beta<\alpha} \mathcal{B}_\beta)_\sigma$ if α is even, and $\mathcal{B}_\alpha = (\bigcup_{\beta<\alpha} \mathcal{B}_\beta)_\delta$ if α is odd; then the class of Borel sets consists of $\bigcup_{\alpha<\omega_1} \mathcal{B}_\alpha$.

On the line or in Euclidean n-space, there are c open sets. Thus $\|\mathcal{G}_o\|_c = c$. Suppose that for each $\beta < \alpha < \omega_1$, $\|\mathcal{G}_\beta\|_c = c$. Since α is a countable ordinal, there at most $c \cdot a$ sets in $\bigcup_{\beta<\alpha} \mathcal{G}_\beta$ and at most $(c \cdot a)^a$ sequences of such sets. By induction, it follows that for each $\alpha < \omega_1$, $c \le \|\mathcal{G}_\alpha\|_c \le (c \cdot a)^a = c$. But then $c \le \|\mathcal{B}\|_c = \|\bigcup_{\alpha<\omega_1} \mathcal{G}_\alpha\|_c \le \aleph_1 \cdot c = c$. Thus there are c Borel subsets of these spaces. A similar argument applies to every infinite separable metric space or, indeed, to any infinite space in which points are closed and there are c open sets in the topology. In such a space, since points are closed, every countable set is an $\mathcal{F}_\sigma$. Since the set of subsets of a countable set has cardinality c, there are at least c Borel sets in such a space. That there are at most c follows from an argument identical to the argument given above. Note that there are 2^c subsets of the line or Euclidean n-space and only c Borel sets. Thus there are 2^c subsets of these spaces which are not Borel sets.

It is worth considering some simple examples of Borel subsets of the line. Every countable subset of the line is an $\mathcal{F}_\sigma$. Thus, the set of rational numbers is an $\mathcal{F}_\sigma$. But the set of rational numbers is not a $\mathcal{G}_\delta$. For if it were, the set of irrational numbers would be an $\mathcal{F}_\sigma$. The closed sets forming this $\mathcal{F}_\sigma$ would necessarily be nowhere dense. Then the line could be written as a countable union of nowhere dense sets which is contrary to the Baire Category Theorem. By a similar argument, using the fact that a perfect set P is a complete metric space, any countable dense subset of a perfect set P is an $\mathcal{F}_\sigma$ but is not a $\mathcal{G}_\delta$. The set of irrational numbers is thus a $\mathcal{G}_\delta$ but not an $\mathcal{F}_\sigma$. The set

$$\{x: \text{if } x \geq 0, \ x \text{ is rational}; \\ \text{if } x < 0, \ x \text{ is irrational}\}$$

is neither an $\mathcal{F}_\sigma$ nor a $\mathcal{G}_\delta$. It is, however, both an $\mathcal{F}_{\sigma\delta}$ and a $\mathcal{G}_{\delta\sigma}$.

One should note further that each of the Borel classes is closed under both finite unions and finite intersections; the sigma classes are also closed under countable unions and the delta classes are closed under countable intersections.

The lower level Borel sets are of particular importance because of the following reason: the quantifiers $\forall$ "for every", and $\exists$ "there exists" have a direct relationship to $\cap$ "intersection", and $\cup$ "union". Specifically, if $\{E_n\}$ is a sequence of sets, then

$$x \in \cap E_n \text{ if and only if } \forall n\ (x \in E_n)$$
$$x \in \cup E_n \text{ if and only if } \exists n\ (x \in E_n).$$

Because of this, it is often possible to take a logical statement about a particular set and translate the statement so that the set can be written using unions and intersections of other sets. Since most logical descriptions involve only a few quantifiers and can frequently be written so that the quantification is over countable indices, the sets where a certain property holds can frequently be shown to be Borel sets. For example, in Chapter 2 it is shown that the set of points of discontinuity of an arbitrary function is a countable union of closed sets, that is, is an $\mathcal{F}_\sigma$. This is accomplished by noting that given a function f the set $E_n = \{x: \omega_f(x) \geq 1/n\}$ is a closed set. Then x is a point of discontinuity of f if and only if there is an n so that $x \in E_n$. In terms of set operations, x is a point of discontinuity of f if and only if $x \in \cup E_n$. The determination of the Borel class of certain collections of sets is referred to as <u>Borel classification</u>.

An important example of the relationship between logic and set operations is as follows: If $\{E_n\}$ is a sequence of sets, the *limit supremum* of the sequence $\{E_n\}$ ("lim sup of E_n") is

$$\overline{\lim}\, E = \{x\colon x \text{ belongs to infinitely many } E_n\}$$

Thus, $x \in \overline{\lim}\, E_n$ iff $\forall n\ \exists k > n\ (x \in E_k)$ which holds iff $x \in \bigcap_n \bigcup_{k>n} E_k$. Consequently, $\overline{\lim}\, E_n = \bigcap_n \bigcup_{k>n} E_k$. The correspondence between $\forall$ and $\cap$, and that between $\exists$ and $\cup$ should be noted. Similarly, the *limit infimum* of the sequence $\{E_n\}$ ("lim inf of E_n") is:

$$\underline{\lim}\, E_n = \{x\colon x \text{ belongs to all but finitely many } E_n\}.$$

Thus, $x \in \underline{\lim}\, E_n$ iff $\exists n\ \forall k > n\ (x \in E_k)$ iff $x \in \bigcup_n \bigcap_{k>n} E_k$. Consequently, $\underline{\lim}\, E_n = \bigcup_n \bigcap_{k>n} E_k$. The *limit of a sequence of sets*, written "$\lim E_n$", is the set $\underline{\lim}\, E_n$ provided $\underline{\lim}\, E_n = \overline{\lim}\, E_n$. Note that $\underline{\lim}\, E_n$ is always contained in $\overline{\lim}\, E_n$.

A sequence of sets is said to be *non-decreasing* provided $E_n \subset E_{n+1}$. It is said to be *non-increasing* if $E_{n+1} \subset E_n$. It is said to be *monotone* if it is either non-decreasing or non-increasing. It is easy to see that monotone sequences always have limits and that if $\{E_n\}$ is non-decreasing, $\lim E_n = \cup E_n$. If $\{E_n\}$ is non-increasing, $\lim E_n = \cap E_n$. Note that $E_n = [1 - 1/n,\ 2 - 1/n]$ has $[1,2)$ as its limit, although the sequence is not monotone. Further note that when $\lim E_n$ exists it can be obtained as either $\bigcup_n \bigcap_{k>n} E_k$ or $\bigcap_n \bigcup_{k>n} E_k$; from this it is easily shown that a limit of a sequence of sets from a given Borel class always belongs to the next largest Borel class. Moreover, since each of the Borel classes is closed under both finite intersections and finite unions, it is always possible to write each element in a sigma class as the increasing union of sets in lower classes; for example, if

$$E = \bigcup_{n=1}^{\infty} E_n, \quad E = \bigcup_{n=1}^{\infty} \left(\bigcup_{k=1}^{n} E_k \right).$$

Likewise, each element of a delta class can be written as a decreasing intersection of sets from lower classes. Thus, the sets in Borel class α are always limits of sequences of sets from lower classes. Also, the lim sup and lim inf of a sequence of sets from a Borel class with index α always belongs to the Borel class with index at most $\alpha + 2$.

The Baire classes of functions on a space X are defined by transfinite induction. The Baire class 0 is defined to consist of the class of continuous functions. Suppose that, for each $\beta < \alpha < \omega_1$, the functions in Baire class β have been defined. Then a function f belongs to Baire class α provided there are $\beta_n < \alpha$ and f_n of Baire class β_n such that at each $x \in X$, $f(x) = \lim f_n(x)$. Such an f is said to be a Baire_α function or to belong to Baire_α. The Baire functions are then equal to $\bigcup_{\alpha<\omega_1} \text{Baire}_\alpha$. To see this, recall that the set of all Baire functions is the smallest complete collection of functions containing the continuous functions. Thus, the class Baire_o is contained in the collection of Baire functions and, by induction, if for every $\beta < \alpha$ the class Baire_β is contained in the set of all Baire functions, then the Baire_α functions are limits of Baire functions and are contained in the collection of all Baire functions. To show that $\bigcup_{\alpha<\omega_1} \text{Baire}_\alpha$ is identical to the collection of all Baire functions, it remains to show that $\bigcup_{\alpha<\omega_1} \text{Baire}_\alpha$ is a complete collection of functions. But, if f is a pointwise limit of functions from $\bigcup_{\alpha<\omega_1} \text{Baire}_\alpha$, there are ordinal numbers α_n and $f_n \in \text{Baire}_{\alpha_n}$ such that at each $x \in X$, $\lim f_n(x) = f(x)$. But there is $\alpha_o < \omega_1$ such that $\alpha_n < \alpha_o$ for each natural number n. Thus $f \in \text{Baire}_{\alpha_o}$. Since f was an arbitrary function among the functions which are pointwise limits of sequences of functions from $\bigcup_{\alpha<\omega_1} \text{Baire}_\alpha$, it follows that $\bigcup_{\alpha<\omega_1} \text{Baire}_\alpha$ is a complete class of functions.

If f and g are two functions in Baire class α and h is a continuous function, then $f + g$, $f - g$, $c \cdot f$, $f \cdot g$, $h \circ f$, $|f|$, $\max(f,g)$ and $\min(f,g)$ are all in Baire class α. This is true for $\alpha = 0$. Suppose it is true for each $\beta < \alpha$ and that $f = \lim f_n$, $g = \lim g_n$ with each f_n and each g_n in Baire classes preceeding Baire_α. Then $f \pm g = \lim(f_n \pm g_n)$, $c \cdot f = \lim c \cdot f_n$, $f \cdot g = \lim f_n \cdot g_n$ and $h \circ f = \lim h \circ f_n$ each belong to Baire_α. If $h(x) = |x|$ it follows that $|f|$ belongs to Baire class α. Since

$$\max(f,g) = \frac{f + g}{2} + \frac{|f - g|}{2}$$

and

$$\min(f,g) = \frac{f+g}{2} - \frac{|f-g|}{2},$$

it also follows that $\max(f,g)$ and $\min(f,g)$ are in Baire class α.

If $\{f_n\}$ is a sequence of functions each in a class preceeding Baire class α, then $\sup f_n$, and $\inf f_n$ are in Baire class α and $\overline{\lim}\, f_n$, and $\underline{\lim}\, f_n$ are in Baire class $\alpha + 1$. Here, if necessary, it is supposed that extended real-valued functions are allowed in the Baire classes. Then $\sup f_n = \lim g_n$ where $g_1(x) = f_1(x)$ and
$g_n(x) = \max(f_1(x), \ldots, f_n(x)) = \max(g_{n-1}(x), f_n(x))$. Similarly, $\inf f_n(x) = \lim h_n(x)$ where $h_n(x) = \min(f_1(x), \ldots, f_n(x))$. Also

$$\overline{\lim}\, f_n(x) = \lim_n \sup_k f_{n+k}(x)$$

and

$$\underline{\lim}\, f_n(x) = \lim_n \inf_k f_{n+k}(x).$$

As with the Borel classes, in the study of real-valued functions of a real variable, the lower level Baire classes occur frequently. An important example is the following:

if $f(x)$ is a derivative,

then $f(x)$ belongs to Baire class 1.

This is because if $f(x) = F'(x)$, then

$$f(x) = \lim_{n\to\infty} n(F(x + 1/n) - F(x))$$

and each of the functions $n(F(x + 1/n) - F(x))$ is continuous. If the derivative of F exists in the extended sense, it belongs to the extended real-valued functions of Baire class 1. Another important example is that of the semi-continuous functions. A function is said to be <u>lower semi-continuous at a point</u> x provided $\underline{\lim}_{t \to x} f(t) \geq f(x)$; it is said to be <u>lower semi-continuous</u> if it is lower semi-continuous at each point. A function is said to be <u>upper semi-continuous at a point</u> x if $\overline{\lim}_{t \to x} f(t) \leq f(x)$; it is <u>upper semi-continuous</u> if upper semi-continuous at each point.

Several observations about the semi-continuous functions are in order. First, a function is continuous at a point x if and only if it is both upper and lower semi-continuous at x. Second, a function f is lower semi-continuous if and only if, for each real number a, $f^{-1}((a,\infty))$ is open; equivalently, each $f^{-1}((-\infty,a])$ is closed; the function f is upper semi-continuous if and only if $-f$ is lower semi-continuous which holds if and only if each $f^{-1}((-\infty,a))$ is open; equivalently, each $f^{-1}([a,\infty))$ is closed. If E is a set, the function which equals 1 if $x \in E$ and 0 if $x \notin E$ is denoted by $C_E(x)$ and is called the <u>characteristic function of</u> E. Note that, if G is an open set, $C_G(x)$ is lower semi-continuous and, if F is a closed set, $C_F(x)$ is upper semi-continuous. Note further that each finite valued lower semi-continuous function defined on a closed interval, (or on a compact space) is bounded below and each finite valued upper semi-continuous function defined on such a space is bounded above. Finally, it is easily shown that the finite sum or the limit of an increasing sequence of lower semi-continuous functions is lower semi-continuous; the finite sum or limit of a decreasing sequence of upper semi-continuous functions is upper semi-continuous.

In particular, we note that the pointwise limit of an increasing sequence of continuous functions is bounded below and lower semi-continuous. That the converse is true for finite valued functions which are lower semi-continuous and bounded below is the content of the next theorem. The proof is given for functions of a real variable although, by replacing $|x - y|$ with $d(x,y)$, it is readily seen to hold for metric spaces. It then follows that each finite valued lower semi-continuous function defined on $[a,b]$ is Baire 1. Moreover, the proof shows how to construct an increasing sequence of continuous functions converging to a given lower semi-continuous one. Of particular interest is the case of $C_G(x)$ where G is open and $C_F(x)$ where F is closed; the theorem implies that each of these functions is in Baire class 1. Indeed, for G open,

$$f_n(x) = \min\left(n\cdot\text{dist}(G^c,x),\ 1\right)$$

is the sequence of functions converging to $C_G(x)$ which the proof specifies. Without loss of generality, it is assumed in the statement of the theorem that the lower semi-continuous function is greater than or equal to 0.

(4.5) <u>Theorem</u>. <u>Let</u> $g(x) > 0$ <u>be a lower semi-continuous function on</u> $[a,b]$. <u>Then there is a non-decreasing sequence of continuous functions</u> $\{f_n\}$ <u>on</u> $[a,b]$ <u>such that</u> $\lim f_n(x) = g(x)$ <u>at every</u> $x \in [a,b]$.

<u>Proof</u>. Let $f_n(x) = \inf_y\{g(y) + n|x - y|\}$. By putting $y = x$, it follows that $f_n(x) \le g(x)$ and since $g(x)$

≥ 0, $f_n(x) \geq 0$. Given x and x', $g(y) + n|x - y| \leq g(y) + n|x' - y| + n|x - x'|$. By taking the infimum with respect to y, it follows that $f_n(x) \leq f_n(x') + n|x - x'|$. By interchanging x and x', $f_n(x') \leq f_n(x) + n|x - x'|$. Hence $|f_n(x) - f_n(x')| \leq n|x - x'|$ and each f_n is continuous. Clearly, if $n < m$, $f_n(x) \leq f_m(x)$ and thus the f_n form a non-decreasing sequence of functions. Thus, at each x, $\lim f_n(x) \leq g(x)$. Fix x. Given $\varepsilon > 0$, choose a sequence of points $\{x_n\}$ so that $f_n(x) > g(x_n) + n|x - x_n| - \varepsilon$. Since $g(x) \geq f_n(x)$ and $g(x_n) \geq 0$, it follows that $g(x) > n|x - x_n| - \varepsilon$. Thus $|x_n - x| \to 0$ as $n \to \infty$. Since g is lower semi-continuous, there is an N so that for $n > N$, $g(x_n) > g(x) - \varepsilon$. Thus, for $n > N$,

$$f_n(x) > g(x_n) + n|x - x_n| - \varepsilon > g(x) - 2\varepsilon.$$

It follows that $\lim f_n(x) \geq g(x)$ and hence that $g(x) = \lim f_n(x)$.

The following theorem concerning the characteristic functions of Borel sets will be needed to delineate the connection between Borel sets and Baire functions.

(4.6) <u>Theorem</u>. <u>For each ordinal number</u> $\alpha < \omega_1$, <u>if</u> E <u>is a Borel set in class</u> $\mathscr{G}_\alpha$ <u>or</u> $\mathscr{F}_\alpha$, <u>then</u> $C_E(x)$ <u>is a Baire function in the Baire class with subscript no more than</u> $\alpha + 1$.

<u>Proof</u>. We have already proved that, when $G \in \mathscr{G}_0 = \mathscr{G}$ or $F \in \mathscr{F}_0 = \mathscr{F}$, then $C_G(x)$ and $C_F(x)$ are Baire 1 functions. Suppose that for each $\beta < \alpha < \omega_1$ the statement of the theorem is true for $\beta < \alpha$. Suppose

E is in the α class which is also a σ-class. Then there are are sets E_n in classes with index β_n so that $\beta_n < \alpha$ and $E = \bigcup_{n=1}^{\infty} E_n$. Then $E = \bigcup_{n=1}^{\infty} E'_n$, where $E'_N = \bigcup_{n=1}^{N} E_n$. Moreover, $\{E'_N\}$ is a non-decreasing sequence of sets and thus $C_E(x) = \lim C_{E'_n}(x)$. Since E'_n belongs to the Borel class with index no more than $\gamma_n = \max(\beta_1, \beta_2, \ldots, \beta_n)$, it follows that $C_{E'_n}$ belongs to Baire class $\gamma_n + 1$ and $C_E(x)$ belongs to Baire class $\alpha + 1$. Similarly, if E is in the α-class which is also a δ-class, there are ordinals $\beta_n < \alpha$ and sets E_n in Borel classes with index β_n and $E = \bigcap_{n=1}^{\infty} E_n$. But, if $E''_N = \bigcap_{n=1}^{N} E_n$, $\{E''_n\}$ is a non-increasing sequence of sets and $C_E(x) = \lim C_{E_n}(x)$. Since E''_N belongs to the Borel class with index no more than $\max(\beta_1, \beta_2, \ldots, \beta_n)$, it again follows that $C_E(x)$ belongs to the Baire class with index no more than $\alpha + 1$. This being true for each $\alpha < \omega_1$, the theorem is proved.

It follows from this theorem that, if f is a real-valued function whose range is finite and if each $f^{-1}(y)$ is a Borel set, then f is a Baire function; indeed, $f(x) = \sum_{i=1}^{n} y_i \cdot C_{E_i}(x)$ where $E_i = f^{-1}(y_i)$ and $\{y_i\}_{i=1}^{n}$ is the range of f. This being so, we are in a position to show that a necessary and sufficient condition for a function f to be a Baire function is that the inverse image under f of each open set is a Borel set.

We first observe a few facts about inverse images of sets. If f is a function taking a set X into a set Y and $\{E_\alpha\}_{\alpha\in I}$ is a collection of subsets of Y, then

$$f^{-1}\left(\bigcup_{\alpha\in I} E_\alpha\right) = \bigcup_{\alpha\in I} f^{-1}(E_\alpha)$$

and

$$f^{-1}\left(\bigcap_{\alpha\in I} E_\alpha\right) = \bigcap_{\alpha\in I} f^{-1}(E_\alpha).$$

This is easily verified by showing that a point x belonging to the left side of each set equality also belongs to the right side and vice versa. Moreover, if $E \subset Y$, $f^{-1}(E^c) = f^{-1}(E)^c$. Thus, inverse images behave nicely with respect to unions, intersections and complements. Incidentally, images of sets under a function do not behave as nicely. While $\cup f(E_\alpha) = f(\cup E_\alpha)$ for collections of sets E_α contained in the domain of f, $f(A \cap B) \subset f(A) \cap f(B)$ and easily obtained examples show that equality need not hold; also, simple examples show that no inclusion relationships need hold between $f(A^c)$ and $f(A)^c$.

We first show that, if the inverse image under a function of each open set is Borel, then the function is a Baire function. Suppose that f is given and that for each open set G, $f^{-1}(G)$ is a Borel set. For each integer m and natural number n, let $E_{n,m} = f^{-1}([\frac{m}{n}, \frac{m+1}{n}))$. Then

$$E_{n,m} = \bigcap_k f^{-1}\left(\left(\frac{m}{n} - \frac{1}{k}, \frac{m+1}{n}\right)\right)$$

and hence each $E_{n,m}$ is a Borel set. Let $f_n(x) = \sum_{m=-n^2}^{n^2} \frac{m}{n} \cdot C_{E_{n,m}}(x)$. Then $f(x) = \lim_n f_n(x)$. This is because, if $|f(x)| < n$, there is an integer m

between $-n^2$ and n^2 so that $f(x) \in [\frac{m}{n}, \frac{m+1}{n})$ and since $f_n(x) = \frac{m}{n}$, $|f_n(x) - f(x)| < 1/n$. It follows that $f_n(x)$ approaches $f(x)$. Since each f_n has a finite range and is the sum of characteristic functions of Borel sets, each f_n is a Baire function. Thus, $f = \lim f_n$ is a Baire function.

Now, suppose that f is a Baire function from Baire class α. If $\alpha = 0$, f is continuous and then $f^{-1}(G)$ is open for each open set G. We proceed by transfinite induction. Suppose that, for every $\beta < \alpha$, g in Baire class β and open set G, $g^{-1}(G)$ is a Borel set. Since f is in Baire class α, $f = \lim f_n$ where each f_n belongs to a Baire class prior to Baire class α.

We first show that for any real number a, $f^{-1}((a,\infty))$ is a Borel set. Note that $x \in f^{-1}((a,\infty))$ iff $\exists N$ such that $\forall n > N$, $x \in f_n^{-1}((a,\infty))$. That is, $x \in f^{-1}((a,\infty))$ iff $x \in \underline{\lim}\, f_n^{-1}((a,\infty))$. Therefore

$$f^{-1}((a,\infty)) = \bigcup_N \bigcap_{n>N} f_n^{-1}((a,\infty))$$

and, since each $f_n^{-1}((a,\infty))$ is a Borel set, $f^{-1}((a,\infty))$ is a Borel set. A similar argument yields the fact that $f^{-1}((-\infty,b))$ is a Borel set. Thus, for each open interval (a,b), $f^{-1}((a,b)) = f^{-1}((-\infty,b)) \cap f^{-1}((a,\infty))$ is a Borel set. Since each open set G is the countable union of open intervals, say, $G = \bigcup_i (a_i,b_i)$, it follows that $f^{-1}(G) = \bigcup_i f^{-1}((a_i,b_i))$ is a Borel set. It is then true that, if $f \in \text{Baire}_\alpha$ and G is open, $f^{-1}(G)$ is a Borel set. By induction this is true for each $\alpha < \omega_1$; that is, for each Baire function. Thus f is a Baire function iff the inverse image under f of each open set is a Borel set.

Because of the behavior of inverse images of unions and of intersections, it is easily seen by induction that f is a Baire function iff the inverse image under f of each Borel set is always a Borel set. From this it also follows that the composition of two Baire functions of a real variable is also a Baire function. Actually, more precise statements about Baire functions hold true. Specifically, f is in Baire class α iff the inverse image of each open set is a Borel set in the σ-class with index α. In addition, if f is a function of a real variable in Baire class α and g is a function in Baire class β, then $f \circ g$ belongs to class $\beta + \alpha$. Proofs of these last two statements require more careful consideration of the logic of the limit process. These proofs are included in the next section of this chapter.

Finally, note that on the line or in Euclidean n-space there are c continuous real-valued functions. Thus $\|Baire_0\|_c = c$. Suppose that for each $\beta < \alpha < \omega_1$, $\|Baire_\beta\|_c = c$. Since α is a countable ordinal, there are at most $c \cdot a$ functions in $\bigcup_{\beta<\alpha} Baire_\beta$ and at most $(c \cdot a)^a$ sequences of such functions. By induction, it follows that for each $\alpha < \omega_1$, $c \le \|Baire_\alpha\|_c \le (c \cdot a)^a = c$. But then, the cardinality of the class of Baire functions is $\|\bigcup_{\alpha<\omega_1} Baire_\alpha\|_c \le \aleph_1 \cdot c = c$. Thus there are c Baire functions on these spaces. Since there are 2^c real-valued functions, and only c

Baire functions, there are 2^c functions which are not Baire functions.

We conclude this section with the following theorem concerning the set of points of continuity of a Baire 1 function. This result is improved in the next section. It will be needed for the forthcoming discussion of measurable functions.

(4.7) Theorem. *If f is a Baire 1 function defined on $[a,b]$ then the set of points of continuity of f is dense in $[a,b]$ and hence this set, belonging to the class $\mathcal{G}_\delta$, is a residual subset of $[a,b]$.*

Proof. Suppose that $f(x) = \lim f_n(x)$ at each $x \in [a,b]$ where each function f_n is continuous on $[a,b]$. Let $I_o \subset [a,b]$ and let $\{\varepsilon_j\}$ be a sequence of positive numbers with $\lim \varepsilon_j = 0$. For each ordered pair (k,n) of natural numbers, let

$$A_{k,n} = \{x \in I_o: |f_k(x) - f_n(x)| \leq \varepsilon_1\}.$$

Note that each set $A_{k,n}$ is closed and hence each $B_k = \bigcap_{n=k}^{\infty} A_{k,n}$ is a closed set. Also $I_o = \bigcup_{k=1}^{\infty} B_k$ because for each $x \in I_o$ there is a natural number k so that if $n \geq k$ then $|f_k(x) - f_n(x)| \leq \varepsilon_1$, and then $x \in B_k$. By the Baire category theorem, there is a natural number K and a closed interval $I_1 \subset \mathring{I}_o$ so that B_K is dense in I_1; since B_K is closed, B_K contains I_1. Moreover, I_1 can be chosen to be so small that for $x, x' \in I_1$, $|f_K(x) - f_K(x')| \leq \varepsilon_1$. Observe that for all $x \in I_1$, and $n \geq K$, $|f_K(x) - f_n(x)| \leq \varepsilon_1$ and, letting n approach ∞, one also has

$$|f_K(x) - f(x)| \leq \varepsilon_1$$

for all $x \in I_1$. But then, for $x, x' \in I_1$, $|f_K(x) - f(x)| \le \varepsilon_1$, $|f_K(x) - f_K(x')| \le \varepsilon_1$ and $|f_K(x') - f(x')| \le \varepsilon_1$ so that $|f(x) - f(x')| \le 3\varepsilon_1$ for all $x, x' \in I_1$. Now, if $I_1, I_2, \ldots, I_j$ have been chosen with $I_{i+1} \subset \mathring{I}_i$ and $|f(x) - f(x')| \le 3\varepsilon_i$ for all $x, x' \in I_{i+1}$, $i = 1, 2, \ldots, j$, then I_{j+1} can be chosen by the above argument so that $I_{j+1} \subset \mathring{I}_j$ and $|f(x) - f(x')| \le 3\varepsilon_{j+1}$ for all $x, x' \in I_{j+1}$. Since $\{I_j\}$ so chosen is a non-increasing sequence of closed intervals, there is a point $x_o \in \bigcap_{j=1}^{\infty} I_j$. Clearly, f is continuous at x_o because given $\varepsilon > 0$ there is $\varepsilon_j < \varepsilon$ and for each $x' \in \mathring{I}_j$,

$$|f(x_o) - f(x')| \le 3\varepsilon_j < 3\varepsilon.$$

Since I_o, an arbitrary subinterval of $[a,b]$, contains a point of continuity of f, the set of points of continuity of f is dense. However, the set of points of continuity of an arbitrary function has been shown to be a $\mathcal{G}_\delta$ set and the points of discontinuity to be an $\mathcal{F}_\sigma$ set. Each of the closed sets making up this $\mathcal{F}_\sigma$ set for f must be nowhere dense and hence the set of points of continuity of a Baire 1 function defined on $[a,b]$ is a residual subset of $[a,b]$.

4.1 Exercises

1. Show that the collection of all finite unions of intervals which are half open on the left forms an algebra of sets.

2. Show that the sets which are both $\mathcal{F}_\sigma$ sets and $\mathcal{G}_\delta$ sets are an algebra.
3. Show that the rational numbers of the form $m/2^n$ are not a $\mathcal{G}_\delta$ subset of $\mathbb{R}$.
4. Show that the set M of all midpoints of component intervals of the complement of the Cantor set is a $\mathcal{G}_\delta$ subset of $\mathbb{R}$.
5. Show that the right hand endpoints of the component intervals of the complement of the Cantor set do not form a $\mathcal{G}_\delta$ subset of $\mathbb{R}$.
6. Show that $C_M(x)$ is a Baire 1 function when M is the set described in Exercise 4.
7. Show that the smallest algebra containing all open sets and all closed sets is properly contained in the sets belonging to $\mathcal{F}_\sigma \cap \mathcal{G}_\delta$.
8. Show that every monotone function is in Baire class 1.
9. If f is continuous on $\mathbb{R}$ and G is open, show that $f(G)$ need not be a $\mathcal{G}_\delta$ set.
10. If f is continuous on $\mathbb{R}$ and E is an $\mathcal{F}_\sigma$, show that $f(E)$ is an $\mathcal{F}_\sigma$.
11. Given any real-valued function f defined on $\mathbb{R}$, show that $\overline{\lim_{t \to x}} f(t)$ is an upper semi-continuous function.
12. A function f taking a topological space X into a space Y is said to be a <u>Borel function</u> if $f^{-1}(G)$ is a Borel set for each open set G in Y. If Y is a metric space, show that the uniform limit of Borel functions is Borel.
13. Show that the only Baire functions from $\mathbb{R} \to \{0,1\}$ are the constant functions. (Thus the Borel

function, defined in the previous exercise, and the Baire function need not coincide.)

4.2 Exact Baire Classes

The determination of the Borel class of a given set or collection of sets is called Borel classification. This term (or sometimes Baire classification) is also used to refer to the determination of the ordinal number which gives the Baire class of a given function or type of functions. For example, the collection of derivatives is contained in Baire class 1; lower semi-continuous and upper semi-continuous functions were also shown to be in Baire class 1. Since such functions need not be continuous, they are not contained in a lower Baire class.

We will be concerned in this section with showing the exact relationship between Baire class α and the Borel sets, giving several characterizations of Baire 1 functions and showing that there are Baire functions and Borel sets in each class α with $\alpha < \omega_1$ which do not belong to previous classes.

We begin by first noting that the argument given in Theorem 4.7 can be applied equally well to any complete metric space; that is, if X is a complete metric space and f is a Baire 1 function defined on X, then the set of points of continuity of f is dense in X. Indeed, the proof of Theorem 2.6 shows that the set of points of discontinuity of a real-valued function defined on a metric space X is always an $\mathcal{F}_\sigma$ and, consequently, if X is a complete metric space, then the set of points of discontinuity of each Baire 1 function must always be a first category subset of X and the points of continuity must be dense.

Suppose that f is a Baire 1 function defined on $[a,b]$. Then each non-empty closed subset $F \subset [a,b]$ is a complete metric space and contains a point x so that f restricted to F is continuous at x. The condition, "f has a point of continuity in each non-empty closed set F with respect to F", turns out to be necessary and sufficient for f to be Baire 1. To show this, we first prove that such a function f must satisfy for each open set G, $f^{-1}(G) \in \mathcal{F}_\sigma$. Then the resulting characterization of Baire 1 functions will follow from the theorem given below which characterizes for each α the functions belonging to Baire class α. This theorem asserts that:

$$f \in \text{Baire } \alpha \quad \text{iff for each open set } G,$$

$$\begin{cases} \text{if } \alpha \text{ is even,} & f^{-1}(G) \in \mathcal{G}_\alpha \\ \text{if } \alpha \text{ is odd,} & f^{-1}(G) \in \mathcal{F}_\alpha. \end{cases}$$

(4.8) <u>Theorem</u>. <u>If</u> f <u>is a real-valued function defined on</u> $[a,b]$ <u>and</u> f <u>has a point of continuity in each non-empty closed subset</u> F <u>of</u> $[a,b]$ <u>with respect to</u> F, <u>then</u> f <u>satisfies for each open set</u> $G \subset \mathbb{R}$, $f^{-1}(G) \in \mathcal{F}_\sigma$.

<u>Proof</u>. Suppose f satisfies the hypotheses of the theorem. It suffices to show that for each real number y, $f^{-1}((y,\infty))$ and $f^{-1}((-\infty,y))$ are $\mathcal{F}_\sigma$ sets. For

then each $f^{-1}((y,y')) = f^{-1}((y,\infty)) \cap f^{-1}((-\infty,y'))$ is an $\mathcal{F}_\sigma$ set and for each open set $G = \cup(y_i, y_i')$, $f^{-1}(G) = \cup f^{-1}((y_i,y_i'))$ is an $\mathcal{F}_\sigma$ set. By symmetry, it suffices to show that each $f^{-1}((y,\infty))$ is an $\mathcal{F}_\sigma$. Fix y and let $\{y_i\}$ be a decreasing sequence of real numbers with $y = \lim y_i$. Let x_o be a point of continuity of f with respect to $F_o = [a,b]$. In the case that $f(x_o) > y$, let G_o be an open set so that $f(G_o \cap F_o) \subset (y,\infty)$; if $f(x_o) \le y$, let G_o be an open set so that $x_o \in G_o$ and $f(G_o \cap F_o) \subset (-\infty,y_1)$. Then $F_1 = F_o \setminus G_o$ is a closed set. Suppose that for each $\beta < \alpha$ a closed set F_β has been defined and, if F_β is not empty, x_β has been chosen in F_β to be a point of continuity of f with respect to F_β. Suppose further that, if $f(x_\beta) > y$, an open set G_β has been chosen so that $f(G_\beta \cap F_\beta) \subset (y,\infty)$ and, if $f(x_\beta) \le y$, G_β has been chosen so that $f(G_\beta \cap F_\beta) < y_1$. Let $F_{\beta+1} = F_\beta \setminus G_\beta$ and for limit ordinals λ let $F_\lambda = \bigcap_{\beta<\lambda} F_\beta$. Then the closed set F_α is defined by the above. By the hypotheses, if $F_\alpha \ne \emptyset$, there is $x_\alpha \in F_\alpha$, a point of continuity of f with respect to F_α and an open set G_α can be chosen with $x_\alpha \in G_\alpha$ so that $f(x_\alpha) > y$ implies $f(G_\alpha \cap F_\alpha) \subset (y,\infty)$ and $f(x_\alpha) \le y$ implies $f(G_\alpha \cap F_\alpha) \subset (-\infty,y_1)$. Then $F_{\alpha+1} = F_\alpha \setminus G_\alpha$. The resulting transfinite sequence of closed sets $\{F_\alpha\}$ is decreasing and hence there is $\alpha_o < \omega_1$ so that $F_{\alpha_o} = \emptyset$. Then $\bigcup_{\alpha<\alpha_o} F_\alpha \cap G_\alpha = [a,b]$. Let $E_1 = \bigcup'_{\alpha<\alpha_o} F_\alpha \cap G_\alpha$ where $\cup'$ is the union over those $\alpha < \alpha_o$ for which $F_\alpha \cap G_\alpha \subset (y,\infty]$. Note that $f^{-1}((y_1,\infty)) \subset E_1 \subset$

$f^{-1}((y,\infty))$. Also, since each open set is an $\mathcal{F}_\sigma$ set, each $F_\alpha \cap G_\alpha$ is an $\mathcal{F}_\sigma$ set and E_1 is an $\mathcal{F}_\sigma$ set. Define the $\mathcal{F}_\sigma$ sets $E_2, E_3, \ldots$ by repeating the above argument using $y_2, y_3, \ldots$ in place of y_1. Then $\bigcup_{i=1}^{\infty} E_i$ is an $\mathcal{F}_\sigma$ set. Since $f^{-1}((y_i,\infty)) \subset E_i \subset f^{-1}((y,\infty))$ and $f^{-1}(y,\infty) = \bigcup_{i=1}^{\infty} f^{-1}((y_i,\infty))$, it follows that $\bigcup_{i=1}^{\infty} E_i = f^{-1}((y,\infty))$. It then follows that for each open subset G of $[a,b]$, $f^{-1}(G)$ is an $\mathcal{F}_\sigma$ set.

The precise characterization of the functions in a given Baire class α requires the next theorem which asserts that each of the Baire classes is closed under uniform limits.

(4.9) Theorem. *For each $\alpha < \omega_1$ if $\{f_n\}$ is a sequence of real-valued functions in Baire class α and f_n approaches f uniformly, then f belongs to Baire class α.*

Proof. Since the theorem is true for $\alpha = 0$, we consider $\alpha > 0$. Suppose that each function f_n belongs to a given Baire class $\alpha > 0$ and that f_n approaches f uniformly. By choosing, if necessary, a subsequence of $\{f_n\}$, we may suppose that $|f_n(x) - f(x)| \le 2^{-n}$ for each x and natural number n. Thus for each x and n, $|f_{n+1}(x) - f_n(x)| \le 2 \cdot 2^{-n}$. We now make use of the fact that if g is in Baire class α and $|g(x)| \le k$, then there is a sequence $\{g_n\}$ with each g_n in a Baire class less than α so that $g(x) = \lim g_n(x)$. Moreover, since each $h_n = \min(g_n, k)$ and each function $\max(h_n, -k)$ belongs to the same Baire class as g_n, the sequence $\{g_n\}$ which approaches g can be chosen so that $|g_n(x)| \le k$. Then, for each natural number n, let $\{g_{n,m}\}_{m=1}^{\infty}$ be sequences of functions from Baire classes with indices less than α so that for each n

$$\lim_{m\to\infty} g_{n,m}(x) = f_{n+1}(x) - f_n(x)$$

and $|g_{n,m}(x)| \le 2\cdot 2^{-n}$. Let $g_m(x) = g_{1,m}(x) + g_{2,m}(x) + \ldots + g_{m,m}(x)$ and note that each $g_m(x)$ belongs to a Baire class β_m with $\beta_m < \alpha$. Since $f(x) = \lim f_n(x) = f_1(x) + \sum_{n=1}^{\infty} f_{n+1}(x) - f_n(x)$, in order to show that f is in Baire class α, it will suffice to show that at each x, $\lim_{m\to 0} g_m(x)$ exists and equals $g(x) = \sum_{n=1}^{\infty} f_{n+1}(x) - f_n(x)$. For then both $g(x)$ and $f(x) = f_1(x) + g(x)$ will be in Baire class α. To do this, fix $\varepsilon > 0$ and choose N so that $4\cdot 2^{-N} < \varepsilon/3$. Hence, for each x,

$$\sum_{n=N+1}^{\infty} |f_{n+1}(x) - f_n(x)| < \varepsilon/3,$$

and also

$$|g(x) - \sum_{n=1}^{N} f_{n+1}(x) - f_n(x)| < \varepsilon/3.$$

Given x, there is $M > N$ so that $n, m \ge M$ implies $|f_{n+1}(x) - f_n(x) - g_{n,m}(x)| < \varepsilon/3N$ because $|f_{n+1}(x) - f_n(x) - g_{n,m}(x)| \le 4\cdot 2^{-n}$. Then, if $m \ge M$,

$$
\begin{aligned}
|g(x) - g_m(x)| &= |g(x) - \sum_{n=1}^{m} g_{n,m}(x)| \\
&\le |g(x) - \sum_{n=1}^{N} f_{n+1}(x) - f_n(x)| \\
&+ |\sum_{n=1}^{N} f_{n+1}(x) - f_n(x) - g_{n,m}(x)| \\
&+ \sum_{n=N+1}^{m} |g_{n,m}(x)| \le \varepsilon.
\end{aligned}
$$

It follows then that f is in Baire class α and that for each $\alpha < \omega_1$ the uniform limit of functions in Baire class α is in Baire class α.

We will need two additional facts about Baire functions in the proof of the next theorem. First, if f and g belong to Baire class α and g is never equal to 0, then f/g belongs to Baire class α. This is true if f and g are continuous and if it is also true for each $\beta < \alpha$ and f and g are in Baire class α, then $f = \lim f_n$, $g = \lim g_n$ and $f/g = \lim f_n \cdot g_n/(g_n^2 + 1/n)$ where f_n and g_n belong to Baire classes with indices less than α and so does $f_n \cdot g_n/(g_n^2 + 1/n)$. Second, f belongs to Baire class α iff for each pair of numbers c,d with $c < d$ the functions $f_c^d = \min(d, \max(f,c))$ belong to Baire class α. Again this is true for continuous functions and follows in general from the fact that $f = \lim f_n$ iff for each c,d with $c < d$ one has $f_c^d = \lim(f_n)_c^d$.

(4.10) Theorem. *Given any ordinal number* $\alpha < \omega_1$, *a real-valued function* f *belongs to Baire class* α *if*

<u>and only if for each open set</u> G, <u>when</u> α <u>is even</u> $f^{-1}(G) \in \mathcal{G}_\alpha$, <u>and when</u> α <u>is odd</u> $f^{-1}(G) \in \mathcal{F}_\alpha$.

<u>Proof</u>. The theorem is true for $\alpha = 0$ because Baire class 0 is by definition the class of continuous functions. Suppose that $\alpha > 0$ and that for every β less than α, for every f in Baire class β and open set G, $f^{-1}(G) \in \mathcal{G}_\beta$ when β is even, and $f^{-1}(G) \in \mathcal{F}_\beta$ when β is odd. Let f belong to Baire class α. Since the Borel classes under consideration are σ-classes, it suffices to show that each $f^{-1}((y,\infty))$ must be in class $\mathcal{G}_\alpha$ when α is even or $\mathcal{F}_\alpha$ when α is odd. Let $f(x) = \lim f_n(x)$ where each f_n belongs to Baire class β_n with $\beta_n < \alpha$. Fix y and let $A_{n,k} = \{x: f_n(x) \geq y + 1/k\}$. Then $\{x: f(x) > y\} = \bigcup_k \bigcup_N \bigcap_{n>N} A_{n,k}$ because $f(x) > y$ iff there is a k and there is an N so that for every $n > N$, $f_n(x) \geq y + 1/k$. Now, if α is odd, each β_n can be chosen to be even. Since, by hypothesis, each $A^c_{n,k} = \{x: f_n(x) < y + 1/k\} \in \mathcal{G}_{\beta_n}$, each $A_{n,k} \in \mathcal{F}_{\beta_n}$, a δ-class, and each $\bigcap_{n>N} A_{n,k} \in \mathcal{F}_{\beta_n}$. Since each $\mathcal{F}_{\beta_n} \subset \mathcal{F}_\alpha$ and $\mathcal{F}_\alpha$ is a σ-class, $\bigcup_k \bigcup_N \bigcap_{n>N} A_{n,k} \in \mathcal{F}_\alpha$. Similarly, if α is even, each β_n can be chosen to be odd. Again, by hypothesis, each $A^c_{n,k} = \{x: f_n(x) < y + 1/k\} \in \mathcal{F}_{\beta_n}$, each $A_{n,k} \in \mathcal{G}_{\beta_n}$, a δ-class, and each $\bigcap_{n>N} A_{n,k} \in \mathcal{G}_{\beta_n}$. Also each $\mathcal{G}_{\beta_n} \subset \mathcal{G}_\alpha$ and $\mathcal{G}_\alpha$ is a σ-class so that $\bigcup_k \bigcup_N \bigcap_{n>N} A_{n,k} \in \mathcal{G}_\alpha$.

It now remains to prove the converse; namely, that if α is even and for each open set G, $f^{-1}(G) \in \mathcal{G}_\alpha$ or if α is odd and for each open set G, $f^{-1}(G) \in \mathcal{F}_\alpha$, then f is in Baire class α. Again noting that this is true for $\alpha = 0$, we fix α with $0 < \alpha < \omega_1$ and f in Baire class α. Because f is in Baire class α iff each f_c^d is in Baire class α we may suppose that f is bounded and, to fix the argument, we will suppose that the range of f is contained in $[0,1]$. Let $0 \le a < b \le 1$, $E^a = \{x: f(x) > a\}$ and $E_b = \{x: f(x) < b\}$. By hypothesis E^a belongs to the σ-class corresponding to α. Thus $E^a = \cup E_n$ where each E_n is in a Borel class β_n with $\beta_n < \alpha$. Then $C_{E_n}(x)$ belongs to Baire class α. Let $g^a(x) = \sum_{n=1}^{\infty} 2^{-n} \cdot C_{E_n}(x)$. Then $g^a(x)$ is the uniform limit of the sequence consisting of $\sum_{n=1}^{N} 2^{-n} \cdot C_{E_n}(x)$ and hence $g^a(x)$ is in Baire class α. Note that $g^a(x) > 0$ iff $x \in E^a$ and $0 \le g^a(x) \le 1$. Similarly, $E_b = \cup E_n'$ where each E_n' is in a Borel class β_n' with $\beta_n' < \alpha$. Thus $g_b(x) = \sum_{n=1}^{\infty} 2^{-n} \cdot C_{E_n'}(x)$ is in Baire class α and satisfies $g_b(x) > 0$ iff $x \in E_b$ and $0 \le g_b(x) \le 1$. Let $h_{a,b}(x) = g^a(x)/(g_b(x) + g^a(x))$. Then $h_{a,b}(x)$ is in Baire class α because $g_b(x) + g^a(x)$ is never 0. Also $h_{a,b}(x) = 1$ if $f(x) > b$, $h_{a,b}(x) = 0$ if $f(x) < a$ and $0 \le h_{a,b}(x) \le 1$ at each x. Now let $h_n(x)$

$= \frac{1}{n} \sum_{m=1}^{n} h_{(m-1)/n,m/n}(x)$. Then each $h_n(x)$ is in Baire class α and if $f(x) \in [(m-1)/n, m/n]$ then $h_n(x) \in [(m-1)/n, m/n]$ and hence $|f(x) - h_n(x)| \le 1/n$ at each x. Thus $h_n(x)$ approaches $f(x)$ uniformly and hence f is in Baire class α.

Note that the above theorem is valid for real-valued functions defined on a metric space. When this theorem is combined with previous results, one obtains the following characterization of functions in Baire class 1, which is also valid when the underlying space is a metric space.

(4.11) Theorem. *A real-valued function f belongs to Baire class 1 iff for every open set G, $f^{-1}(G) \in \mathcal{F}_\sigma$ and this holds if and only if every closed set F contains a point of continuity of the function f restricted to F.*

Proof. Theorem 4.7 shows that each function f in Baire class 1 has on each closed set F a point of continuity with respect to F. Theorem 4.8 states that functions f with this property satisfy for each open set G, $f^{-1}(G) \in \mathcal{F}_\sigma$. Finally, the last theorem implies that f is in Baire 1 iff for each open set G, $f^{-1}(G) \in \mathcal{F}_\sigma$.

We return to the general case of functions in a given Baire class. For each $\alpha < \omega_1$, let $\mathcal{M}_\alpha$ denote the σ-class with ordinal number α of Borel sets and

$\mathcal{N}_\alpha$ denote the δ-class; that is, $\mathcal{M}_\alpha = \mathcal{G}_\alpha$ and $\mathcal{N}_\alpha = \mathcal{F}_\alpha$ if α is even, $\mathcal{M}_\alpha = \mathcal{F}_\alpha$ and $\mathcal{N}_\alpha = \mathcal{G}_\alpha$ if α is odd. This notation is convenient for the following theorem regarding inverse images of sets and compositions of functions.

(4.12) *Theorem. If* f *is a real-valued function of a real variable in Baire class* α, g *a function in Baire class* β *and* $E \subset \mathbb{R}$ *belongs to* $\mathcal{M}_\gamma$ (*resp.* $\mathcal{N}_\gamma$), *then* $g^{-1}(E) \in \mathcal{M}_{\beta+\gamma}$ (*resp.* $\mathcal{N}_{\beta+\gamma}$) *and* $f \circ g$ *is in Baire class* $\beta + \alpha$.

Proof. By Theorem 4.10, if g is in Baire class β and $E \in \mathcal{M}_0$ (resp. $\mathcal{N}_0$), then $g^{-1}(E) \in \mathcal{M}_\beta$ (resp. $\mathcal{N}_\beta$). By induction, since $g^{-1}(\cup E_n) = \cup g^{-1}(E_n)$ and $g^{-1}(\cap E_n) = \cap g^{-1}(E_n)$, it follows that if $E \in \mathcal{M}_\gamma$ (resp. $\mathcal{N}_\gamma$) then $g^{-1}(E) \in \mathcal{M}_{\beta+\gamma}$ (resp. $\mathcal{N}_{\beta+\gamma}$). Now let f be a function of a real variable in Baire class α. Let $G \subset \mathbb{R}$ be an open set. Then $f^{-1}(G) \in \mathcal{M}_\alpha$ and $g^{-1}(f^{-1}(G)) \in \mathcal{M}_{\beta+\alpha}$. Since G was an arbitrary open subset of the real numbers and $(f \circ g)^{-1}(G) = g^{-1}(f^{-1}(G)) \in \mathcal{M}_{\beta+\alpha}$, it follows that $f \circ g$ belongs to Baire class $\beta + \alpha$.

We turn now to the proof that, for functions of a real variable defined on $[0,1]$, each Baire class α with $0 < \alpha < \omega_1$ contains functions which are not in any previous class. It will then follow that there are also Borel sets in each given class $\mathcal{G}_\alpha$ or $\mathcal{F}_\alpha$, $\alpha < \omega_1$, which do not belong to any previous class. It is easily seen that this result does not hold for functions which merely have a metric space for their domain. For example, if X is a discrete metric space, every real-valued function on X is continuous; if $X = \{0\} \cup \{1/n\}_{n=1}^{\infty}$, then every real-valued function defined on this space X is in Baire class 1. The result for functions defined on $[0,1]$ will follow from the next theorem, due to Lebesgue, which asserts that for each α with $0 < \alpha < \omega_1$ there exists a Baire function $F(x,y)$ defined on the unit square so that, if f belongs to Baire class α on $[0,1]$, there is a $y \in [0,1]$ such that $f(x) = F(x,y)$. Such a function F is sometimes called a <u>universal function for Baire class</u> α.

(4.12) <u>Theorem</u>. <u>For each</u> α <u>with</u> $0 < \alpha < \omega_1$ <u>there is a real-valued function</u> $F_\alpha(x,y)$ <u>which is a Baire function on</u> $[0,1] \times [0,1]$ <u>such that for each</u> $f(x)$ <u>in Baire class</u> α <u>there is</u> $y \in [0,1]$ <u>so that</u> $f(x) = F_\alpha(x,y)$.

<u>Proof</u>. According to Weierstrass' approximation theorem, each continuous function is the uniform limit of a sequence of polynomials. By varying the coefficients of these polynomials to nearby rational numbers, it is clear that the coefficients in the sequence of polynomials can be chosen to be rational numbers; that is, the polynomials with rational coefficients restricted to $[0,1]$ are dense in the space of continuous functions defined on $[0,1]$. Since the collection of polynomials of degree n with rational coefficients are cardinally equivalent to the set of ordered n-tuples of rational numbers, a countable set, the set of all polynomials with rational

coefficients, being the countable union of countable sets, is countable. Let $\{P_n(x)\}_{n=1}^{\infty}$ be an enumeration of the polynomials with rational coefficients. Let $F_o(x,y) = P_n(x)$ if $y = 1/n$; $F_o(x,y) = 0$, otherwise. Then $F_o(x,y)$ is a Baire function defined on $[0,1] \times [0,1]$. Indeed, $E_n = \{(x,y): y = 1/n\}$ is a closed set and $F_o(x,y) = \sum_{n=1}^{\infty} P_n(x) \cdot C_{E_n}(x,y)$ is in Baire class 2 since each function $C_{E_n}(x,y)$ is in Baire class 1 as is each function $P_n(x) \cdot C_{E_n}(x,y)$. Now, for each $y \in [0,1]$, let $y = \Sigma a_n/2^n$ where $a_n = 0$ or 1 and, when there are two expansions of y, the one ending in 0's is chosen. The function $a_n(y) = a_n$ when $y = \Sigma a_n/2^n$ is a Baire 1 function since it is the characteristic function of the union of a finite collection of intervals. Let $g_1(y) = \sum_{i=1}^{\infty} a_{2i-1}(y)/2^i$ and, in general, let $g_j(y) = \sum_{i=1}^{\infty} a_{n_{i,j}}(y)/2^i$ where $n_{i,j} = (2i - 1) \cdot 2^{j-1}$. Then each $g_j(y)$ is in Baire class 1 since it is the uniform limit of Baire 1 functions. Note that the image of each g_j is $[0,1]$ and that for each sequence $\{y_n\} \subset [0,1]$ there is $y \in [0,1]$ with $g_n(y) = y_n$. We now defined the function $F_\alpha(x,y)$ by induction. Suppose that for every $\beta < \alpha$, $F_\beta(x,y)$ has been defined. Select $\{\beta_n\}$, a non-decreasing sequence of ordinals, so that α is the least ordinal greater than all the β_n. Let

$$F_\alpha(x,y) = \overline{\lim} F_{\beta_n}(x, g_n(y)).$$

Then F_α is a Baire function. Now suppose $f(x)$ is in Baire class α. Then f_n can be chosen in Baire classes β_n so that $f(x) = \lim f_n(x)$. By hypothesis for each f_n there is y_n so that $f_n(x) = F_{\beta_n}(x,y_n)$. Let y_n determine the point y_o so that, for each natural number n, $y_n = g_n(y_o)$. But then

$$F_\alpha(x, y_o) = \overline{\lim} F_{\beta_n}(x, g_n(y_o))$$
$$= \overline{\lim} f_n(x) = f(x).$$

Thus $F_\alpha(x,y)$ is a universal Baire function for functions f in Baire class α.

We are now in a position to show that the Baire classes of functions defined on $[0,1]$ are strictly increasing for $0 < \alpha < \omega_1$ as are the classes of Borel subsets of $[0,1]$.

(4.13) Theorem. *For each $\alpha < \omega_1$ there are Baire functions in Baire class α which are in no lower class and Borel sets in $\mathcal{G}_\alpha$ and $\mathcal{F}_\alpha$ which are in no lower Borel classes.*

Proof. If not, there would be a least ordinal $\alpha_o < \omega_1$ so that Baire class $\alpha_o + 1$ was the same as Baire class α_o. Likewise, by considering characteristic functions of Borel sets, Theorem 4.10 implies that $\mathcal{G}_{\alpha_o+1}$ (resp. $\mathcal{F}_{\alpha_o+1}$) would be the same as $\mathcal{G}_{\alpha_o}$ (resp. $\mathcal{F}_{\alpha_o}$). Let $F(x,y)$ be the universal Baire function for class α_o. Let $E = \{(x,x): F(x,y) = 0\}$ and $E' = \{x: (x,x) \in E\}$. Then E is a Borel set since it is the intersection of the line $y = x$ with $\{(x,y): F(x,y) = 0\}$. Also E' is a Borel set since it could be constructed with unions and intersections parallel to those involved in the construction of E.

However, $C_{E'}(x)$ cannot be in Baire class α_o since there is no $y \in [0,1]$ so that $C_{E'}(x)$ equals $F(x,y)$. This contradiction proves the theorem. Again, Theorem 4.10 implies that for each $\alpha < \omega_1$ there must be Borel sets in $\mathcal{G}_\alpha$ and $\mathcal{F}_\alpha$ which are in no lower Borel class.

4.2 Exercises

1. Suppose $[0,1] = \bigcup_{n=1}^{\infty} F_n$ where each F_n is closed. Suppose further that f is continuous when restricted to each set F_n. Show that $f^{-1}(G)$ is both an $\mathcal{F}_\sigma$ and a $\mathcal{G}_\delta$ for each open set G.
2. Suppose f is defined on $[0,1]$ and for every open set G, $f^{-1}(G)$ is both an $\mathcal{F}_\sigma$ and a $\mathcal{G}_\delta$. Show that $[0,1] = \cup F_n$ where each F_n is closed and f restricted to F_n is continuous.
3. If f is a Baire function on $[0,1]$, show using transfinite induction that there is a residual subset E of $[0,1]$ so that f restricted to E is continuous on E.
4. Show that given any real-valued function f there is a countable dense set A so that f restricted to A is continuous on A.
5. Show that every $\mathcal{G}_\delta$ subset of $\mathbb{R}$ is either at most countable or contains a nonempty perfect set.

Chapter Five

5.1 Measure and Measurable Sets

The idea behind the concept of measure is that of assigning a non-negative, possibly infinite number to many of the subsets of a given set in such a fashion that the number represents a quantity associated with the set. For example, $b - a$ is the length of the interval $[a,b]$ and it seems natural to assign $\Sigma(b_i - a_i)$ as the length of an open set $G \subset \mathbb{R}$ when (a_i,b_i) $i = 1, 2, \ldots$ are the component intervals of G. A closed set $F \subset (a,b)$ can be assigned $(b - a) - \Sigma(b_i - a_i)$ where (a_i,b_i) $i = 1, 2, \ldots$ are the component intervals of $(a,b)\setminus F$. Historically, the definition of length was extended by Borel using transfinite induction to a measurement of the length of each Borel subset of the line. Clearly, many other measures are desirable; for example, area of subsets of the plane, volume in space, length and lower dimensional measures in higher dimensional spaces, measurements of other physical quantities, and the measurements of probability on the collection of "events"; that is, on a collection of subsets of a probability space consisting of all possible "outcomes of an experiment".

In these situations, it is desirable to have a measure function which assigns to as many subsets E of a set X as possible a number $m(E)$ so that $m(\emptyset) = 0$ and, whenever $\{E_n\}_{n=1}^{\infty}$ is a sequence of pairwise disjoint sets, $m(\cup E_n) = \Sigma m(E_n)$. It turns out that such a function is not in general attainable for all the subsets of a set X, but is frequently obtainable on a large collection of subsets of X.

Rather than following strictly the historical development of this subject, the direct approach to defining a measure as developed by Caratheodory will be taken here. This approach will be shown to be completely general.

An <u>outer measure</u> m^* is a non-negative, possibly infinite, real-valued function defined on each subset E of a given set X and satisfying:

(1) $m^*(\emptyset) = 0$

(2) if $A \subset B$, then $m^*(A) \le m^*(B)$

(3) for any sequence of sets $\{E_n\}$, $m^*(\cup E_n) \le \Sigma m^*(E_n)$.

Property (2) means that the function m^* is <u>non-decreasing</u> and property (3) is called <u>countable</u>

<u>sub</u>-<u>additivity</u>. Outer measures are readily constructed on the set of all subsets of a set X as follows: Let $\mathscr{C}$ be a collection of subsets of X such that $\emptyset \in \mathscr{C}$ and $X \in \mathscr{C}$ and let τ be any non-negative, possibly infinite function defined on the sets in $\mathscr{C}$ with $\tau(\emptyset) = 0$. Then $m^*(E)$ is defined as

$$\inf\{\sum_{i=1}^{\infty} \tau(E_i): E \subset \bigcup_{i=1}^{\infty} E_i, \quad E_i \in \mathscr{C}\}.$$

The possibility that there only be finitely many elements to the sum is obtained by letting all but finitely many of the E_i be empty.

Clearly $m^*(\emptyset) = 0$ and if $A \subset B$,

$$\begin{aligned} m^*(A) &= \inf\{\Sigma\tau(E_i): A \subset \cup E_i\} \\ &\le \inf\{\Sigma\tau(E_i): B \subset \cup E_i\} = m^*(B). \end{aligned}$$

If $\{E_i\}$ is a sequence of subsets of X and $\Sigma m^*(E_i) = \infty$, (3) is automatically satisfied. On the other hand, if $\Sigma m^*(E_i) < \infty$, then, given $\varepsilon > 0$, for each natural number i there are sets $E_{i,n} \in \mathscr{C}$ so that $E_i \subset \bigcup_n E_{i,n}$ and $m^*(E_i) + \varepsilon/2^i \ge \sum_n \tau(E_{i,n})$. Then $\bigcup_i E_i \subset \bigcup_{i,n} E_{i,n}$ and $m^*(\bigcup_i E_i) \le \sum_{i,n} \tau(E_{i,n}) \le \sum_i m^*(E_i) + \varepsilon$. Since this holds for all $\varepsilon > 0$, $m^*(\cup E_i) \le \Sigma m^*(E_i)$ and (3) is satisfied. It follows that m^* defined from such a τ is an outer measure.

If $X = \mathbb{R}^n$ and

$$\mathscr{C} = \{(a_1,b_1) \times (a_2,b_2) \times \dots \times (a_n,b_n): -\infty \le a_i < b_i \le \infty\}$$

and

$$\begin{aligned} &\tau\big((a_1,b_1) \times (a_2,b_2) \times \dots \times (a_n,b_n)\big) \\ &\qquad = (b_1 - a_1)(b_2 - a_2) \dots (b_n - a_n), \end{aligned}$$

the resulting m^* is <u>Lebesgue</u> n-<u>dimensional</u> <u>outer</u> <u>measure</u>. In particular, <u>Lebesgue</u> <u>outer</u> <u>measure</u> m^* on the real line is given by

$$m^*(E) = \inf\{\Sigma(b_i - a_i): E \subset \cup(a_i,b_i)\}.$$

This concept of outer measure originated with Lebesgue.

Recall that a <u>σ-algebra</u> $\mathcal{A}$ of subsets of a set X is a collection of subsets of X which contains both $\emptyset$ and X, and is closed under countable unions, countable intersections and complements. A <u>measure</u> m on a set X is traditionally considered to involve a triple $(X,\mathcal{A},m)$ which is sometimes called a <u>measure space</u>. Here $\mathcal{A}$ is a σ-algebra of subsets of X and m is a non-negative, possibly infinite function which satisfies $m(\emptyset) = 0$ and, for $\{E_i\}$ any sequence of pairwise disjoint elements of $\mathcal{A}$, $m(\cup E_i) = \Sigma m(E_i)$. This last property is called <u>countable additivity</u> of m. The sets belonging to $\mathcal{A}$ are called <u>measurable</u>.

Given finitely many pairwise disjoint sets E_1, E_2, ..., E_N, $m(\bigcup_{i=1}^{N} E_i) = \sum_{i=1}^{N} m(E_i)$ follows by considering $E_i = \emptyset$ for $i > N$. Note further that, if $A \subset B$ with A and B measurable, then $m(A) + m(B\backslash A) = m(B)$ and since m is non-negative, $m(B\backslash A) \geq 0$ implies $m(A) \leq m(B)$. Since this is true for all measurable sets A and B, m is non-decreasing. The following theorem gives several additional important properties satisfied by a measure on the collection of measurable sets.

(5.1) <u>Theorem</u>. <u>If</u> $(X,\mathcal{A},m)$ <u>is a measure space and</u> $\{E_n\}$ <u>is a sequence of measurable sets, then</u>

i) <u>if</u> $\{E_n\}$ <u>is non-decreasing</u>, $m(\cup E_n) = \lim m(E_n)$

ii) <u>if</u> $\{E_n\}$ <u>is non-increasing and</u> $m(E_1) < \infty$,

$$m(\cap E_n) = \lim m(E_n)$$

iii) $m(\underline{\lim}\, E_n) \le \underline{\lim}\, m(E_n)$

iv) <u>if</u> $\cup E_n \subset E$ <u>and</u> $m(E) < \infty$,

$$\overline{\lim}\, m(E_n) \le m(\overline{\lim}\, E_n)$$

v) <u>if</u> $\cup E_n \subset E$ <u>and</u> $m(E) < \infty$ <u>and</u> $\lim E_n$ <u>exists</u>, $m(\lim E_n) = \lim m(E_n)$.

<u>Proof</u>.

i) If $\{E_n\}$ is a non-decreasing sequence of measurable sets, letting $E_o = \emptyset$, we have $E_n \setminus E_{n-1}$ is a pairwise disjoint sequence of sets. Thus

$$m\Big(\bigcup_{n=1}^{\infty} E_n\Big) = m\Big(\bigcup_{n=1}^{\infty} (E_n \setminus E_{n-1})\Big) = \sum_{n=1}^{\infty} m\big(E_n \setminus E_{n-1}\big).$$

But

$$\sum_{n=1}^{\infty} m(E_n \setminus E_{n-1}) = \lim \sum_{n=1}^{N} m(E_n \setminus E_{n-1}) = \lim m(E_N).$$

Hence $m(\cup E_n) = \lim m(E_n)$ when $\{E_n\}$ is non-decreasing.

ii) Suppose $\{E_n\}$ is a non-increasing sequence of measurable sets and $m(E_1) < \infty$. Then $\{E_1 \setminus E_n\}$ is non-decreasing. Since $E_n \subset E_1$ and $m(E_1) < \infty$, $m(E_1 \setminus E_n) = m(E_1) - m(E_n)$ and by i)

$$\begin{aligned} m\big(\cup(E_1 \setminus E_n)\big) &= \lim m(E_1 \setminus E_n) \\ &= \lim \big(m(E_1) - m(E_n)\big) = m(E_1) - \lim m(E_n). \end{aligned}$$

Since $\cap E_n \subset E_1$ and $m(E_1) < \infty$, $m(E_1 \setminus \cap E_n) = m(E_1) - m(\cap E_n)$ and

$$m\big(\cup(E_1 \setminus E_n)\big) = m(E_1 \setminus \cap E_n) = m(E_1) - m(\cap E_n).$$

From the equality of the right hand sides of the equations beginning with $m\big(\cup(E_1 \setminus E_n)\big)$, it follows that $m(\cap E_n) = \lim m(E_n)$.

iii) Recall that $\underline{\lim}\, E_n = \bigcup_k \bigcap_{n \ge k} E_n$. Let $A_k = \bigcap_{n \ge k} E_n$. Then $\{A_k\}$ is non-decreasing and $m(\bigcup_k A_k) =$

$\lim m(A_k)$. But for each k and $n > k$, $m(E_n) \geq m(A_k)$ because $E_n \supset A_k$ when $n \geq k$. Hence, for each k, $\underline{\lim}\, m(E_n) \geq m(A_k)$ and thus, since $\underline{\lim}\, E_n = \bigcup_k A_k$,

$$m(\underline{\lim}\, E_n) = \lim m(A_k) \leq \underline{\lim}\, m(E_n).$$

iv) Suppose $\cup E_n \subset E$ with $m(E) < \infty$. Note that $\overline{\lim}\, E_n = \bigcap_k \bigcup_{n \geq k} E_n = E \backslash \underline{\lim}(E \backslash E_n)$. From this it follows that $E \backslash \overline{\lim}\, E_n = \underline{\lim}\, E \backslash E_n$ and since $m(E) < \infty$, $m(E) - m(\overline{\lim}\, E_n) = m(\underline{\lim}\, E \backslash E_n \leq \underline{\lim}\, m(E \backslash E_n)$ by iii). But $\underline{\lim}\, m(E \backslash E_n) = \underline{\lim}\,(m(E) - m(E_n)) = m(E) - \overline{\lim}\, m(E_n)$ and subtracting $m(E)$ from both sides of the resulting inequality yields

$$\overline{\lim}\, m(E_n) \leq m(\overline{\lim}\,(E_n)).$$

v) If $\cup E_n \subset E$ with $m(E) < \infty$ and $\lim E_n$ exists, then by iii) and iv),

$$m(\underline{\lim}\, E_n) \leq \underline{\lim}\, m(E_n) \leq \overline{\lim}\, m(E_n) \leq m(\overline{\lim}\, E_n).$$

Since $\overline{\lim}\, E_n = \underline{\lim}\, E_n$, the limit of the measures exists and $m(\lim E_n) = \lim m(E_n)$.

The possibility of obtaining countable additivity for a measure function motivated the development of measure theory. Lebesgue defined $m^*(E)$ for $E \subset [0,1]$ as described above, defined the <u>inner measure</u> $m_*(E)$ as $1 - m^*(E^c)$ and called a set $E \subset [0,1] = X$ measurable if $m^*(E) = m_*(E)$. He then proved that the measurable sets form a σ-algebra and that m^* is countably additive on this σ-algebra. Caratheodory took an approach which holds in a completely general setting. He called a set E <u>measurable</u> if, for any set $A \subset X$, $m^*(E \cap A) + m^*(E^c \cap A) = m^*(A)$. Heuristically speaking, if E is considered to divide each set A into two parts $A \cap E$ and $A \cap E^c$, E is measurable if m^* adds on each of these two parts no matter which $A \subset X$ is considered. If E is a measurable set, for brevity, we will write $E \in \mathcal{M}$. That the measurable sets in this general setting satisfy the properties desired of them is shown by Caratheodory's theorem:

(5.2) _Theorem_. (Caratheodory) _Given an outer measure m^* on the subsets of a set X, the measurable subsets of X form a σ-algebra and m^* restricted to this σ-algebra is a measure._

Proof. Given an outer measure m^* on a set X, let

$$\mathcal{M} = \{E \subset X: \text{for each } A \subset X, \; m^*(A) = m^*(A \cap E) + m^*(A \cap E^c)\}.$$

To show that a set E is measurable, that is, that $E \in \mathcal{M}$, it is sufficient to show that for any set A,

$$m^*(A) \geq m^*(A \cap E) + m^*(A \cap E^c)$$

because the other inequality is given by the subadditivity of m^*. Since only this inequality needs to be checked, it is also sufficient to consider only sets A with $m^*(A) < \infty$; for if $m^*(A) = \infty$, the inequality is clearly satisfied. The remainder of the proof proceeds in a sequence of six steps some of which have additional significance of their own.

i) If $m^*(E) = 0$, $E \in \mathcal{M}$. (That is, all sets of outer measure zero are measurable.) This is because, if $m^*(E) = 0$, then for any set A, $m^*(A \cap E) = 0$ by property (2) of outer measures and

$$m^*(A) \geq m^*(A \cap E^c) = m^*(A \cap E) + m^*(A \cap E^c)$$

again by property (2) of outer measures. Hence, E is measurable.

ii) If $E \in \mathcal{M}$, $E^c \in \mathcal{M}$. (That is, the complements of measurable sets are measurable.) This is because, given $E \in \mathcal{M}$ and $A \subset X$, $m^*(A) = m^*(A \cap E) + m^*(A \cap E^c)$ by the measurability of E. But then $m^*(A) = m^*(A \cap E^c) + m^*(A \cap E)$ and thus E^c is also measurable.

iii) If $E_1, E_2, \ldots, E_n$ are measurable, so is $\bigcup_1^n E_i$ measurable. (That is, finite unions of measurable sets are measurable.) To see this, let E_1 and E_2 be measurable sets and let $A \subset X$. Then $m^*(A) = m^*(A \cap E_1) + m^*(A \cap E_1^c)$. Using the set $A \cap E_1^c$ along with the fact that E_2 is measurable yields

$$m^*(A \cap E_1^c) = m^*(A \cap E_1^c \cap E_2) + m^*(A \cap E_1^c \cap E_2^c).$$

By combining these two formulas, we have:

$$m^*(A) = m^*(A \cap E_1) + m^*(A \cap E_1^c \cap E_2) + m^*(A \cap E_1^c \cap E_2^c).$$

Then, by the subadditivity of m^* and the identities

$$A \cap (E_1 \cup E_2) = (A \cap E_1) \cup (A \cap E_1^c \cap E_2)$$

and

$$A \cap E_1^c \cap E_2^c = A \cap (E_1 \cup E_2)^c,$$

we have $m^*(A) \geq m^*[A \cap (E_1 \cup E_2)] + m^*[A \cap (E_1 \cup E_2)^c]$. As noted above, this inequality implies the measurability of $E_1 \cup E_2$ and the measurability of finite unions of measurable sets follows by induction.

iv) If $E_1, E_2, \ldots, E_n$ are pairwise disjoint measurable sets and $S_n = \bigcup_1^n E_i$, then for $A \subset X$, $m^*(A \cap S_n) = \Sigma\, m^*(A \cap E_i)$. (That is, if a subset is chosen from each set belonging to a finite collection of pairwise disjoint measurable sets, then outer measure adds on the finite collection of subsets.) This is clearly true if $n = 1$ for then, trivially, $m^*(E_1 \cap A) = m^*(E_1 \cap A)$. Assume it is true for n. Given A and S_{n+1}, by the measurability of S_n it follows that

$$\begin{aligned} m^*(A \cap S_{n+1}) &= m^*(A \cap S_{n+1} \cap S_n) + m^*(A \cap S_{n+1} \cap S_n^c) \\ &= m^*(A \cap S_n) + m^*(A \cap E_{n+1}) \\ &= \sum_{i=1}^{n+1} m^*(A \cap E_i) \end{aligned}$$

since by the induction hypothesis $m^*(A \cap S_n) = \sum_{i=1}^{n} m^*(A \cap E_i)$. Thus iv) is true for any finite collection of pairwise disjoint sets.

v) If $\{E_n\}$ is a sequence of pairwise disjoint measurable sets and $S = \bigcup_{n=1}^{\infty} E_n$, then for $A \subset X$, $m^*(A \cap S) = \sum_{n=1}^{\infty} m^*(A \cap E_n)$. (That is, the additivity described in iv) holds for countable collections as well.) To see this, let E_n and A be given as above. Let $S_n = \bigcup_{i=1}^{n} E_i$. By the non-decreasing property of m^* and by iv) above,

$$m^*(A \cap S) \geq m^*(A \cap S_n) = \sum_{i=1}^{n} m^*(A \cap E_i).$$

By letting n aproach ∞, we have $m^*(A \cap S) \geq \sum_{i=1}^{\infty} m^*(A \cap E_i)$. Since the opposite inequality follows from the subadditivity of m^*, we have $m^*(A \cap S) = \sum_{n=1}^{\infty} m^*(A \cap E_n)$.

vi) If $\{E_n\}$ is a sequence of measurable sets, then $\bigcup_{n=1}^{\infty} E_n$ is measurable. To see this, given $\{E_n\}$, let $H_1 = E_1$ and for $n > 1$ let $H_n = E_n \setminus \bigcup_1^{n-1} E_i$ and let $S_n = \bigcup_{i=1}^{n} E_i$. If the E_n are measurable, by iii) and ii) each H_n and S_n are measurable. Thus, given $A \subset X$, $m^*(A) = m^*(A \cap S_n) + m^*(A \cap S_n^c)$. By iv), $m^*(A) = \sum_{i=1}^{n} m^*(A \cap H_i) + m^*(A \cap S_n^c)$ and, by the non-decreasing property of m^*, $m^*(A) \geq \sum_{i=1}^{n} m^*(A \cap H_i) + m^*(A \cap S^c)$. Letting n approach ∞ in this inequality and applying v) yields $m^*(A) \geq m^*(A \cap S) + m^*(A \cap S^c)$. Hence $S = \bigcup_{n=1}^{\infty} E_n$ is measurable.

To complete the proof of the theorem, note that ii) and vi) imply that the collection of measurable sets is closed under countable unions, countable intersections and complements and thus forms a σ-algebra. Given a sequence $\{E_n\}$ of pairwise disjoint measurable sets, using v) with $A = \cup E_n$, it follows that $m^*(\cup E_n) = \sum_{n=1}^{\infty} m^*(E_n)$. Thus m^* restricted to $\mathcal{M}$ is a measure. These observations complete the proof of the theorem.

The outer measure m^* gives rise to a σ-algebra $\mathcal{M}$ of sets and, if E is a set which is known to be measurable, we write $m(E)$ instead of $m^*(E)$. To see that the approach given above is completely general, first note that no structure is required on the set X. All that is required is a non-negative function τ or a non-decreasing subadditive function m^* defined on the subsets of X and equal to 0 on $\emptyset$. The measure which results is said to be <u>complete</u>; that is, if $Z \subset X$ and $Z \subset E$ with $m(E) = 0$, then Z is measurable and $m(Z) = 0$. A measure space $(X,\mathcal{A}_1,m_1)$ is said to be the <u>completion</u> of a measure m if m_1 is complete, if each m-measurable set E is m_1-measurable with $m_1(E) = m(E)$ and if $E_1 \subset \mathcal{A}_1$, then $E_1 = E \cup Z$ where $E \in \mathcal{A}$ and Z is a subset of a set of m-measure 0. That every measure has a completion follows from the observation that, if $(X,\mathcal{A},m)$ is given,

$$\mathcal{A}_1 = \{H \subset X: H = E \cup Z, \quad E \in \mathcal{A}, \quad Z \subset Z_1 \in \mathcal{A} \text{ with } m(Z_1) = 0\}$$

is a σ-algebra and $(X,\mathcal{A}_1,m_1)$ a measure space, where if $E \subset H$ and $H \setminus E$ is a subset of a set of m-measure 0, $m_1(H) = m(E)$.

An outer measure is said to be <u>regular</u> provided that for each set $H \subset X$ there is a set $E \in \mathcal{M}$ with $H \subset E$ and $m^*(H) = m(E)$. The following theorem shows that every measure $(X,\mathcal{A},m)$ can be obtained from a regular outer measure. The theorem shows the generality of the Caratheodory construction; specifically, every measure can be constructed using the procedure with an appropriate τ function.

(5.3) <u>Theorem</u>. <u>If</u> $(X,\mathcal{A},m)$ <u>is a measure space and</u> $\tau(E) = m(E)$ <u>for each</u> $E \in \mathcal{A}$, <u>the resulting outer</u>

<u>measure</u> m_1^* <u>is regular; the measure</u> m_1 <u>is the completion of</u> m <u>and thus agrees with</u> m <u>on each</u> $E \in \mathcal{A}$.

<u>Proof</u>. That m_1 is complete is a consequence of i) of the previous theorem. Now

$$\begin{aligned} m_1^*(E) &= \inf\{\Sigma m(H_i): E \subset \bigcup_{i=1}^{\infty} H_i \in \mathcal{A}\} \\ &= \inf\{m(H): E \subset H,\ H \in \mathcal{A}\} \end{aligned}$$

since for $H = \cup H_i$, $m(H) \le \Sigma m(H_i)$. For $E \in \mathcal{A}$, since $m(E)$ is non-decreasing on $\mathcal{A}$, letting $H = E$ yields $m_1^*(E) \le m(E)$. Since $m(E)$ is countably additive and hence subadditive on $\mathcal{A}$, $m_1^*(E) \ge m(E)$. Thus, for $E \in \mathcal{A}$, $m_1^*(E) = m(E)$ and $m_1^*(E)$ agrees with m on $\mathcal{A}$. Note that, if $A \subset X$ and $m_1^*(A) < \infty$, then there are sets $H_n \in \mathcal{A}$ with $A \subset H_n$ and $m_1^*(A) \le m(H_n) + 1/n$. It follows that $A \subset H = \cap H_n$ and $m_1^*(A) = m(H)$. Since $H \in \mathcal{A}$, it follows that each set A is contained in an m-measurable set H with $m_1^*(A) = m(H)$. Finally, if $E \in \mathcal{A}$ and $A \subset X$ is any set with finite m_1^* outer measure, there exist H, H_1, H_2 in $\mathcal{A}$ such that $A \subset H$ and $m_1^*(A) = m(H)$, $A \cap E \subset H_1$ and $m_1^*(A \cap E) = m(H_1)$, $A \cap E^c \subset H_2$ and $m_1^*(A \cap E^c) = m(H_2)$. Then

$$\begin{aligned} m_1^*(A) = m(H) &\ge m(H \cap H_1 \cap E) + m(H \cap H_2 \cap E^c) \\ &\ge m_1^*(A \cap E) + m_1^*(A \cap E^c). \end{aligned}$$

Thus, if E is m-measurable, it is also m_1-measurable and from the above $m(E) = m_1(E)$. Thus m_1^* is a regular outer measure. Moreover, $(X, \mathcal{M}, m_1)$ is the completion of m. This is because, if $E = H \backslash Z$ with $Z \subset Z_1$, $m(Z_1) = 0$, and $H \in \mathcal{A}$, then $m_1(Z) = 0$ and E is m_1-measurable and $m_1(E) = m(H) - m_1(Z) = m(H)$. Conversely, if E is m_1-measurable, then $E \subset H$

where $m_1(E) = m(H)$ and, $H\setminus E$ being a set of m_1-measure 0, by the regularity of m_1, $H\setminus E$ is contained in a set of m-measure 0. Thus m_1^* is the completion of m.

For regular outer measures on a space X, where the measure of X is finite, Caratheodory's definition of measurability and Lebesgue's are equivalent as is shown by the following theorem.

(5.4) Theorem. *If m^* is a regular outer measure on a set X with $m(X) < \infty$, then a set $E \subset X$ is measurable iff $m(X) = m^*(E) + m^*(E^c)$; that is, if and only if $m^*(E) = m_*(E)$ where $m_*(E) = m(X) - m^*(E^c)$.*

Proof. Clearly, if E is measurable, then $m(X) = m(E) + m(E^c) = m^*(E) + m^*(E^c)$. On the other hand, suppose $m^*(E) + m^*(E^c) = m(X)$. Let A_1 and A_2 be two measurable sets with $E \subset A_1$, $E^c \subset A_2$ and $m(A_1) = m^*(E)$, $m(A_2) = m^*(E^c)$. Since A_1 and A_2 are measurable and $A_1 \cup A_2 = X$,

$$m(A_1) + m(A_2\setminus A_1) = m(X),$$
$$m(A_2) + m(A_1\setminus A_2) = m(X),$$

and

$$m(A_1) + m(A_2) = m^*(E) + m^*(E^c) = m(X).$$

From these three equations and the fact that $m(X) < \infty$, it follows that $m(A_2\setminus A_1) + m(A_1\setminus A_2) = m(X)$. But since $m(A_2\setminus A_1) + m(A_1\setminus A_2) + m(A_1 \cap A_2) = m(X)$, it follows that $m(A_1 \cap A_2) = 0$. Now let A be any set. Then $m^*(A \cap A_1^c) \leq m^*(A \cap A_2) \leq m^*(A \cap (A_2\setminus A_1)) + m^*(A \cap A_1 \cap A_2) \leq m^*(A \cap A_1^c)$ since $m^*(A \cap A_1 \cap A_2) =$

0. Thus $m^*(A \cap A_1^c) = m^*(A \cap A_2)$.

Now

$$\begin{aligned} m^*(A \cap E) + m^*(A \cap E^c) &\le m^*(A \cap A_1) + m^*(A \cap A_2) \\ &\le m^*(A \cap A_1) + m^*(A \cap A_1^c) \\ &= m^*(A) \end{aligned}$$

because A_1 is measurable. But this shows that E is also measurable which completes the proof of the theorem.

Note that if m^* was generated from τ and $\mathscr{C}$ and if $\mathscr{C}_\sigma$ denotes the collection of all countable unions of sets from $\mathscr{C}$ and if $\mathscr{C}_{\sigma\delta}$ denotes the collection of all countable intersections of sets from $\mathscr{C}_\sigma$, then each subset E of X is contained in a set from $\mathscr{C}_{\sigma\delta}$ which has the same outer measure as E. Hence, if the sets in $\mathscr{C}$ can be shown to be measurable, the measure is regular. Further, if $m(X) < \infty$, then the criteria given in the previous theorem for measurability can be used for all sets $E \subset X$.

We next present some simple examples of measures on sets.

<u>Example</u> 1. Let $X = \{a,b\}$. Let $\tau(X) = 1$, $\tau(\emptyset) = 0$, $\tau(\{a\}) = \tau(\{b\}) = 3/4$. Then $m^*(A) = \tau(A)$, $\emptyset$ and X are the measurable sets, $\{a\}$ and $\{b\}$ are not measurable. If m_1 is generated as above, $m_1(\emptyset) = 0$, $m_1(X) = m_1^*(\{a\}) = m_1^*(\{b\}) = 1$.

<u>Example</u> 2. Let X be any set, $A \subset X$ be a countable set and $f(x) = 0$ if $x \notin A$, $f(x) \ge 0$ if $x \in A$ (f

possibly taking on $+\infty$). Then $m(E) = \sum_{x \in E \cap A} f(x)$ is a measure and every subset of X is measurable. Such a measure is called a discrete measure.

Example 3. Let X be a set and for $A \subset X$ let $m(A) = 0$ if A is at most countable, $m(A) = \infty$ if A is uncountable. Then m is a measure and again every subset of X is measurable.

We are of course interested in Lebesgue measure on the line and in Euclidean n-space. Recall that if $I = (a,b)$, then $\tau(I) = b - a$ and if $E \subset \mathbb{R}$, $m^*(E) = \inf\{\Sigma(b_i - a_i): E \subset \bigcup_{i=1}^{\infty} (a_i,b_i)\}$. The following theorem on the measurable sets is in order:

(5.5) Theorem. Given Lebesgue outer measure on $\mathbb{R}$ as defined above, every Borel subset of $\mathbb{R}$ is measurable and the Lebesgue measure of (a,b) is $b - a$.

Proof. In order to show that Borel sets are measurable it will suffice to show that each interval (a,∞) or $(-\infty,a)$ is measurable. Then, since the measurable sets form a σ-algebra, it follows that open intervals (a,b) are measurable; open sets, being the countable union of open intervals, are measurable; Borel sets, making up the smallest σ-algebra which contains the open sets, are also measurable. Consider the interval (a,∞) and A any subset of $\mathbb{R}$ with $m^*(A) < \infty$. Given $\varepsilon > 0$, let $\{(c_i,d_i)\}_{i=1}^{\infty}$ be a collection of intervals such that $A \subset \cup(c_i,d_i)$ and $m^*(A) + \varepsilon \geq \sum_{i=1}^{\infty} (d_i - c_i)$.

Without any loss of generality, assume that no interval (c_i, d_i) is contained in the union of all the others. If an interval (c_i, d_i) were so contained, it could be removed from the collection with the resulting collection still covering A and, proceeding inductively, all such intervals could be removed. This being so, each point x, and in particular a, belongs to at most two intervals (c_i, d_i). If (c_i, d_i) contains a, replace (c_i, d_i) with three intervals, (c_i, a), $(a-\delta, a+\delta)$ and (a, d_i) where $\delta < \varepsilon$. Reorder the resulting intervals into a sequence $\{(a_i, b_i)\}$. Then

$$m^*(A \cap (a,\infty)) \le \Sigma'(b_i - a_i)$$

and

$$m^*(A \cap (-\infty,a]) \le \Sigma''(b_i - a_i)$$

where Σ' is over all (a_i, b_i) which meet (a,∞) and Σ'' is over all (a_i, b_i) which meet $(-\infty,a]$. Since at most two replaced intervals meet both sets and these have length less than 2ε,

$$m^*(A) \ge \Sigma(d_i - c_i) - \varepsilon \ge \Sigma'(b_i - a_i) + \Sigma''(b_i - a_i) - \varepsilon - 4\varepsilon$$

and

$$m^*(A) \ge m^*(A \cap (-\infty,a]) + m^*(A \cap (a,\infty)) - 5\varepsilon.$$

Since $\varepsilon > 0$ is arbitrary, $m^*(A) \ge m^*(A \cap (-\infty,a]) + m^*(A \cap (a,\infty))$. Since A was an arbitrary set with $m^*(A) < \infty$, (a,∞) is a measurable set. Also, $(-\infty,a)$ is measurable by a symmetric argument. To see that $m((a,b)) = b - a$, first note that $m((a,b)) \le b - a$ since (a,b) covers itself. Given $\varepsilon > 0$, let $\{(c_i, d_i)\}$ be a countable collection of intervals whose union contains (a,b) so that $m((a,b)) \ge \Sigma(d_i - c_i) - \varepsilon$. Adjoin $(a-\varepsilon, a+\varepsilon)$ and $(b-\varepsilon, b+\varepsilon)$ to this cover. The resulting collection covers $[a,b]$ and hence contains a finite subcover of $[a,b]$. Without loss of generality, we may again assume that no interval in the subcover is contained in the union of the others. Then each $x \in [a,b]$ is covered at most twice by intervals in the subcover and the subcover can be arranged in order (a_1,b_1), (a_2,b_2), ..., (a_N,b_N) where $a_1 < a_2 < \dots < a_N$. But then, because no interval is contained in the union of the others, $b_1 > a_2$ and $b_i > a_{i+1}$ for each $i < N$. Thus $\sum_{i=1}^{N}(b_i - a_i) \geq \sum_{i=1}^{N-1}(a_{i+1} - a_i) + b_N - a_N = b_N - a_1$. Thus $m(a,b) \geq \Sigma(d_i - c_i) - \varepsilon \geq \Sigma(b_i - a_i) - 5\varepsilon \geq b_N - a_1 - 5\varepsilon \geq b - a - 5\varepsilon$. Here the additional 4ε comes from the intervals $(a-\varepsilon, a+\varepsilon)$ and $(b-\varepsilon, b+\varepsilon)$ adjoined to the cover. Since $\varepsilon > 0$ is arbitrary, $m\big((a,b)\big) = b - a$.

Some basic facts regarding Lebesgue measure on the line are worth noting. First of all, every finite or at most countable set has Lebesgue measure equal to 0. To see this, let $\varepsilon > 0$ be given and let $\{x_n\}_{n=1}^{\infty}$ be an enumeration of the at most countable set. Then, if $I_n = (x_n - \varepsilon/2^n, x_n + \varepsilon/2^n)$, $\{x_n\} \subset \cup I_n$ and $\Sigma|I_n| = \sum_{n=1}^{\infty} 2(\varepsilon/2^n) = 2\varepsilon$. Since $\varepsilon > 0$ is arbitrary, the measure of the at most countable set is 0.

The measure of the Cantor ternary set is also 0. This is because for each natural number n the Cantor ternary set is contained in 2^n closed intervals each of length 3^{-n}. It follows that the measure of the

Cantor set is less than $(2/3)^n$ and thus its measure is 0. By noting that each of the 2^c subsets of the Cantor set, being subsets of a set of measure 0, are also measurable, one obtains the fact that there are more Lebesgue measurable sets on the line than Borel sets and thus that there are measurable sets which are not Borel. However, the following theorem indicates a relationship between the Borel sets and the measurable ones. A measure on a topological space is said to be *Borel regular* provided every measurable set is contained in a Borel set which has the same measure as the measurable one, a measure space is said to be of *σ-finite measure* if it is the countable union of measurable sets of finite measure.

(5.6) *Theorem. Given any Lebesgue measurable set* E *and* $\varepsilon > 0$

i) *there is an open set* G *such that* $E \subset G$ *and* $m(G \setminus E) < \varepsilon$,

ii) *there is a* $\mathcal{G}_\delta$ *set* A *such that* $E \subset A$ *and* $m(A \setminus E) = 0$,

iii) *there is a closed set* F *such that* $F \subset E$ *and* $m(E \setminus F) < \varepsilon$,

iv) *there is an* $\mathcal{F}_\sigma$ *set* H *such that* $H \subset E$ *and* $m(E \setminus H) = 0$,

v) *if* $m(E) < \infty$, *there is a compact set* $K \subset E$ *so that* $m(E \setminus K) < \varepsilon$.

Moreover, the statements i) - iv) *are true for any measure defined on a topological space which is generated by Caratheodory's methods using a* τ *defined on a covering class* $\mathcal{C}$ *which consists of open sets,*

<u>providing the space is of σ-finite measure and open sets are measurable</u>. <u>Furthermore</u>, v) <u>is true provided the space is also the countable union of a sequence of compact sets</u>.

<u>Proof</u>. We will prove i) - iv) in the general situation. Note that the line is the union of $[n, n+1)$ where n ranges over the integers and, in general, if X is σ-finite, $X = \bigcup_{n=1}^{\infty} X_n$ where each X_n is measurable and $m(X_n) < \infty$; if $X_o = \emptyset$ and $E_n = X_n \setminus \bigcup_{1}^{n-1} X_i$, then $X = \bigcup_{n=1}^{\infty} E_n$ and the E_n are both measurable and pairwise disjoint.

i) Given a measurable set E and $\varepsilon > 0$, if $X = \cup E_n$ as above, there are open sets $G_{i,n} \in \mathscr{C}$ such that $E \cap E_n \subset \bigcup_i G_{i,n}$, and $m(E \cap E_n) \geq \sum_i \tau(G_{i,n}) - \varepsilon/2^n$. Thus, if $G_n = \bigcup_i G_{i,n}$, $m(G_n) \leq \Sigma\, \tau(G_{i,n})$ and $m(E \cap E_n) \geq m(G_n) - \varepsilon/2^n$. Since $m(G_n)$ and $m(E \cap E_n)$ are finite, $m(G_n \setminus (E \cap E_n)) \leq \varepsilon/2^n$. Now, if $G = \cup G_n$, G is open and $m(G \setminus E) \leq m(\bigcup_n G_n \setminus (E \cap E_n)) \leq \Sigma\, m(G_n \setminus (E \cap E_n)) \leq \varepsilon$.

ii) Let G_k be open sets with $E \subset G_k$ and $m(G_k \setminus E) < 1/k$. Such sets are guaranteed by i). If $A = \cap G_k$ then $m(A \setminus E) \leq m(G_k \setminus E) \leq 1/k$ and since k is arbitrary, $m(A \setminus E) = 0$. Also, A is a $\mathscr{G}_\delta$ set containing E.

iii) Again, using i), let $G \supset E^c$ with $m(G\setminus E^c) < \varepsilon$. Let $F = G^c$. Since open sets are measurable, so are closed sets. Now $F \subset E$, $E\setminus F = G\setminus E^c$ and hence $m(E\setminus F) < \varepsilon$.

iv) Now using ii), let A be a $\mathcal{G}_\delta$ with $E^c \subset A$ and $m(A\setminus E^c) = 0$. Then, if $H = A^c$, H is an $\mathcal{F}_\sigma$ and $H \subset E$, $E\setminus H = A\setminus E^c$ and thus $m(E\setminus H) = 0$.

v) Let K_n be compact and $X = \cup K_n$. Clearly, the K_n can be chosen to be nondecreasing. If $m(E) < \infty$, since $m(E \cap K_n)$ approaches $m(E)$ as n approaches ∞, there is an N so that $m(E) - m(E \cap K_N) < \varepsilon/2$. Then $m(E\setminus K_N) < \varepsilon/2$. There is, according to iii), a closed set $K \subset E \cap K_N$ so that $m[(E \cap K_N)\setminus K] < \varepsilon/2$. Because K is a closed subset of K_N, it is compact. Then $m(E\setminus K) = m(E\setminus K_N) + m[(E \cap K_n)\setminus K] < \varepsilon$.

Theorem 5.6 leads to the existence of perfect nowhere dense sets of positive measure. For example, let E be the set of irrational numbers in $[0,1]$. Then $m(E) = 1$ because $m(E^c) = 0$. There is a closed set $F \subset E$ with $m(F) \geq 1/2$. Clearly, F must be nowhere dense. For if F were dense in an interval I, F being closed, F would contain I; but F is contained in E and contains no rational numbers. Now $F = P \cup D$ where P is perfect and D is at most countable. Since $m(D) = 0$, $m(P) = m(F) \geq 1/2$. Actually, F can be constructed by enumerating the rational numbers in $[0,1]$ in a sequence $\{r_n\}$. Let $G = \underset{n}{\cup}(r_n-5^{-n}, r_n+5^{-n})$. Then $m(G) \leq 2\cdot\Sigma 5^{-n} = 1/2$ and $F = G^c$ has measure greater than or equal to $1/2$. Numerical examples of closed nowhere dense sets of positive measure can also be explicitly constructed.

Theorem 5.6 also leads to the existence of non-measurable subsets of the reals. Let S be the totally imperfect set (Example 3.11) constructed in Chapter 3. It is thus not possible to find a closed subset of positive measure in either S or S^c. For such a set F would be the union of a perfect set P and a denumerable set D. Since there is no such P, every closed subset of S or S^c is at most countable and hence of measure 0. Theorem 5.6 implies that if S were measurable, then $m(S) = 0$. For the same reason $m(S^c)$ would equal 0 contrary to the fact that $m(R) = \infty$ and $R = S \cup S^c$. The same construction used to produce S can be performed on any measurable set E of positive measure. That is, there is a set $T \subset E$ so that neither T nor $E \setminus T$ contains a perfect set. If T were measurable, both T and $E \setminus T$ would have measure 0. Then $m(E)$ would equal $m(T) + m(E \setminus T)$ and since this is not the case, it follows that T is not measurable. In fact $m^*(T) = m^*(E \setminus T) = m(E)$ follows from the fact that, if $m^*(T) < m(E)$, then there would be an open set G with $T \subset G$ and $m(G) < m(E)$. But then $E \setminus G$, being of positive measure, would contain a perfect set and it does not.

Lebesgue measure in Euclidean n-space is a special case of what is called a product measure. If $(X, \mathcal{A}, m_1)$ and $(Y, \mathcal{B}, m_2)$ are two measure spaces, a natural measure can be defined on $X \times Y = \{(x,y): x \in X, y \in Y\}$ by defining $\tau(A \times B) = m_1(A) \cdot m_2(B)$ for each pair of sets $A \in \mathcal{A}$ and $B \in \mathcal{B}$. The resulting measure is sometimes written as $m_1 \times m_2$. This can clearly be done with the cross product of more than two spaces.

In the case where $X = Y = \mathbb{R}$ and m_1 and m_2 are Lebesgue measure, this definition yields the same values for two dimensional Lebesgue measure as does the definition using the covering class

$$\mathscr{C} = \{(a,b) \times (c,d) : a < b, \ c < d\}.$$

Each pair of measurable sets A and B of finite measure can be covered so that, given $\varepsilon > 0$, $A \subset \cup(a_i, b_i)$ and $m(A) > \Sigma(b_i - a_i) - \varepsilon$ and $B \subset \cup(c_j, d_j)$ and $m(B) > \Sigma(d_j - c_j) - \varepsilon$. Then, it follows that $\tau(A \times B) = m(A)\cdot m(B) \le (\Sigma(b_i - a_i) - \varepsilon)(\Sigma(d_j - c_j) - \varepsilon)$. Then for each set E in the plane, one obtains

$$\inf\{\Sigma m(A_n)\cdot m(B_n) : E \subset \bigcup_n A_n \times B_n\}$$
$$= \inf\{\Sigma(b_k - a_k)(d_k - c_k) : E \subset \cup(a_k, b_k) \times (c_k, d_k)\}$$

Thus the measure determined by cross products of measurable sets is the same as that determined by cross products of open intervals.

If A and B are measurable in X and Y respectively, two questions arise concerning the cross product measures. Is $A \times B$ an $(m_1 \times m_2)$ measurable set? Is $(m_1 \times m_2)(A \times B) = m_1(A)\cdot m_2(B)$? Both questions have affirmative answers. The measurability of $A \times B$ is shown easily. For let E be any subset of $X \times Y$ with $(m_1 \times m_2)^*(E) < \infty$. Let A_i, B_i satisfy $E \subset \cup(A_i \times B_i)$ and $m^*(E) \ge \Sigma m_1(A_i)\cdot m_2(B_i) - \varepsilon$. Then

$$\Sigma m_1(A_i)\cdot m_2(B_i)$$
$$= \Sigma m_1(A_i \cap A)\cdot m_2(B_i) + \Sigma m_1(A_i \cap A^c)\cdot m_2(B_i)$$
$$= \Sigma m_1(A_i \cap A)\cdot m_2(B_i \cap B)$$
$$+ \Sigma m_1(A_i \cap A^c)\cdot m_2(B_i \cap B)$$
$$+ \Sigma m_1(A_i \cap A)\cdot m_2(B_i \cap B^c)$$

$$+ \quad \Sigma m_1(A_i \cap A^c)\cdot m_2(B_i \cap B^c)$$

$$\geq \quad (m_1\times m_2)^*\big(E \cap (A\times B)\big) + (m_1\times m_2)^*\big(E \cap (A\times B)^c\big).$$

The last inequality is due to the fact that

$$E \cap (A \times B) \subset \cup(A_i \cap A) \times (B_i \cap B)$$

and

$$E \cap (A \times B)^c \subset$$
$$\bigcup_i \big[(A_i \cap A^c) \times (B_i \cap B) \cup (A_i \cap A) \times (B_i \cap B^c) \cup (A_i \cap A^c) \times (B_i \cap B^c)\big].$$

Since $\varepsilon > 0$ was arbitrary,

$$(m_1 \times m_2)^*(E) \geq$$
$$(m_1 \times m_2)^*\big(E \cap (A \times B)\big) + (m_1 \times m_2)^*\big(E \cap (A \times B)^c\big).$$

It follows that $A \times B$ is $(m_1 \times m_2)$ measurable.

The proof that $(m_1 \times m_2)(A \times B)$ equals $m_1(A)\cdot m_2(B)$ for all measurable sets A and B will be postponed until Chapter 7. Such sets $A \times B$ are sometimes called <u>measurable rectangles</u>. A direct argument can be given that the Lebesgue measure of $(a,b) \times (c,d)$ in the plane is $(b - a)(d - c)$. It is as follows:

First of all, each of the boundary line segments of $(a,b) \times (c,d)$ are of area measure 0. For example

$$\{(x,y): x = a, \quad y \in [c,d]\}$$

can be covered with $(a-\varepsilon,a+\varepsilon) \times (c-\varepsilon,d+\varepsilon)$, a rectangle of area $2\varepsilon(d - c + 2\varepsilon)$. Thus it suffices to show that $S = [a,b] \times [c,d]$ has measure $(b - a)(d - c)$. Let $\bigcup_{i=1}^{\infty} (a_i,b_i) \times (c_i,d_i)$ contain S. Since S is compact there is a finite collection of the $(a_i,b_i) \times (c_i,d_i)$ which also covers S. Let $\{x_j\}_{j=1}^{N}$ be an enumeration of all the a_i and b_i from this collection and let $\{y_k\}_{k=1}^{M}$ be an enumeration of all the c_i and d_i. Assume the x_j and y_k are listed in order of smallest to largest. Fix i. Since

$$\{[x_{j-1},x_j]\times[y_{k-1},y_k]: a_i \le x_{j-1} \le b_i,\ c_i \le y_{k-1} \le d_i\}$$

covers $[a_i,b_i] \times [c_i,d_i]$ and, for the summation over all pairs j, k for which $[x_{j-1},x_j] \times [y_{k-1},y_k]$ are in this set, we have

$$\begin{aligned}\sum_{j,k} (y_k - y_{k-1})(x_j - x_{j-1}) &= \sum_k \Big(\sum_j y_k(x_j - x_{j-1}) - \sum_j y_{k-1}(x_j - x_{j-1})\Big) \\ &= \sum_k y_k(b_i - a_i) - \sum_k y_{k-1}(b_i - a_i) \\ &= (b_i - a_i)(d_i - c_i).\end{aligned}$$

Thus all the rectangles $[x_{j-1},x_j] \times [y_{k-1},y_k]$ together form an at least equally efficient cover of $[a,b] \times [c,d]$ as that of the collection of $[a_i,b_i] \times [c_i,d_i]$. However,

$$\begin{aligned}\sum_{j,k} (y_k - y_{k-1})(x_j - x_{j-1}) &\ge \sum_k (y_k - y_{k-1})(b - a) \\ &\ge (b - a)(d - c).\end{aligned}$$

Now the measure of $(a,b) \times (c,d)$ is no more than $(b - a)(d - c)$, because the rectangle $(a,b) \times (c,d)$ covers itself. It follows that $(b - a)(d - c)$ is the measure of the rectangle. Similar considerations hold for Euclidean k-dimensional space. Since k-cells are measurable with the appropriate measure, so are all the Borel sets, the Borel sets being the smallest σ-algebra which contains the k-cells in k-dimensional space.

We turn now to two geometric theorems involving Lebesgue measure. The first, the Vitali Covering

Theorem will be proved for n-dimensional Lebesgue measure in Euclidean n-space. This is the key theorem for making the connection between integration and differentiation. The second theorem, the Lebesgue Density Theorem, is proved only for Lebesgue measure on the line. Similar and, indeed, stronger analogous results hold in Euclidean n-space. In what follows, if a set E is measurable, where the measure is Lebesgue measure, either $|E|$ or $m(E)$ will be used to denote the measure of E. The letters I and J will frequently be used to denote an interval on the line or a k-cell in k-dimensional space. The k-dimensional measure of a k-cell I will most often be written $|I|$.

A collection of closed sets $\mathcal{V}$ is said to be a <u>Vitali cover</u> of a set $E \subset \mathbb{R}^k$ or to <u>cover</u> E <u>in the sense of Vitali</u> provided that for each $x \in E$ there is a number $r(x) > 0$, a sequence of k-cells I_n with equal length sides centered at x and sets from the cover $A_n \subset I_n$ so that $m(A_n)/|I_n| > r(x)$ and $|I_n| \to 0$. The number $r > 0$ is called a <u>parameter of regularity</u> of a set A if A is contained in a k-cell I with equal length sides and $|A|/|I| \geq r$. Note that, if a set E is covered by $\mathcal{V}$ in the sense of Vitali, it is covered quite densely. Thus, if G is an open set and $E \subset G$ and $\mathcal{V}_G = \{A \in \mathcal{V}: A \subset G\}$, then $\mathcal{V}_G$ is also a Vitali cover of E. Alternately, if for $\varepsilon > 0$, $\mathcal{V}_\varepsilon = \{A \in \mathcal{V}: \text{diam } A < \varepsilon\}$, then $\mathcal{V}_\varepsilon$ is also a Vitali cover of E. Restricting the cover in such a fashion is sometimes referred to as "pruning".

What follows is a general version of the Vitali Covering Theorem. The proof is given for sets $E \subset \mathbb{R}^k$

with m equal to k-dimensional Lebesgue measure.

(5.7) Theorem. (The Vitali Covering Theorem) *If $E \subset \mathbb{R}^k$ is any set which is covered by $\mathcal{V}$ in the sense of Vitali, there is a finite or countable sequence $\{A_n\}$ of pairwise disjoint sets belonging to $\mathcal{V}$ which satisfy $m(E \setminus \cup A_n) = 0$.*

Proof. First consider a closed k-cell I with equal sides of length s. Suppose I' is a larger closed k-cell with equal length sides and same center as I and that for $x \in I$, $x' \notin I'$, we have $d(x,x') > 2(\operatorname{diam} I)$. For this to happen an easy calculation shows that it is sufficient that $\operatorname{diam} I' \geq a(\operatorname{diam} I)$ where $(as - s)/2 = 2s\sqrt{k}$. That is, $a = 4\sqrt{k} + 1$. This number a is required later in the proof. To begin, fix $E \subset \mathbb{R}^k$ and $\mathcal{V}$ a Vitali cover of E. For each natural number n, let

$$E_n = \{x \in E: r(x) > 1/n \text{ and } \|x\| < n\}.$$

Fix n. Let

$$\mathcal{V}_{n,1} = \{A \in \mathcal{V}: A \subset N_n(0) \text{ and there is } I \text{ with } A \subset I \text{ and } m(A)/m(I) > 1/n\}.$$

Then $\mathcal{V}_{n,1}$ is a Vitali cover of E_n. Choose $A_1 \in \mathcal{V}_{n,1}$ satisfying $\operatorname{diam} A_1 > (1/2)\sup\{\operatorname{diam} A: A \in \mathcal{V}_{n,1}\}$. Then, since A_1 is closed, either $E_n \setminus A_1 = \emptyset$ or $\mathcal{V}_{n,2} = \{A \in \mathcal{V}_{n,1}: A \cap A_1 = \emptyset\}$ is a Vitali cover of $E_n \setminus A_1$. If sets A_i and covers $\mathcal{V}_{n,i+1}$ of $E_n \setminus \bigcup_1^i A_j$ have been determined, choose $A_{i+1} \in \mathcal{V}_{n,i+1}$ so that $\operatorname{diam} A_{i+1} > (1/2)\sup\{\operatorname{diam} A: A \in \mathcal{V}_{n,i+1}\}$. Then either

$E_n \setminus \bigcup_1^{i+1} A_j = \emptyset$ or $\mathcal{V}_{n,i+2} = \{A \in \mathcal{V}_{n,i+1}: A \cap A_{i+1} = \emptyset\}$ is a Vitali cover of $E_n \setminus \bigcup_1^{i+1} A_j$. Let $B_n = E_n \setminus \bigcup_1^{\infty} A_i$ and suppose it is possible that $m(B_n) > 0$. Each A_i is contained in a k-cell I_i with equal length sides and satisfying $m(A_i)/|I_i| > 1/n$. Let I_i' be the k-cell with the same center as I_i, equal length sides so that $\operatorname{diam} I_i' = a(\operatorname{diam} I_i)$. Since all the sets A_i are contained in $N_n(0)$, it follows that

$$\Sigma|I_i'| = a^k\Sigma|I_i| < a^k\cdot n\cdot\Sigma m(A_i) < a^k\cdot n\cdot(2n)^k < \infty.$$

This implies that $\operatorname{diam} I_i' \to 0$, $\operatorname{diam} A_i \to 0$ and that there is a natural number M so that $\sum_{i=M}^{\infty} |I_i'| < m(B_n)$. Thus there is $x_o \in B_n$ so that $x_o \notin I_j'$ for $j > M$. There is a set $A_o \in \mathcal{V}_{n,M}$ such that $A_o \cap \bigcup_1^{M-1} A_i = \emptyset$, $A_o \subset I_o$ where I_o is a k-cell centered at x_o with equal length sides and $m(A_o)/|I_o| > 1/n$. In effect, this set A_o should have been chosen for one of the A_i. This is because $A_o \cap A_i = \emptyset$ for $i = 1, 2, \ldots, M-1$ due to the fact that $A_o \in \mathcal{V}_{n,M}$. Furthermore, A_o can have no points in common with A_i, for $i \geq M$. Indeed, if A_o did have points in common with a given A_i with $i \geq M$, then $\operatorname{diam} A_o$ would be larger than $2(\operatorname{diam} I_i)$ and A_i could not have been chosen rather than A_o. Thus A_o meets no A_i. But $\operatorname{diam} A_i \to 0$. Thus some A_i has diameter less than $(1/2)\operatorname{diam} A_o$ and that A_i could not have been chosen. This contradiction implies that $m(E_n \setminus \cup A_i) = 0$. Now, let ε_n decrease to 0. There is M_1 so that $m^*(E_1 \setminus \bigcup_1^{M_1} A_i) < \varepsilon_1$. By pruning $\mathcal{V}$ to $(\bigcup_1^{M_1} A_i)^c$, there are A_i, $i = M_1 + 1, \ldots, M_2$ so that $m^*(E_2 \setminus \bigcup_1^{M_2} A_i) < \varepsilon_2$. Continuing in this fashion, there are A_i, $i = M_n + 1, \ldots, M_{n+1}$ so that $m^*(E_N \setminus \bigcup_1^{M_{n+1}} A_i) < \varepsilon_n$. Thus for each n, $m(E_n \setminus \bigcup_1^{\infty} A_i) = 0$ and, since $E \setminus \bigcup_1^{\infty} A_i = \bigcup_n E_n \setminus \bigcup_1^{\infty} A_i$, it follows that $m(E \setminus \cup A_i) = 0$. This completes the proof of the theorem.

The Lebesgue Density Theorem, which follows, is a natural application of the Vitali Covering Theorem. If E is a subset of the line and $E \in \mathcal{M}$ and x is any point (not necessarily in E), several density concepts are defined indicating the denseness of E near x. We write

$$\overline{D}_E(x) = \overline{\lim_{\varepsilon \to 0^+}} \{m(E \cap I)/|I| : x \in I \text{ and } |I| < \epsilon\}$$

and call this the <u>upper density</u> of E at x. This limit supremum is sometimes written $\overline{\lim_{I \to x}} |E \cap I|/|I|$. Similarly, the <u>lower density</u> of E at x is

$$\underline{D}_E(x) = \underline{\lim_{I \to x}} \frac{|E \cap I|}{|I|} = \underline{\lim_{\varepsilon \to 0^+}} \{|E \cap I|/|I| : x \in I, \ |I| < \varepsilon\}.$$

The <u>density</u> of E at x, $D_E(x) = \lim_{I \to x} |E \cap I|/|I|$ is the common value of the lim inf and lim sup when they are equal. If E is arbitrary; that is, not necessarily measurable, then four densities are defined.

The <u>upper outer density</u> is

$$\overline{D}_E^+(x) = \overline{\lim_{I \to x}}\, m^*(E \cap I)/|I|;$$

the <u>upper inner density</u> is

$$\overline{D}_E^-(x) = \overline{\lim_{I \to x}}\, m_*(E \cap I)/|I|;$$

the <u>lower outer density</u> is

$$\underline{D}_E^+(x) = \underline{\lim_{I \to x}}\, m^*(E \cap I)/|I|;$$

and the <u>lower inner density</u> is

$$\underline{D}_E^-(x) = \underline{\lim_{I \to x}}\, m_*(E \cap I)/|I|.$$

If E is measurable, then $\overline{D}_E^+(x) = \overline{D}_E^-(x)$ and $\underline{D}_E^+(x) = \underline{D}_E^-(x)$. A point x is called a <u>point of density</u> of a set E if $\underline{D}_E^+(x) = 1$ and thus if $E \in \mathcal{M}$ and x is a point of density of E, all the densities of E at x equal 1.

To illustrate the concept of density, let G be an open subset of the reals. Then every point $x \in G$ is a point of density of G. A point x is called a <u>point of dispersion</u> of E if $\overline{D}_E^+(x) = 0$; that is, if all the densities of E at x are 0.

The Lebesgue Density Theorem asserts that every point x which belongs to a measurable set E, except for a set Z of points of E with $m(Z) = 0$, are points of density of E. A property which holds at all the points of a set E except for those in a subset of measure 0 is said to hold <u>almost everywhere</u> on E or at almost every point of E. This is abbreviated a.e.

(5.8) Theorem. (Lebesgue Density Theorem) *Given any set E, almost every point of E is a point of density of E. If, in addition E is measurable, almost every point of E^c is a point of dispersion of E.*

Proof. Suppose not. That is, suppose there is a set E such that the set $B = \{x \in E: \underline{D}_E^+(x) < 1\}$ has positive outer measure. Then if

$$B_n = \{x \in E \cap (-n,n): \underline{D}_E^+(x) < 1 - 1/n\},$$

$B = \cup B_n$ and there is a natural number N so that $m^*(B_N) > 0$. Now let $\mathcal{V}$ be the collection of closed intervals I which satisfy the condition that there is a point of B_N in I and $m^*(E \cap I)/|I| < 1 - 1/N$. Then $\mathcal{V}$ covers B_N in the sense of Vitali. Let G be an open set with $B_N \subset G$ such that $m(G) < (1 + 1/N)\cdot m^*(B_N)$. Then $\mathcal{V}_G = \{I \in \mathcal{V}: I \subset G\}$ also covers B_N in the sense of Vitali. Thus, there is a sequence of pairwise disjoint closed intervals $\{I_j\}$ contained in $\mathcal{V}_G$ such that $m\left(B_N \setminus \bigcup_{j=1}^{\infty} I_j\right) = 0$. But then

$$m^*(B_N) \leq \sum_j m^*(B_N \cap I_j) \leq \Sigma(1 - 1/N)|I_j|$$
$$\leq (1 - 1/N)|G| \leq (1 - 1/N)(1 + 1/N)\cdot m^*(B_N)$$

Thus $m^*(B_N) \leq (1 - 1/N^2)\cdot m^*(B_N)$ contradicts the fact that $m^*(B_N) > 0$. Thus $m(B_N) = 0$ and $m(B) = 0$ and almost every point of E is a point of density of E. Now, if $E \in \mathcal{M}$, then for each interval I, $|E \cap I| + |E^c \cap I| = |I|$. Thus $\dfrac{|E \cap I|}{|I|} + \dfrac{|E^c \cap I|}{|I|} = 1$. Since almost every point of E^c is a point of density of E^c, if x is such a point,

$$\lim_{I\to x} \frac{|E^c \cap I|}{|I|} = 1$$

and hence

$$\lim_{I\to x} \frac{|E \cap I|}{|I|} = \lim_{I\to x}\left(1 - \frac{|E^c \cap I|}{|I|}\right) = 0$$

and x is a point of dispersion of E. Thus for a measurable set E, a.e. point of E is a point of density of E and a.e. point of E^c is a point of dispersion of E.

5.1 Exercises

1. Suppose $m(X) < \infty$ and for $A, B \subset X$, $A \sim B$ if $m(A\backslash B) + m(B\backslash A) = 0$. Show that $d(A,B) = m(A\backslash B) + m(B\backslash A)$ is a metric on these equivalence classes of measurable subsets of X. (It is called the <u>Hausdörff metric</u>.)
2. Show that a set $E \subset \mathbb{R}$ is Lebesgue measurable iff for every $\varepsilon > 0$ there are open sets G_1 and G_2 so that $E \subset G_1$, $E^c \subset G_2$ and $m(G_1 \cap G_2) < \varepsilon$.
3. Show that there is an $\mathcal{F}_\sigma$ set $E \subset [0,1]$ such that for each $I \subset [0,1]$, $m(E \cap I) > 0$ and $m(E^c \cap I) > 0$.
4. Show that the real numbers can be written as a union of c non-measurable sets.
5. Prove that there does not exist a measurable set $E \subset \mathbb{R}$ with the property that $m(E \cap I) = (1/2)\cdot m(I)$ for each interval I.
6. Show that $\{x: x = \sum_{i=2}^{\infty} (a_i/2^i + a_{i-1}/8^i),\ a_i = 0,1\}$ is a nowhere dense perfect set of measure $1/2$.

7. Given $x \in [0,1]$, let x_n be the nth digit in the decimal expansion of x. (If two expansions of x exist, choose the one ending in 0's.) Let $\{n_k\}$ be an increasing sequence of natural numbers. Show that $\{x\colon \text{for every } k,\ n_k \neq 5\}$ has measure 0.
8. Using the notation of Exercise 7, show
$$\{x \in [0,1]\colon\ n_k = 5,\ k = 1, 2, \ldots\}$$
has measure 0.
9. A number x is said to be a <u>normal</u> <u>number</u> (with respect to the decimal system) if
$$x = \Sigma a_i/10^i, \quad a_i = 0, 1, \ldots, 9$$
and
$$\lim_{n\to\infty} \|\{a_i\colon a_i = k,\ i \le n\}\|_c/n = 1/10$$
for $k = 0, 1, \ldots, 9$. Show that almost every number is a normal number.
10. Show that if $E \subset \mathbb{R}$, $E \in \mathcal{M}$ and $m(E) > 0$, then then for each natural number n, E contains n equispaced points; i.e., points $x_1, \ldots, x_n$ with each $d(x_{i-1}, x_i)$ equal.
11. Show that $Z \subset [0,1]$ is of measure 0 iff there is a sequence of intervals $\{I_n\}$ with $\Sigma m(I_n) < \infty$ and for each N, $Z \subset \bigcup_{n=N}^{\infty} I_n$.
12. Show that if E is a measurable set contained in $[0,1]$, then for any $\varepsilon > 0$ there is a set A which is a finite union of intervals so that $m(E\backslash A) + m(A\backslash E) < \varepsilon$.
13. A set $X \subset \mathbb{R}^k$ has porosity p at a point x if p is the infimum of the number p' such that if for each $\varepsilon > 0$ there are r, r' with $0 < r' < r < \varepsilon$ and $y \in N_r(x)$ so that $N_{r'}(y) \subset N_r(x)$,

$N_{r'}(x) \cap X = \emptyset$ and $r'/r \geq p'$. Then X is said to be porous at x if it has porosity $p > 0$ at x; X is porous if it is porous at each of its points; X is σ-porous if it is a countable union of porous sets. Show that every σ-porous set is of first category and of measure 0.

5.2 Additive Functions of a Set

Before considering a variety of measures which have proven useful, we consider a generalization of the concept of measure in which both positive and negative values are allowed. A function Φ is called a completely additive function of a set provided Φ is defined on a σ-algebra $\mathcal{A}$ of subsets of a set X and satisfies $\Phi(\emptyset) = 0$ and, if $\{E_n\}$ is a sequence of pairwise disjoint subsets of $\mathcal{A}$, then $\Phi(\cup E_n) = \Sigma\, \Phi(E_n)$. Such a function Φ is sometimes called a signed measure.

A function Φ which is defined on an algebra of subsets of X is called finitely additive if $\Phi(\cup E_n) = \Sigma\Phi(E_n)$ for any finite, pairwise disjoint collection of sets E_n from the algebra and $\Phi(\emptyset) = 0$. (Recall that an algebra is a collection of subsets of a set X which is closed under finite unions, finite intersections and complements and contains $\emptyset$.) Such functions Φ are allowed to take on infinite values but, because of the impossibility of defining $\infty + (-\infty)$, even a finitely additive function can take on only one of the two values $+\infty$ or $-\infty$. To see this, suppose $\Phi(A) = \infty$ and $\Phi(B) = -\infty$. If $\Phi(A \cap B)$ were finite, then $\Phi(A \setminus B) = \infty$ and $\Phi(B \setminus A) = -\infty$ would be required by the fact that $\Phi(A) = \Phi(A \cap B) + \Phi(A \setminus B)$ and $\Phi(B) = \Phi(A \cap B) + \Phi(B \setminus A)$. But it is then impossible for Φ to be defined on $(A \setminus B) \cup (B \setminus A)$. On the other hand, if $\Phi(A \cap B)$ were infinite, say $+\infty$, then it would be impossible to define Φ on $B \setminus A$ so that $\Phi(B) = -\infty = \Phi(A \cap B) + \Phi(B \setminus A)$. Similarly, if $\Phi(A \cap B)$ were $-\infty$, it would be impossible to define Φ on $A \setminus B$ so that $\Phi(A) = \infty = \Phi(A \cap B) + \Phi(A \setminus B)$.

Note that if $\Phi(A)$ happens to be infinite and $A \subset B$, then $\Phi(B) = \Phi(A)$ is required in order that $\Phi(B) = \Phi(A) + \Phi(B \setminus A)$. Also, if $\Phi(A)$ is finite and $B \subset A$, $\Phi(B)$ is finite.

An easy example of a completely additive function of a set is obtained by considering the difference of two measures m_1 and m_2 defined on the same σ-algebra $\mathcal{A}$ of subsets of a set X where one of the two measures is finite valued. Then $\Phi(E) = m_1(E) - m_2(E)$ is clearly countably additive and the difficulty of defining $\infty + (-\infty)$ is avoided by the requirement that one of the two measures be finite. While completely additive functions of a set play an important role in analysis, we will see that every such function is the difference of two measures. Then completely additive functions of a set can usually be dealt with by considering each of the two measure functions separately.

We begin by stating two properties which completely additive functions of a set have in common with measures.

(5.9) Theorem. Let Φ be a completely additive function of a set defined on a σ-algebra $\mathcal{A}$ of subsets of a set X. Let $\{E_n\}$ be a sequence of sets each belonging to $\mathcal{A}$. Then

i) if $\{E_n\}$ is non-decreasing, $\Phi(\cup E_n) = \lim \Phi(E_n)$

ii) if $\{E_n\}$ is non-increasing and $\Phi(E_1)$ is finite, $\Phi(\cap E_n) = \lim \Phi(E_n)$.

Proof.

i) Suppose $\{E_n\}$ is non-decreasing. Let $E_o = \emptyset$ and note that $\{E_n \backslash E_{n-1}\}$ is a pairwise disjoint sequence and $\Phi(E_n) = \sum_{i=1}^{n} \Phi(E_i \backslash E_{i-1})$ and $\Phi(\bigcup_{n=1}^{\infty} E_n) = \sum_{i=1}^{\infty} \Phi(E_i \backslash E_{i-1})$ because $E_n = \bigcup_{i=1}^{n} E_i \backslash E_{i-1}$ and $\bigcup_{n=1}^{\infty} E_n = \bigcup_{i=1}^{\infty} E_i \backslash E_{i-1}$. Thus $\Phi(\bigcup_{n=1}^{\infty} E_n) = \sum_{i=1}^{\infty} \Phi(E_i \backslash E_{i-1}) = \lim_{n\to\infty} \sum_{i=1}^{n} \Phi(E_i \backslash E_{i-1}) = \lim_{n\to\infty} \Phi(E_n)$.

ii) Suppose $\{E_n\}$ is non-increasing and $\Phi(E_1)$ is finite. Then $\{E_1 \backslash E_n\}$ is non-decreasing. Since each E_n and $\cap E_n$ are contained in E_1, each number $\Phi(E_n)$ and $\Phi(\cap E_n)$ is finite. Thus

$$\Phi(E_1 \backslash E_n) = \Phi(E_1) - \Phi(E_n)$$

and

$$\Phi(E_1 \backslash \cap E_n) = \Phi(E_1) - \Phi(\cap E_n).$$

Then $\Phi(\bigcup_{n=1}^{\infty} E_1 \backslash E_n) = \lim_{n\to\infty} \Phi(E_1 \backslash E_n) = \Phi(E_1) - \lim_{n\to\infty} \Phi(E_n)$.

Since $\bigcup_{n=1}^{\infty} E_1 \backslash E_n = E_1 \backslash \bigcap_{n=1}^{\infty} E_n$, $\Phi(\bigcup_{n=1}^{\infty} E_1 \backslash E_n) = \Phi(E_1) -$

$\Phi(\bigcap_{n=1}^{\infty} E_n)$. We thus have that $\Phi(E_1) - \lim_{n\to\infty} \Phi(E_n) = \Phi(E_1) - \Phi(\bigcap_{n=1}^{\infty} E_n)$ and, since $\Phi(E_1)$ is finite, $\Phi(\bigcap_{n=1}^{\infty} E_n) = \lim_{n\to\infty} \Phi(E_n)$.

Given a completely additive function Φ defined on $\mathcal{A}$ which does not take on the value $+\infty$, it is shown below that there is a number M such that for each $E \in \mathcal{A}$, $\Phi(E) \le M$; that is, Φ is bounded above. A similar argument yields that, if Φ does not take on the value $-\infty$, Φ is bounded below. Since Φ never has both $+\infty$ and $-\infty$ in its range of values, it will follow that a completely additive function of a set must be either bounded above or bounded below. To see this, suppose Φ is defined on an algebra $\mathcal{A}$ of subsets of X and $\Phi(X) \ne \infty$. Suppose, if possible, that Φ is not bounded above. Then there is $E_1 \subset X$ with $\Phi(E_1) > 1$. Also Φ is not bounded above on at least one of E_1 or $X \backslash E_1$; for otherwise, Φ would be bounded above on X. Let X_1 be either E_1 or $X \backslash E_1$ so that Φ is not bounded above on X_1. Proceed inductively to define E_{n+1} in X_n with $\Phi(E_{n+1}) > n + 1$ and let X_{n+1} be either E_{n+1} or $X_n \backslash E_{n+1}$ so that Φ is not bounded above on X_{n+1}. It is then not possible that $X_n = E_n$ for infinitely many natural numbers n. For if $X_{n_k} = E_{n_k}$, $k = 1, 2, \ldots,$ then E_{n_k} is a non-increasing sequence of sets, $\Phi(E_{n_1})$ is finite and by ii) of the previous theorem, $\Phi(\bigcap_{k=1}^{\infty} E_{n_k})$

$= \lim \Phi(E_{n_k}) = \infty$. Thus there must be a natural number N so that for $n > N$, $X_{n+1} = X_n \setminus E_{n+1}$. But then $\{E_n\}_{n=N}^{\infty}$ is a pairwise disjoint collection of sets. Hence $\Phi\left(\bigcup_{n=N}^{\infty} E_n\right) = \sum_{n=N}^{\infty} \Phi(E_n) = \infty$ and it follows that $\Phi(X) = \infty$, contrary to the initial assumption. It follows that, if $\Phi(X) \neq \infty$, Φ is bounded above.

Given an additive function of a set Φ, for each set E belonging to the σ-algebra $\mathcal{A}$, the <u>upper variation of Φ on E</u> is defined to be $\overline{V}_\Phi(E) = \sup\{\Phi(A): A \subset E, A \in \mathcal{A}\}$ and the <u>lower variation of Φ on E</u> to be $\underline{V}_\Phi(E) = \inf\{\Phi(A): A \subset E, A \in \mathcal{A}\}$. Since, by the above discussion, one of $\overline{V}_\Phi(E)$ or $\underline{V}_\Phi(E)$ is finite, the <u>variation of Φ on E</u> can be defined as $V_\Phi(E) = \overline{V}_\Phi(E) - \underline{V}_\Phi(E)$. Note that, since $\emptyset \subset E$, $\overline{V}_\Phi(E) \geq 0 \geq \underline{V}_\Phi(E)$.

The following theorem relates these variations to Φ and shows that every completely additive function of a set is the difference of two measures one of which is finite.

(5.10) <u>Theorem</u>. (Jordan Decomposition Theorem) <u>Given a completely additive function of a set Φ defined on a σ-algebra $\mathcal{A}$, each of $\overline{V}_\Phi(E)$ and $-\underline{V}_\Phi(E)$ are measures and for each $E \in \mathcal{A}$, $\Phi(E) = \overline{V}_\Phi(E) + \underline{V}_\Phi(E)$.</u>

<u>Proof</u>. Clearly $\overline{V}_\Phi(\emptyset) = 0$. We first consider $E \subset X$ with $\overline{V}_\Phi(E) < \infty$. Given a sequence $\{E_n\}$ of pairwise disjoint sets from $\mathcal{A}$ with $E = \cup E_n$ and given $\varepsilon > 0$, there are sets $A_n \subset E_n$ such that $\overline{V}_\Phi(E_n) < \Phi(A_n) + \varepsilon/2^n$. Thus, if $A = \cup A_n$,

$\bar{V}_\Phi(E) \geq \Phi(A) = \Sigma\Phi(A_n) \geq \Sigma\bar{V}_\Phi(E_n) - \varepsilon/2^n = \Sigma\bar{V}_\Phi(E_n) - \varepsilon$. If $B \subset E$ and $B \in \mathscr{A}$ and $\Phi(B) + \varepsilon > \bar{V}_\Phi(E)$, letting $B_n = B \cap E_n$ yields

$$\bar{V}_\Phi(E) \leq \Phi(B) + \varepsilon = \Sigma\Phi(B_n) + \varepsilon \leq \Sigma\bar{V}_\Phi(E_n) + \varepsilon.$$

Since $\varepsilon > 0$ was arbitrary, it follows that $\bar{V}_\Phi(E) = \Sigma\bar{V}_\Phi(E_n)$. Now consider the case where $\bar{V}_\Phi(E) = \infty$ and $E = \cup E_n$ where the $E_n \in \mathscr{A}$ are pairwise disjoint. Let $C \subset E$, $C \in \mathscr{A}$ and let $C_n = E_n \cap C$. Then $\bar{V}_\Phi(E_n) \geq \Phi(C_n)$ and $\Sigma\bar{V}_\Phi(E_n) \geq \Sigma\Phi(C_n) = \Phi(C)$. Since C is an arbitrary element of $\mathscr{A}$ contained in E, $\Sigma\bar{V}_\Phi(E_n) \geq \bar{V}_\Phi(E) = \infty$. That is, $\Sigma\bar{V}_\Phi(E_n) = \bar{V}_\Phi(E) = \infty$. It follows that $\bar{V}_\Phi$ is a measure. Similar consideration applied to $\underline{V}_\Phi$ shows that $-\underline{V}_\Phi$ is a measure. Indeed, this fact can be obtained by noting that $-\Phi$ is a completely additive function of a set and $-\underline{V}_\Phi = \bar{V}_{(-\Phi)}$. Now if $E \subset X$, $E \in \mathscr{A}$ and $\Phi(E) = \infty$, then $\bar{V}_\Phi(E) = \infty$, Φ must be bounded below and $\underline{V}_\Phi(E) \neq -\infty$ and hence $\infty = \Phi(E) = \bar{V}_\Phi(E) + \underline{V}_\Phi(E)$. Likewise, if $\Phi(E) = -\infty$, $-\infty = \Phi(E) = \bar{V}_\Phi(E) + \underline{V}_\Phi(E)$ because $\underline{V}_\Phi(E) = -\infty$ and $\bar{V}_\Phi(E) \neq \infty$. Finally, if $\Phi(E)$ is finite, let $A \subset E$ with $A \in \mathscr{A}$. Then $\underline{V}_\Phi(E) \leq \Phi(E\backslash A) \leq \bar{V}_\Phi(E)$. Adding $\Phi(A)$ to this inequality yields $\underline{V}_\Phi(E) + \Phi(A) \leq \Phi(E) \leq \bar{V}_\Phi(E) + \Phi(A)$. Since this is true for all $A \subset E$ with $A \in \mathscr{A}$, taking the supremum on the left side of the inequality and the infimum on the right yields

$$\underline{V}_\Phi(E) + \bar{V}_\Phi(E) \leq \Phi(E) \leq \bar{V}_\Phi(E) + \underline{V}_\Phi(E)$$

and $\Phi(E) = \bar{V}_\Phi(E) + \underline{V}_\Phi(E)$.

We are now in a position to prove a second decomposition theorem for completely additive functions of a set.

(5.11) Theorem. (Hahn Decomposition Theorem) *If* Φ *is a completely additive function of a set defined on a* σ*-algebra* $\mathcal{A}$ *of subsets of a set* X, *then there is a set* $A \subset X$ *with* $A \in \mathcal{A}$ *such that for each* $B \in \mathcal{A}$, *if* $B \subset A$, $\Phi(B) \geq 0$ *and if* $B \subset X \setminus A$, $\Phi(B) \leq 0$. *Also, if* A' *is any other set with the above property,* $\Phi(A \setminus A') = 0 = \Phi(A' \setminus A)$. *Moreover, for any set* $E \in \mathcal{A}$, $\bar{V}_\Phi(E) = \Phi(E \cap A)$ *and* $\underline{V}_\Phi(E) = \Phi[E \cap (X \setminus A)]$.

Proof. Since Φ is either bounded above or bounded below, without loss of generality, we may assume Φ is bounded above. Let $\bar{V}_\Phi(X) = M < \infty$. Then for each natural number n there is a set $E_n \subset X$ with $E_n \in \mathcal{A}$ and $\Phi(E_n) > M - 1/2^n$. Given such a sequence $\{E_n\}$, note that for each set $E \in \mathcal{A}$ with $E \subset E_{n+1} \setminus E_n$, $\Phi(E) > -1/2^n$; for otherwise, $\Phi(E_{n+1} \setminus E)$ would be greater than M because $\Phi(E_{n+1}) > M - 1/2^{n+1}$. Thus $\Phi(E_1) > M - 1/2$, $\Phi(E_2 \setminus E_1) > -1/2$, $\Phi[E_3 \setminus (E_1 \cup E_2)] > -1/4$ and $\Phi(E_{n+1} \setminus \bigcup_{i=1}^{n} E_i) > -1/2^n$. It follows that $\Phi[\bigcup_{i=1}^{\infty} E_i] > M - 1/2 - 1 = M - 3/2$. Also note that $\Phi(E_n) > M - 1/2^n$, $\Phi(E_{n+1} \setminus E_n) > -1/2^n$, $\Phi[E_{n+2} \setminus (E_n \cup E_{n+1})] > -1/2^{n+1}$ and $\Phi(E_{n+k+1} \setminus \bigcup_{n}^{n+1} E_i) > -1/2^{n+k}$. It follows that $\Phi[\bigcup_{i=n}^{\infty} E_i] > M - 1/2^n - 1/2^{n-1} = M - 3/2^n$. Now let $A = \overline{\lim} E_n = \bigcap_{k=1}^{\infty} A_k$ where $A_k = \bigcup_{n=k}^{\infty} E_n$. Since $\{A_k\}$ is a non-increasing sequence of sets and $\Phi(A_1)$ belongs to the interval $[M - 3/2, M]$, $\Phi(A_1)$ is finite and $\Phi(A) = \lim \Phi(A_n)$; that is, $\Phi(A) = M$. Now if $B \in \mathcal{A}$ and B

$\subset A$, it is not possible that $\Phi(B) < 0$; for otherwise, $\Phi(A\setminus B) = \Phi(A) - \Phi(B) > \Phi(A) = M$. Similarly, if $B \subset X\setminus A$ it is not possible that $\Phi(B) > 0$; for otherwise, $\Phi(A\cup B) = \Phi(A) + \Phi(B) > M$. Suppose, $A' \in \mathcal{A}$ is another set with the same property as A. Then, since $A'\setminus A$ is contained in $X\setminus A$, $\Phi(A'\setminus A) \leq 0$; since $A'\setminus A$ is contained in A', $\Phi(A'\setminus A) \geq 0$ and thus $\Phi(A'\setminus A) = 0$. The same reasoning shows that $\Phi(A\setminus A') = 0$. Finally, if $E \in \mathcal{A}$, then $\bar{V}_\Phi(E) \geq \Phi(E\cap A)$. But if $\bar{V}_\Phi(E)$ were greater than $\Phi(E\cap A)$ there would be a set $B \subset E\setminus A$ so that $\Phi(B) > 0$. Since $E\setminus A \subset X\setminus A$, this is impossible. It follows that $\bar{V}_\Phi(E) = \Phi(E\cap A)$. The identical argument applied to $\underline{V}_\Phi$ shows that $\underline{V}_\Phi(E) = \Phi[E\cap(X\setminus A)]$.

Due to the Hahn and Jordan decomposition theorems a completely additive function of a set Φ can always be considered as the difference of two measures either by writing $\Phi(E)$ as $\bar{V}_\Phi(E) - [-\underline{V}_\Phi(E)]$ or as $\Phi(E\cap A) - [-\Phi(E\cap A^c)]$. The former is a decomposition into two measures defined on X; the latter a decomposition of X into two sets A and A^c so that Φ is a measure on A and $-\Phi$ is a measure on A^c. Having reduced the consideration of completely additive functions of a set to that of measures we now turn our attention to examples of measures that have proven useful.

Recall the method of constructing an outer measure using a non-negative, extended real-valued function τ defined on a collection $\mathcal{C}$ of subsets of a set X with $X \in \mathcal{C}$, $\emptyset \in \mathcal{C}$ and $\tau(\emptyset) = 0$. While this method is completely general, it is sometimes referred to as method I. Method II consists of using a sequence of classes $\{\mathscr{C}_n\}$ with $\mathscr{C}_{n+1} \subset \mathscr{C}_n$ and for each n, $\emptyset \in \mathscr{C}_n$ and the set X is contained in a countable union of sets from $\mathscr{C}_n$. If τ is defined on $\mathscr{C}_1$, then for each $E \subset X$,

$$m_n^*(E) = \inf\{\Sigma\, \tau(E_i): E \subset \bigcup_{i=1}^{\infty} E_i \text{ and each } E_i \in \mathscr{C}_n\}$$

is an outer measure. If $E \subset X$ and $n \geq N$ then $m_N^*(E) \leq m_n^*(E)$. This is because the infimum used to calculate m_n^* is taken over a subset of the collection of sets which are used to calculate m_N^*. Thus for each $E \subset X$, $\{m_n^*(E)\}$ is a non-decreasing sequence and has a limit (possibly ∞) which defines $m^*(E)$. That m^* is an outer measure is shown as follows:

i) $m^*(\emptyset) = \lim m_n^*(\emptyset) = 0$;

ii) If $E_1 \subset E_2$, for each natural number n, $m_n^*(E_1) \leq m_n^*(E_2)$ and hence $\lim m_n^*(E_1) = m^*(E_1) \leq m^*(E_2) = \lim m_n^*(E_2)$.

iii) Given a sequence of sets E_i, for each natural number n, $m_n^*(\cup E_i) \leq \Sigma m_n^*(E_i)$, and letting n approach ∞ yields $m^*(\cup E_i) \leq \Sigma m^*(E_i)$.

Indeed, the above shows that the limit of a non-decreasing sequence of outer measures is an outer measure. It can be shown similarly that this also holds for a non-increasing sequence. A slight variation of this method allows for classes $\mathscr{C}_\delta$ for $\delta > 0$ which satisfy $\mathscr{C}_\delta \subset \mathscr{C}_{\delta'}$ whenever $\delta < \delta'$. One then lets $m^*(E) = \lim_{\delta \to 0^+} m_\delta^*(E)$.

An important collection of geometrical measures is defined in this way; namely, the Hausdörff s-dimensional measures with s a fixed number greater than or equal to 0. The motivation for these measures is that of generalizing length in the plane and n-dimensional volume in k-dimensional space with $k \geq n$. It was Hausdörff who recognized the possibility of defining dimensional measures with any non-negative real number as the dimension.

Note that if I is a k-cell, $c \geq 0$ and $c*I$ is the k-cell of all points cx with $x \in I$, then the k-dimensional Lebesgue measure of $c*I$ is c^k times the measure of I. Any outer measure m^* on $\mathbb{R}^k$ with the property that $m^*(E) = m^*(E+x)$ for all $E \subset \mathbb{R}^k$, $x \in \mathbb{R}^k$ is called <u>translation invariant</u>. If m^* is translation invariant and also satisfies $m^*(c*E) = c^s \cdot m^*(E)$ for each E in $\mathbb{R}^k$ and $c > 0$, then m^* is called an s-<u>dimensional outer measure</u> on $\mathbb{R}^k$. Given any metric space X, and $E \subset X$, let $\tau(E) = \text{diam}(E)^s$ and $s\text{-}m_\delta^*(E) =$

$$\inf\{\Sigma \text{diam}(E_i)^s\colon E \subset \cup E_i \text{ and each } \text{diam}(E_i) < \delta\}.$$

Then, by applying method II, one has

$$s\text{-}m^*(E) = \lim_{\delta \to 0^+} s\text{-}m_\delta^*(E).$$

When $X = \mathbb{R}^k$, then $s\text{-}m^*(E)$ is the <u>Hausdörff</u> s-<u>dimensional outer measure</u> of E. It is easily seen in $\mathbb{R}^k$ that $s\text{-}m^*(E)$ is translation invariant and satisfies $s\text{-}m^*(c*E) = c^s \cdot s\text{-}m^*(E)$ for each $c > 0$.

When $s = 0$, the resulting measure is the counting measure. When $s = 1$, the resulting measure agrees in $\mathbb{R}^k$ with length on every set E which is an arc, where by arc is meant a continuous one to one image of $[0,1]$. The k-dimensional Hausdörff measure in $\mathbb{R}^k$ is a constant multiple of Lebesgue k-dimensional measure; the constant is such that the sphere of diameter 1 has unit Hausdörff measure.

A generalization of s-measure which is also studied is obtained by letting $h(x)$ be defined on the non-negative real numbers with h non-decreasing and $h(x) = \lim_{t \to x^+} h(t)$. Then, if $\tau(E) = h[\text{diam}(E)]$, method II yields the outer measure $h\text{-}m^*_\delta(E)$ and $h\text{-}m^*(E)$. The case where $h(x) = x^s$ is s-dimensional Hausdörff measure.

We note in passing that, if $s\text{-}m(E) < \infty$ and $s' > s$, then $s'\text{-}m(E) = 0$. This is because, if $\text{diam}(E_i) < \delta$ and $E \subset \cup E_i$, then $\Sigma \text{diam}(E_i)^{s'} \le \Sigma \text{diam}(E_i)^s \cdot \delta^{s'/s}$ and hence, by taking the infimum over the countable collection $\{E_i\}$ with $E \subset \cup E_i$ and $\text{diam}(E_i) < \delta$, it follows that $s'\text{-}m^*_\delta(E) \le s\text{-}m^*_\delta(E) \cdot \delta^{s'/s}$. Letting $\delta \to 0^+$ yields for $s' > s$ that $s'\text{-}m(E) = 0$ whenever $s\text{-}m(E) < \infty$. The <u>Hausdörff dimension</u> of a set E can then be defined to be $\inf\{s: s\text{-}m^*(E) = 0\}$ or ∞ if $\{s: s\text{-}m^*(E) = 0\}$ is empty. It follows from the above that, if $s'' < \dim(E)$, then $s''\text{-}m^*(E) = \infty$.

The problem of identifying the measurable sets arises naturally here; in particular, we wish to show that the Borel subsets of a metric space are measurable with respect to each s measure. Following Caratheodory, the proof of this will be accomplished in a more general setting.

An outer measure m^* defined on a metric space is said to be a <u>metric outer measure</u> provided that whenever two sets A and B satisfy

$$d(A,B) = \inf\{d(x,y): x \in A,\ y \in B\} > 0,$$

then $m^*(A \cup B) = m^*(A) + m^*(B)$. It is easy to see that Hausdörff s-measure in a metric space is a metric outer measure. For if $d(A,B) > \delta > 0$, then each countable cover $\mathcal{U}$ of $A \cup B$ with sets of diameter less than $\delta/2$ contains two covers $\mathcal{U}_1$ and $\mathcal{U}_2$ of A and B where $\mathcal{U}_1 = \{E \in \mathcal{U}: E \cap A \neq \emptyset\}$ and

$$\mathcal{U}_2 = \{E \in \mathcal{U}: E \cap B \neq \emptyset\}.$$

Now, if $E_1 \in \mathcal{U}_1$ and $E_2 \in \mathcal{U}_2$, $d(E_1,E_2) > 0$ and $E_1 \cap E_2 = \emptyset$. Hence $s\text{-}m^*_{\delta/2}(A \cup B) \geq s\text{-}m^*_{\delta/2}(A) + s\text{-}m^*_{\delta/2}(B)$ and thus $s\text{-}m^*_{\delta/2}(A \cup B) = s\text{-}m^*_{\delta/2}(A) + s\text{-}m^*_{\delta/2}(B)$ for each δ with $\delta < d(A,B)$. It follows that $s\text{-}m^*(A \cup B) = s\text{-}m^*(A) + s\text{-}m^*(B)$. Since this holds for any two sets A and B with $d(A,B) > 0$, it follows that s-measure is a metric outer measure.

It is often easier, as above, to determine that one is dealing with a metric outer measure than to determine that the Borel subsets of the space are measurable. That these two assertions are equivalent is the content of the next theorem.

(5.12) <u>Theorem</u>. <u>Given an outer measure</u> m^* <u>defined on a metric space</u> X, <u>the Borel subsets of</u> X <u>are measurable if and only if the measure is a metric outer measure</u>.

<u>Proof</u>. Suppose the Borel sets of a metric space X are measurable with respect to m. Let A and B be two sets with $d(A,B) > 0$. Then the open set $G = \{x: d(x,A) < d(A,B)\}$ does not intersect B. Since G is measurable, $B \subset G^c$ and $A \subset G$, testing G with $A \cup B$ yields $m^*(A \cup B) = m^*[(A \cup B) \cap G] + m^*[(A \cup B) \setminus G] = m^*(A) + m^*(B)$. It follows that m^* is a metric outer measure. Now suppose that F is a closed subset of X

and m^* is a metric outer measure on X. Let A be any set with $m^*(A) < \infty$ and let

$$A_n = \{x \in A: d(x,F) \ge 1/n\}.$$

Let $D_n = A_{n+1} \setminus A_n$. If $D_{n+1} \ne \emptyset$ and $A_n \ne \emptyset$, then $d(D_{n+1}, A_n) \ge 1/n - 1/(n+1) > 0$. Using the fact that m^* is a metric outer measure, it follows by induction that for each natural number n, $m^*(A_{2n+1}) \ge \sum_{i=1}^{n} m^*(D_{2i}) = m^*\left(\bigcup_{i=1}^{n} D_{2i}\right)$ and $m^*(A_{2n}) \ge \sum_{i=1}^{n} m^*(D_{2i-1}) = m^*\left(\bigcup_{i=1}^{n} D_{2i-1}\right)$. Since $A \setminus F = \bigcup_{n=1}^{\infty} A_n$, $\infty > m^*(A) \ge m^*(A \setminus F) \ge m^*(A_{2n}) \ge \sum_{i=1}^{n} m^*(D_{2i-1})$ and also $m^*(A \setminus F) \ge \sum_{i=1}^{n} m^*(D_{2i})$. It follows that $\sum_{n=1}^{\infty} m^*(D_{2n-1}) < \infty$ and $\sum_{n=1}^{\infty} m^*(D_{2n}) < \infty$. But then

$$m^*(A \setminus F) \le m^*(A_{2N}) + \sum_{n=N}^{\infty} m^*(D_{2n}) + \sum_{n=N}^{\infty} m^*(D_{2n-1})$$

and since both of the sums on the right hand side tend to 0 as N approaches ∞, $m^*(A \setminus F) \le \overline{\lim}\, m^*(A_{2N}) \le \overline{\lim}\, m^*(A_n)$. Since each $A_n \subset A \setminus F$, $m^*(A_n) \le m^*(A \setminus F)$ and since $A_n \subset A_{n+1}$, $m^*(A \setminus F) = \lim m^*(A_n)$. But then $m^*(A) \ge m^*\left((A \cap F) \cup A_n\right) = m^*(A \cap F) + m^*(A_n)$ and letting n approach ∞ yields $m^*(A) \ge m^*(A \cap F) + m^*(A \setminus F)$. It follows that F is measurable and, since F was an arbitrary closed subset of X, it also follows that each closed subset and hence each Borel subset of X is measurable.

An outer measure m^* on a topological space X is said to be Borel regular if the Borel sets are measurable and if each subset E of X is contained in a Borel set E' with $m^*(E) = m^*(E')$. The next theorem guarantees the Borel regularity of metric outer measures constructed by method II.

(5.13) Theorem. *If method II is used to construct m^* on X using covering classes $\mathscr{C}_n$ with $\mathscr{C}_{n+1} \subset \mathscr{C}_n$ and $\mathscr{C} = \mathscr{C}_1$, then each set $E \subset X$ is contained in a set $E' \in \mathscr{C}_{\sigma\delta}$ with $m^*(E') = m^*(E)$. In particular, if $\mathscr{C}$ consists of open sets of a topological space X, then each $E \subset X$ is contained in a $\mathscr{G}_\delta$ set E' with $m^*(E') = m^*(E)$; if $\mathscr{C}$ consists of Borel sets, then each set $E \subset X$ is contained in a Borel set E' with $m^*(E) = m^*(E')$.*

Proof. Note first that, if $m^*(E) = \infty$, let $E' = X$. Then for each n, by the definition of method II, X is contained in a countable union of sets from $\mathscr{C}_n$ and hence $X \in \mathscr{C}_{\sigma\delta}$. On the other hand, if $E \subset X$ and $m^*(E) < \infty$, then for each natural number n, there is a sequence of sets $\{E_{n,k}\}_{k=1}^{\infty}$ so that each $E_{n,k} \in \mathscr{C}_n$ and $\sum_k \tau(E_{n,k}) \le m_n^*(E) + 2^{-n}$ and $E \subset \bigcup_k E_{n,k}$. If $E' = \bigcap_{n=1}^{\infty} \bigcup_k E_{n,k}$, then $E \subset E'$ and for each n,

$$\sum_{k=1}^{\infty} \tau(E_{n,k}) - 2^{-n} \le m_n^*(E) \le m_n^*(E') \le \sum_{k=1}^{\infty} \tau(E_{n,k})$$

because for each n, $E' \subset \bigcup_{k=1}^{\infty} E_{n,k}$. Letting n approach ∞, $m^*(E) = \lim m_n^*(E) = \lim m_n^*(E') = m^*(E')$ because

$m_n^*(E') - 2^{-n} \le m_n^*(E) \le m_n^*(E')$. Thus each $E \subset X$ is contained in a set $E' \in \mathscr{C}_{\sigma\delta}$. Furthermore, if each set $E_{n,k}$ is an open set, then $\cup_k E_{n,k}$ is an open set and $E' = \cap_n \cup_k E_{n,k}$ is a $\mathscr{G}_\delta$; if each $E_{n,k}$ is a Borel set then $E' = \cap_n \cup_k E_{n,k}$ is also a Borel set. If $\mathscr{C}$ consists of measurable Borel subsets of a topological space X, m* is Borel regular.

Note that for each subset E of a metric space X and $s \ge 0$, $s\text{-}m^*(E) = \lim s\text{-}m^*_{1/n}(E)$. Thus, when $\mathscr{C}_n$ is the collection of open subsets of X of diameter less than 1/n, the fact that s-measure is Borel regular may be obtained from the theorem above. To do this note that each set of diameter less than 1/n is contained in an open set whose diameter is as close to that of the original set as desired. Thus letting $\mathscr{C}$ consist of the open subsets of X and $\mathscr{C}_n$ consist of the open subsets of diameter less than 1/n, method II applied to $\mathscr{C}_n$ with $\tau(E) = \text{diam}(E)^s$ yields the Hausdörff outer s-measure of each set E. Thus each E is contained in a set $E' \in \mathscr{G}_\delta$ so that $s\text{-}m(E') = s\text{-}m^*(E)$ and, in particular, if E is measurable and of finite measure, then the set E' so chosen satisfies $s\text{-}m(E' \setminus E) = 0$.

Note that each set Z with $s\text{-}m(Z) = 0$ also has $s\text{-}m_\delta(Z) = 0$ for every $\delta > 0$ and hence is $s\text{-}m_\delta$ measurable. However, the Borel subsets of X are not in general measurable with respect to $s\text{-}m^*_\delta$ outer measure. Indeed, if $E \subset \mathbb{R}^k$, $s < k$, $0 < s\text{-}m(E) < \infty$ and $\delta > 0$, it is normally the case that E is not

s-m_δ measurable. A simple example illustrates why this occurs.

Let $E_1 \subset \mathbb{R}^2$ be $\{(x,y): 0 \le x \le 1, \ y = 0\}$ and let $s = 1$ and $\delta > 0$. If E_1 were 1-m_δ measurable, since s-m_δ is clearly translation invariant, so would the set $E_2 = \{(x,y): 0 \le x \le 1, \ y = \delta/2\sqrt{2}\}$ be 1-m_δ measurable and so would the union of E_1 and E_2 be measurable. But $E_1 \cup E_2$ is contained in squares of side length $\delta/2\sqrt{2}$ and diameter $\delta/2$ and at most $[\![2\sqrt{2}/\delta]\!]$ of these are needed to cover this set. (Here $[\![x]\!]$ is the least integer greater than or equal to x.) Thus $1\text{-}m_\delta^*(E_1 \cup E_2) \le \delta/2\sqrt{2}(1 + 2\sqrt{2}/\delta) = 1 + \delta/2\sqrt{2}$. Since $1\text{-}m_\delta^*(E_1) = 1\text{-}m_\delta^*(E_2) = 1$, neither E_1 nor E_2 are 1-m_δ measurable. For if they were, $m_\delta^*(E_1 \cup E_2)$ would have to equal $m_\delta^*(E_1) + m_\delta^*(E_2)$. Similar considerations hold for sets of dimension s of finite, non-zero outer s-measure in $\mathbb{R}^k$ with $k > s$.

Additional measures in $\mathbb{R}^k$ can be defined using a function τ defined on a collection $\mathscr{C}$ which consists of intervals contained in $\mathbb{R}^k$. Without loss of generality, we will consider intervals which are half open on the left; that is, intervals of the form $(a_1,b_1;\ a_2,b_2;\ \ldots;\ a_k,b_k] =$

$$\{(x_1,x_2,\ldots,x_k): a_1<x_1\le b_1,\ a_2<x_2\le b_2,\ \ldots,\ a_k<x_k\le b_k\}.$$

Such intervals will be written as $(a,b]$ where $a = (a_1,a_2,\ldots,a_k)$ and $b = (b_1,\ b_2,\ \ldots,\ b_k)$. Note that for any $\delta > 0$, $\mathbb{R}^k$ can be written as a countable union of such intervals and any interval $(a,b]$ and the difference $(a,b]\setminus(c,d]$ of two such intervals can be written as a finite union of such intervals and this may be done so that each interval in the union is of diameter less than δ. Moreover, these intervals fit together in such a way that the intervals in the union can be chosen to be pairwise disjoint.

A collection of intervals of the form $(a,b]$ of $\mathbb{R}^k$ is called a <u>net</u> providing the intervals in the collection are pairwise disjoint and their union covers $\mathbb{R}^k$. Given a net N, let

$$d(N) = \sup\{\operatorname{diam}(a,b]: (a,b] \in N\}.$$

A sequence $\{N_n\}$ of nets is said to be <u>regular</u> if for each natural number n each interval of N_n is the union of finitely many intervals from N_{n+1} and $\lim d(N_n) = 0$. A frequently used regular sequence of nets in $\mathbb{R}^k$ is the <u>sequence of binary nets</u> $\{N_n\}$ where N_n consists of all intervals of the form

$$\left(\frac{m_1}{2^n}, \frac{m_1 + 1}{2^n}; \frac{m_2}{2^n}, \frac{m_2 + 1}{2^n}; \ldots; \frac{m_k}{2^n}, \frac{m_k + 1}{2^n} \right]$$

Either all the intervals $(a,b]$ of $\mathbb{R}^k$ or those from a regular sequence of nets can be used as a covering class to generate measures. A real-valued function F of an interval which is non-negative and defined on the intervals of the covering class yields a measure by letting $\tau(I) = F(I)$ and $\tau(\emptyset) = 0$ and applying either method I or II. Here, it is clear that method II with $\mathcal{C}_\delta = \{I: \operatorname{diam} I \le \delta\}$ or $\mathcal{C}_n = \{I: I \in \bigcup_{i=n}^{\infty} N_i\}$ will always yield a metric outer measure. Frequently, it is easier to deal with a sequence of nets than with all intervals or with other covering classes. For example, Besicovich showed that, if the sequence of binary nets in $\mathbb{R}^k$ is used with $F(I) = \operatorname{diam}(I)^s$, then the resulting net measures

$$Nh\text{-}m^*_\delta(E) = \inf\{\Sigma h(I_m): E \subset \cup I_m \text{ with } I_m \in N_n \text{ and each } \operatorname{diam}(I_m) \le \delta\}$$

and $Nh\text{-}m^*(E) = \lim_{\delta\to 0} Nh\text{-}m^*_\delta(E)$ are comparable to $h\text{-}m^*_\delta(E)$ and $h\text{-}m^*(E)$ respectively. His exact result is contained in the following theorem.

(5.14) Theorem. *If h is a non-decreasing function defined on $[0,\infty)$ with $h(0) = 0$ and $h(x) = \lim_{t\to x^+} h(t)$ then for each set $E \subset \mathbb{R}^k$ and δ with $0 < \delta < 1$, $h\text{-}m^*_\delta(E) \le Nh\text{-}m^*_\delta(E) \le 3^k.2^{k(k+1)}h\text{-}m^*_\delta(E)$ and $h\text{-}m^*(E) \le Nh\text{-}m^*(E) \le 3^k.2^{k(k+1)}h\text{-}m^*(E)$. Thus $h\text{-}m^*(E)$ is zero, finite and non-zero, or infinite iff $Nh\text{-}m^*(E)$ is respectively zero, finite and non-zero, or infinite.*

Proof. Let E be any subset of $\mathbb{R}^k$ and $\delta \in (0,1)$. Note that the diameter of each interval of N_n is $2^{-n}\sqrt{k}$. Since it is clear that $h\text{-}m^*_\delta(E) \le Nh\text{-}m^*_\delta(E)$ we may assume that $h\text{-}m^*_\delta(E) < \infty$. Then, given $\varepsilon > 0$, there is a sequence of sets $\{E_i\}$ with each $\operatorname{diam}(E_i) < \delta$, $E \subset \cup E_i$ and $h\text{-}m^*_\delta(E) > \Sigma h(\operatorname{diam} E_i) - \varepsilon$. Then each E_i meets one of the intervals from each N_n. Let s_i be the side length of an interval $I_i \in \cup N_n$ which meets E_i and satisfies $s_i/2 \le \operatorname{diam} E_i < s_i$. Then E_i is contained in the 3^k intervals of $\cup N_n$ whose boundary meets that of I_i and in the $3^k.2^{k(k+1)}$ smaller intervals from $\cup N_n$ which make up these 3^k intervals. Each of these $3^k.2^{k(k+1)}$ intervals has side length $s_i \cdot 2^{-k-1}$ and diameter

$s_i \cdot 2^{-k-1}\sqrt{k} \le s_i/2 \le \text{diam } E_i$. Hence

$$3^k \cdot 2^{k(k+1)}\left(h\text{-}m_\delta^*(E) + \varepsilon\right) > \sum_i 3^k \cdot 2^{k(k+1)} h(\text{diam } E_i)$$
$$\ge Nh\text{-}m_\delta^*(E).$$

Since the number $\varepsilon > 0$ is arbitrary, $Nh\text{-}m_\delta^*(E) \le 3^k \cdot 2^{k(k+1)} h\text{-}m_\delta^*(E)$. By letting δ approach 0, one also has $h\text{-}m^*(E) \le Nh\text{-}m^*(E) \le 3^k \cdot 2^{k(k+1)} h\text{-}m^*(E)$. Thus $h\text{-}m^*(E)$ is zero, finite and non-zero, or infinite iff $Nh\text{-}m^*(E)$ is respectively zero, finite and non-zero, or infinite.

An important class of measures in $\mathbb{R}^k$ consists of the Borel regular measures which are finite valued on each bounded subset of $\mathbb{R}^k$. These measures are called Lebesgue-Stieltjes measures because of the way in which they can be obtained from functions of an interval. In $\mathbb{R}^k$ a finite valued function of an interval is called an <u>additive function of an interval</u> (on intervals of the form $I = (a,b]$) if $F(I) = F(I_1) + F(I_2)$ whenever $I = I_1 \cup I_2$; F is said to be <u>non-decreasing</u> if, whenever $I_1 \subset I_2$, then $F(I_1) \le F(I_2)$. Clearly, each non-negative additive function of an interval is necessarily non-decreasing. We will see that in generating Lebesgue-Stieltjes measures there is no loss in generality if one begins with functions F which are non-negative, additive and also continuous on the right; that is, whenever $(a,b]$ is the intersection of a decreasing sequence $\{(a,b_n]\}$ of intervals, then $F\left((a,b]\right) = \lim F\left((a,b_n]\right)$. One then applies method I or II with $\tau(I) = F(I)$ and $\tau(\emptyset) = 0$. For $b =$

$(b_1, b_2, \ldots, b_k)$, let $b_h = (b_1+h, b_2+h, \ldots, b_k+h)$. Then to assert that a non-negative additive function F is continuous on the right, it is enough to assert that for each $(a,b]$ one has $\lim_{h\to 0^+} F((a,b_h]) = F((a,b])$. Such a function F must also have the property that, for each interval $(a,b]$, $\lim_{h\to 0^+} F((a_h,b]) = F((a,b])$ and $\lim_{h\to 0^+} F((a,a_h]) = 0$. To show this we consider only the case $\mathbb{R}^k = \mathbb{R}^2$; the general situation in other $\mathbb{R}^k$ is shown similarly. Given $(a,b] = (a_1,b_1;\ a_2,b_2]$ let $c_1 < a_1$, $c_2 < a_2$. Then $(c_1,b_1;\ c_2,b_2] \subset (a_h,b] \cup (c_1,a_1;c_2,a_2] \cup (a_1,c_2;b_1+h,a_2+h] \cup (c_1,a_2;a_1+h,b_2+h]$. By letting $h \to 0^+$ and using the fact that F is non-decreasing and continuous on the right, one obtains $F((c,b]) \le \overline{\lim}_{h\to 0^+} F((a_h,b]) + F((c_1,a_1;\ c_2,a_2]) + F((a_1,c_2;\ b_1,a_2]) + F((c_1,a_2;\ a_1,b_2])$. By the fact that F is additive, $F((a,b]) \le \overline{\lim}\, F((a_h,b])$ and since for each $h > 0$, $F((a,b]) \ge F((a_h,b])$ it follows that $\lim_{h\to 0^+} F((a_h,b]) = F((a,b])$. Also $\overline{\lim}_{h\to 0^+} F((a,a_h]) \le \overline{\lim}\, F((c,a_h]) - F((c,a]) = 0$ and then $\lim_{h\to 0^+} F((a,a_h]) = 0$.

The theorems which follow show that each function F defined on half open intervals which is non-negative, additive and continuous on the right generates a Borel regular measure which is finite on bounded sets and that each such measure can be generated from such a function F.

(5.15) *Theorem. Let F be a non-negative additive function of an interval defined on the intervals of the form $(a,b]$ in $\mathbb{R}^k$. If F is continuous on the right, then the measures generated either by method I or method II using the covering class consisting of all such $(a,b]$ are the same. This measure, denoted by m_F, satisfies $m_F(I) = F(I)$ for each $I = (a,b]$. (Hence m_F is finite valued on bounded subsets of $\mathbb{R}^k$.)*

Proof. Given $\delta > 0$, the additivity of F on half-open intervals guarantees that each cover $\{I_n\}$ of a set E can be replaced with a cover $\{I'_j\}$ of E consisting of intervals of diameter less than δ which has the value $\Sigma F(I'_j)$ equal to $\Sigma F(I_n)$. This is done simply by subdividing the intervals I_n into a finite number of disjoint intervals of diameter less than δ. Thus, for each set E, the outer measure of E as determined by method II is less than or equal to that determined by method I; since method II always generates a measure greater than or equal to that of method I, the two outer measures are the same. Denoting this measure by m_F, we have for each $I = (a,b]$ that $m_F(I) \le F(I)$. Suppose, if possible, that some $m_F(I) < F(I)$. Then there is $\varepsilon > 0$ so that $m_F(I) < F(I) - \varepsilon$. Since $F((a_h,b])$ approaches $F((a,b])$ as $h \to 0^+$, there is $I' = (a_h,b]$ with $m_F(I) < F(I') - \varepsilon$. Since there is a sequence of intervals $\{I_n\}$ with $I \subset \cup I_n$ and $\Sigma F(I_n) < F(I') - \varepsilon$, by the continuity of F on the right each of these intervals $I_n = (a_n,b_n]$ can be replaced by intervals

$I'_n = (a_n, b_{n,h_n}]$ so that $\Sigma F(I'_n) < F(I') - \varepsilon$. But now $\bar{I}'$ is contained in the union of the interiors of the I'_n and hence there is a finite subcover of the I'_n which covers I', say $I'_{n_1}, I'_{n_2}, \ldots, I'_{n_m}$. Then each $I'_{n_j} \setminus \cup_{i<j} I'_{n_i}$ can be written as a finite union of pairwise disjoint half-open intervals. The union of this finite collection $\{I''_j\}$ of pairwise disjoint intervals contains $\bar{I}'$. Thus

$$F(I') \le \Sigma F(I''_j) \le \sum_{i=1}^{m} F(I'_{n_i}) < F(I') - \varepsilon$$

and this contradiction shows that $m_F(I) = F(I)$.

Note that the covering class of all intervals half open on the right is made up of $\mathcal{F}_\sigma$ sets; thus the resulting measure m_F is Borel regular. (Borel sets are measurable because m_F is a metric outer measure which is generated by method II with each $\mathcal{C}_\delta$ consisting of Borel sets of diameter less than δ). Indeed, by the continuity of F on the right, the measure m_F can also be generated by method I with $\mathcal{C} = \{\mathring{I}: I = (a,b)\}$ and $\tau((a,b)) = F((a,b])$. To see this let $E \subset \mathbb{R}^k$ and $\varepsilon > 0$; let $E \subset \cup I_n$ with $m_F^*(E) < \Sigma F(I_n) + \varepsilon$ and $I_n \subset \mathring{I}'_n$ with $F(I'_n) < F(I_n) + \varepsilon/2^n$. Then, if m^* is the outer measure generated by $\mathcal{C}$ and τ, clearly, $m^*(E) \le m_F^*(E)$ and thus

$$m^*(E) \le m_F^*(E) \le \varepsilon + \sum_n F(I_n)$$

$$\le \varepsilon + \sum_n F(I'_n) + \varepsilon/2^n \le m^*(E) + 2\varepsilon$$

and since $\varepsilon > 0$ was arbitrary, $m^*(E) = m_F^*(E)$ and the measure generated in this way agrees with m_F. Now it follows from Theorem 5.6 that for any set E which is measurable with respect to m_F and $\varepsilon > 0$

i) there is an open set G with $E \subset G$ and $m_F(G\setminus E) < \varepsilon$

ii) there is a $\mathcal{G}_\delta$ set A with $E \subset A$ and $m_F(A\setminus E) = 0$

iii) there is a closed set $F \subset E$ with $m_F(E\setminus F) < \varepsilon$

iv) there is an $\mathcal{F}_\sigma$ set $H \subset E$ with $m_F(E\setminus H) = 0$ and if $m_F(E) < \infty$, there is a compact set $K \subset E$ with $m^*(E\setminus K) < \varepsilon$.

To show that each Borel regular measure which is finite on bounded sets in $\mathbb{R}^k$ is an m_F, we need the following result.

(5.16) Theorem. *Each open set $G \subset \mathbb{R}^k$ can be written as a countable union of pairwise disjoint half-open intervals. Indeed if $\{N_n\}$ is a regular sequence of nets in $\mathbb{R}^k$, G can be written as a countable union of pairwise disjoint intervals of $\cup N_n$.*

Proof. It clearly suffices to show that each G can be written as a union of pairwise disjoint intervals from $\cup N_n$ where $\{N_n\}$ is a given regular sequence of nets. Recall that if $I_1, I_2 \in \cup N_n$ then either I_1 and I_2 are pairwise disjoint (if they belong to the same N_n) or one of them is contained in the other. Also, because $\lim d(N_n) = 0$, if x is any point of $\mathbb{R}^k$ and $\delta > 0$, then there is $I \in \cup N_n$ with $x \in I$ and $\operatorname{diam}(I) < \delta$. Thus G is the union of all I in $A = \{I\colon I \subset G$ and $I \in \cup N\}$. Consider

$$B = \{I \in A\colon \text{ for each } I' \in A, \quad I' \cap I \neq \emptyset \text{ implies } I' \subset I\}.$$

That is, B consists of those $I \in A$ which are not contained in any larger interval $I' \in A$; alternately, given $x \in G$, if n is the least natural number for which there is an $I \in N_n$ with $x \in I$, then that I belongs to B. Consequently, $G = \bigcup_{I \in B} I$ and the intervals of B are pairwise disjoint.

(5.17) <u>Theorem</u>. <u>If m^* is a Borel regular outer measure defined on $\mathbb{R}^k$ which is finite valued on bounded sets, then there is a non-negative additive function of an interval F defined on the half-open intervals of $\mathbb{R}^k$ which is continuous on the right and</u> $m_F^* = m^*$.

<u>Proof</u>. Given such an m^* defined on $\mathbb{R}^k$, let $F(I) = m^*(I)$ on each $I = (a,b]$. Since m is a measure on the Borel sets of $\mathbb{R}^k$, each I is measurable and $F(I)$ is non-negative and is an additive function of an interval. Since m^* is finite valued on bounded sets, for each $I = (a,b]$, $F(I) = m(I) = \lim_{h\to 0^+} m((a,b_h]) = \lim_{h\to 0^+} F((a,b_h])$ and F is continuous on the right. But then $m_F((a,b]) = F((a,b])$ and, since by the previous theorem each open set G is the union of a sequence of pairwise disjoint intervals $\{I_n\}$, then $m(G) = \Sigma m(I_n) = \Sigma m_F(I_n) = m_F(G)$. Thus m and m_F agree on open sets. Now each Borel class $\mathcal{G}_\alpha$ is closed under finite unions and intersections; thus each bounded Borel set in class $\mathcal{G}_\alpha$, α even, is the union of a non-decreasing sequence of bounded Borel sets from lower

classes and each bounded Borel set in class $\mathcal{G}_\alpha$, α odd, is the intersection of a non-increasing sequence of bounded Borel sets from lower classes. Thus, by transfinite induction, using the fact that $m(\lim E_n) = \lim m(E_n) = \lim m_F(E_n) = m_F(\lim E_n)$ for monotone sequences of sets of finite measure, it follows that m agrees with m_F on each bounded Borel set. Since each Borel set is the union of a non-decreasing sequence of bounded Borel sets, m and m_F agree on every Borel set. Finally, since m^* and m_F are Borel regular and agree on each Borel set, for each set $E \subset \mathbb{R}^k$, $m^*(E) = m_F^*(E)$.

On the line, non-negative additive functions F of an interval correspond to non-decreasing functions of a point. That is, given $F(I)$, define F at 0 to be any real number c and let $F(x) = c + F\big((0,x]\big)$ for $x > 0$ and $F(x) = c - F\big((x,0]\big)$ for $x < 0$. One easily checks that $F(x)$ is continuous on the right iff $F(I)$ is and that for each $I = (a,b]$, $F(I) = F(b) - F(a)$. Moreover, any $G(x)$ with the property that $F(I) = G(b) - G(a)$ when $I = (a,b]$ must differ from $F(x)$ by a constant. If the restriction of right continuity is dropped, it is traditional to consider additive functions on closed intervals I; that is, $F(I) = F(I_1) + F(I_2)$ whenever $I = I_1 \cup I_2$ and $\mathring{I}_1 \cap \mathring{I}_2 = \emptyset$. It is also traditional to define $F^*(E)$ to be $\inf\{\Sigma F(I_n)\colon E \subset \cup \mathring{I}_n\}$. The resulting outer measure F^* is Borel regular and $F^*(I) = F(I)$ on each interval $I = [a,b]$ for which $F(x)$ is continuous at both a and b; in general, $F^*(I) = F(b+) - F(a-) \geq F(I)$. Also, if $F(x)$ is a non-decreasing function defined on the line, then $G(x) = \lim_{t \to x^+} F(x)$ is non-decreasing and continuous on the right. However, the measures m_G, G^* and F^* do not in general coincide; of course they do coincide on each interval for which F is continuous at both a and b.

In $\mathbb{R}^k$ with $k > 1$, the situation is more complicated. Non-negative additive functions F of an interval which are continuous on the right do correspond to functions of a point which are continuous on the right and non-decreasing in each variable but the reverse correspondence does not hold without considerable restrictions on the function of a point. To see the correspondence, given $F(I)$, define $F(0) = c$ and for $b = (b_1, b_2, \ldots, b_k)$ with each $b_i > 0$, define $F(b) = c + F((0,b])$ and for $a = (a_1, a_2, \ldots, a_k)$, define $F(a) = c + F((a,b])$ where b is chosen with $b_i > a_i$, $i = 1, 2, \ldots, k$. This $F(x)$ is well-defined, non-decreasing in each variable and continuous on the right in each variable. To recapture $F(I)$ from $F(x)$, define for $h_i > 0$,

$$\begin{aligned} Dh_i \circ F(x) &= F(x_1, x_2, \ldots, x_i + h_i, \ldots, x_k) \\ &\quad - F(x_1, x_2, \ldots, x_i, \ldots, x_k). \end{aligned}$$

If $(a,b] = (a_1, a_1+h_1; \ldots; a_k, a_k+h_k]$, then $F((a,b]) = Dh_1 \circ Dh_2 \circ \cdots \circ Dh_k(F(a))$. The motivation for this definition can be easily visualized in the plane. For suppose F is a non-negative function of an interval in $\mathbb{R}^2$ and $I = (a,b] = (a_1, a_1+h_1; a_2, a_2+h_2]$ is in the first quadrant. Let $I_1 = (0,b]$, $I_2 = (0, a_1+h_1; 0, a_2]$, $I_3 = (0, a_1; 0, a_2+h_2]$ and $I_4 = (0,a]$. Then $I = ((I_1 \setminus I_2) \cup I_4) \setminus I_3$ and hence

$$F(I) = F(I_1) - F(I_2) + F(I_4) - F(I_3)$$

$$= F(a_1+h_1, a_2+h_2) - F(a_1+h_1, a_2) - F(a_1, a_2+h_2) + F(a_1, a_2)$$
$$= Dh_1 \circ Dh_2 \circ F(a).$$

Similar considerations show that this also holds for other intervals in $\mathbb{R}^2$ and also for intervals in higher dimensional spaces.

In order to obtain the reverse correspondence from a function F of a point to a non-negative additive function of an interval which is continuous on the right, one must require that $F(x_1, x_2, \ldots, x_n)$ satisfy

i) $Dh_1 \circ Dh_2 \circ \ldots \circ Dh_k \circ F(x) \geq 0$ for each x and $h_1, h_2, \ldots, h_k > 0$

Note that for any $F(x)$ the numbers $Dh_1 \circ \ldots \circ Dh_k \circ F(a)$ are the same no matter what order is placed on the Dh_i. Furthermore, if F satisfies i) at each $x \in \mathbb{R}^k$, the resulting function of an interval is additive and non-negative. However, it is convenient to consider only those $F(x)$ that also satisfy:

ii) F is continuous on the right in each variable x_i

iii) F is non-decreasing in each variable x_i.

Condition ii) means that for each variable x_i and fixed values x_j, $j \neq i$, $F(x_1, \ldots, x_i, \ldots, x_k)$ is a real-valued function which is continuous on the right; iii) means that each such real-valued function is non-decreasing. We will see that when determining non-negative additive functions of an interval $I = (a,b]$ which are continuous on the right in $\mathbb{R}^k$ there is no loss of generality in beginning with functions $F(x)$ which satisfy i), ii) and iii).

(5.18) Theorem. If $F(x)$ is defined on $\mathbb{R}^k$ and satisfies i), ii), and iii) above, then $F((a,b]) = Dh_1 \circ \ldots \circ Dh_k \circ F(a)$ where $h_i = b_i - a_i$ is an additive non-negative function of an interval which is continuous on the right on each interval $I = (a,b]$. (Thus, for each I, $m_F(I) = F(I)$.)

Proof. For any finite valued function F of a point whatsoever in $\mathbb{R}^k$, $F((a,b]) = Dh_1 \circ \ldots \circ Dh_k \circ F(a)$ where $h_i = b_i - a_i$ is an additive function of an interval and i) guarantees that $F(I)$ is non-negative. It remains to see that i), ii) and iii) together guarantee that $F(I)$ is continuous on the right on each $I \subset \mathbb{R}^k$. Fix $I = (a,b]$ and $\varepsilon > 0$. Given $x = (x_1, \ldots, x_k)$, let δ_1 be chosen so that

$$F(x_1+\delta_1, x_2, \ldots, x_k) - F(x_1, x_2, \ldots, x_k) < \varepsilon/k.$$

Choose δ_2 so that $F(x_1+\delta_1, x_2+\delta_2, \ldots, x_k) - F(x_1+\delta_1, x_2, \ldots, x_k) < \varepsilon/k$. Continue in this fashion until δ_k is determined so that

$$F(x_1+\delta_1, x_2+\delta_2, \ldots, x_{k-1}+\delta_{k-1}, x_k+\delta_k) - F(x_1+\delta_1, x_2+\delta_2, \ldots, x_{k-1}+\delta_{k-1}, x_k) < \varepsilon/k.$$

Then, by iii), if $h_i < \delta_i$ then $F(x_1+h_1, \ldots, x_k+h_k) - F(x_1, \ldots, x_k) < \varepsilon$. A similar argument applies to each corner of $(a,b]$. It then follows that, if h is less than each of the values of δ determined for the corners of $(a,b]$, then $F((a,b_h]) - F((a,b]) < 2^k\varepsilon$. Thus F is a non-negative additive function of an interval which is continuous on the right on each $I \subset \mathbb{R}^k$.

Note that the fact that ii) and iii) are not absolutely required follows from the observation that for any F satisfying i), ii), and iii), $F_o(x_1, \ldots, x_k) = F(x_1, \ldots, x_k) + F_1(x_1) + F_2(x_2) + \cdots + F_k(x_k)$ gives rise to the same function of an interval whenever $F_i(x_i)$ are any functions whatsoever of the single variables x_i. For example, for any real function G and H, $F(x,y) = G(x) + H(y)$ produces $F(I)$ with $F(I) = 0$ for each $I \subset \mathbb{R}^2$. Note also that i) is not implied by ii) and iii). For example, $F(x,y) = (x + y)^{1/2}$ if $x + y > 0$ and $F(x,y) = 0$ if $x + y \le 0$ satisfies ii) and iii) but $F((0,1; 0,1]) = \sqrt{2} - 2 < 0$. Condition i), while necessary and sufficient in order that $F(I)$ be non-negative, does not seem to be guaranteed by any simpler hypothesis.

Important examples of such functions $F(x)$ are $F(x_1, \ldots, x_k) = \Pi F_i(x_i)$ where each F_i is a monotone non-decreasing function of a real variable which is continuous on the right. In this case, m_F is the product measure in $\mathbb{R}^k$ of each m_{F_i} in $\mathbb{R}^1$. For the particular case where each $F_i(x) = x$, the resulting measure m_F is Lebesgue measure in $\mathbb{R}^k$.

As on the line, in $\mathbb{R}^k$ a measure can be generated from any non-negative additive function of an interval (not necessarily continuous on the right). Such an F is said to be additive on closed intervals if $F(I) = F(I_1) + F(I_2)$ whenever $I = I_1 \cup I_2$ and $\mathring{I}_1 \cap \mathring{I}_2 = \emptyset$. The resulting measure, denoted by F^* is given by

$$F^*(E) = \inf\{\Sigma F(I_n) : E \subset \cup \mathring{I}_n\},$$

is a metric outer measure, Borel regular, finite valued on bounded sets and hence is a σ-finite measure. Of the hyperplanes in $\mathbb{R}^k$ (a <u>hyperplane</u> in $\mathbb{R}^k$ is a subset of $\mathbb{R}^k$ congruent to $\mathbb{R}^{k-1}$) given by $\{x \in \mathbb{R}^k : x_i = c\}$, at most countably many can have positive measure with respect to F^*; these are called <u>hyperplanes of discontinuity</u> of F. If no face of $\bar{I}$ lies in a hyperplane of discontinuity then $F(I) = F^*(I)$. Finally, obtaining a right continuous function G which generates the same measure as F can perhaps be most easily accomplished by defining $G(I) = F^*(I)$ for each $I = (a,b]$.

We turn now to a consideration of some probability measures where by a <u>probability measure</u> is meant any measure on a space X so that $m(X) = 1$. When a measure on X is a probability measure, $Pr(A)$ is usually written for the measure of sets $A \subset X$. Given any set $B \subset X$ with $Pr(B) > 0$, the relative measure of subsets of B, $Pr_B(A) = Pr(A \cap B)/Pr(B)$, is written $Pr(A|B)$, read "the conditional probability of A given B" and is clearly also a probability measure on X. The simplest examples of such measures are obtained by letting $\{x_i\}$ be a finite or infinite sequence of elements of a set X and defining m at each x_i so that $0 \le m(x_i)$ and $\Sigma m(x_i) = 1$. Then $m(A) = \sum_{x_i \in A} m(x_i)$ is clearly a measure and every subset of X is measurable with respect to m. While these are important measures, they present no real technical difficulties for analysis.

If a probability measure is defined on $\mathbb{R}$ or on $\mathbb{R}^k$, it is traditional to define the <u>distribution function</u> F by $F(x) = m((-\infty,x])$ or $F(x_1, \ldots, x_k) = m((-\infty,x_1] \times \ldots \times (-\infty,x_k])$. Then such F satisfy i), ii) and iii) of the previous theorem. For such F

defined on $\mathbb{R}^k$,

iv) when F is defined on $\mathbb{R}^1$, $\lim_{x\to-\infty} F(x) = 0$ and $\lim_{x\to\infty} F(x) = 1$ and, when F is defined on $\mathbb{R}^k$, $k > 1$, $F(x_1, x_2, \ldots, x_k) \to 0$ as $(x_1, x_2, \ldots, x_k) \to (-\infty, -\infty, \ldots, -\infty)$ and $F(x_1, x_2, \ldots, x_k) \to 1$ as $(x_1, x_2, \ldots, x_k) \to (\infty, \infty, \ldots, \infty)$.

A function F satisfying i), ii), iii) and iv) is necessarily the distribution function of the resulting probability measure on $\mathbb{R}^k$ determined by $F(I)$. The measure is Borel regular and, in the context of probability theory, one frequently considers the measure to be a function defined only on the Borel sets.

From given distribution functions, other useful probability measures on the line or in Euclidean space can be generated. For example, given a distribution function $F(x) = F(x_1, \ldots, x_k)$ and its associated probability measure Pr, the <u>marginal distribution functions</u> F_i are given by

$$F_i(x) = \Pr(\{(x_i, \ldots, x_k): x_i \le x\}) = \Pr(x_i \le x).$$

Given j indices $i_1, i_2, \ldots, i_j$ with $j < k$, the function $F_{i_1,\ldots,i_j}$ of j variables, are distribution functions when

$$\begin{aligned} F_{i_1,\ldots,i_j}(y_1, \ldots, y_j) &= \Pr(\{(x_1, \ldots, x_k): x_{i_1} \le y_1, \ldots, x_{i_j} \le y_j\}) \\ &= \Pr(x_{i_1} \le y_1, \ldots, x_{i_j} \le y_j). \end{aligned}$$

The distribution G of averages of the x_i is given by

$$G(x) = \Pr\left(\frac{1}{k} \sum_{i=1}^{k} x_i \le x\right).$$

Distribution functions which arise due to changes of variable or from a function h taking $\mathbb{R}^k$ into $\mathbb{R}^{k'}$ are $F_h(y_1, \ldots, y_{k'}) =$

$$Pr(h_1(x_1, \ldots, x_k) \leq y_1, \ldots, h_{k'}(x_1, \ldots, x_k) \leq y_{k'})$$

provided all of these sets are measurable. All of these measures are of the same type as those which were considered above in that they are Borel regular measures defined on some Euclidean space.

We consider at this point several examples of probability measures which are defined on larger spaces, spaces of sequences or spaces of functions. Two problems for analysis arise in this context. The measure Pr is determined from a function τ defined on a covering class $\mathscr{C}$. The first problem consists of showing that the sets in $\mathscr{C}$ are measurable; the second consists in determining the measure of each set in $\mathscr{C}$, in particular cases this is often the value of τ on the sets in $\mathscr{C}$. Computing the probability of other sets, while potentially complicated, can frequently be done using the logic which describes a given set X_o along with the fact that the probability space, having finite measure, satisfies $m(X_o) = m(\lim X_n) = \lim m(X_n)$ whenever the sequence $\{X_n\}$ has the set X_o for its limit.

Some probability spaces can be reduced to the already considered examples of probability measures on $\mathbb{R}^k$. Such an example on a space of sequences is that of a random walk. A <u>random walk</u> (on the line) is a sequence $\{x_n\}_{n=0}^{\infty}$ with $x_o = 0$ and $(x_{n+1} - x_n) = 1$. The probability that $x_{n+1} = x_n + 1$ is P_n and the probability that $x_{n+1} = x_n - 1$ is $1 - P_n$. The problem is to construct a probability measure on all such sequences. Here one desires a probability measure on this space which would allow the answer to easy questions of probability such as determining $\Pr(x_4 = x_2)$ and also to more difficult ones such as

$$\Pr(x_n = 0 \quad \text{for some} \quad n > N)$$

or $\Pr($ for all n, $x_n \le \sqrt{n})$, etc. In the case of a random walk on the line, one can represent each sequence $\{x_n\}$ as a sequence of 0's and 1's where 1 indicates progression to the right and 0 to the left. On removing the set of measure 0 of sequences which have only finitely many 0's, the resulting space can be represented by the interval $[0,1)$ with each x written in its binary expansion as $.x_1x_2\cdots$. For each I in the sequence of binary nets let $\tau(I) = P'_1 \cdot \ldots \cdot P'_n$ when $I = \{x\colon x_1 = c_1, \ldots, x_n = c_n\}$ where $c_i = 0, 1$ are specified and $P'_i = P_i$ if $c_i = 1$, $P'_i = 1 - P_i$ if $c_i = 0$. Then τ is additive on these intervals and using the collection N_n of such intervals with n specified x_i's and applying either method I or II will yield a measure Pr. Moreover, $\Pr(I) = \tau(I)$ when $I \in N_n$ and each such I in each N_n is measurable.

Another example of a probability space consisting of sequences consists of assigning, for each natural number n, a distribution function $F_n(x)$ defined on $\mathbb{R}$. Assume that the space of all sequences $\{x_n\}$ of real numbers is to be given a measure Pr so that

$$\Pr[x_1 \in (a_1,b_1],\ldots,x_n \in (a_n,b_n]] = \prod_{i=1}^{n} (F_i(b_i)-F_i(a_i)).$$

This represents the assumption that the choice of each x_i is independent of other values in the sequence and distributed with probability given by F_i. The case where all the F_i are identically equal to some F can represent the repeated running of an experiment whose outcome is a real number with the value of the outcome distributed by F. The symbol Pr has been written above; this, of course, should be read as τ until it is shown that each set $(a_1,b_1] \times \cdots \times (a_n,b_n]$ is measurable and that its measure is given by its τ value.

Actually, we now examine a more general situation than the above. Suppose $F_1(x)$ is a distribution function on $\mathbb{R}$ and Pr_1 is its probability measure. Suppose $F_2(x,y)$ with Pr_2 its probability measure satisfies $F_1(x) = Pr_2(\{(t,y): t \le x\})$. Suppose, in general, that $F_{k+1}(x_1, \ldots, x_{k+1})$ is a distribution function on $\mathbb{R}^{k+1}$ with Pr_{k+1} its probability measure and that

$$F_k(x_1,\ldots,x_k) = Pr_{k+1}(\{(y_1,\ldots,y_{k+1}): y_1 \le x_1, \ldots, y_k \le x_k\}).$$

Let the space X consist of all sequences $\{x_n\}$ and the covering class $\mathscr{C}$ be all sets of the form $\{\{x_i\}: a_i < x_i \le b_i \quad i = 1, 2, \ldots, k\}$ which, for brevity, will be written as $(a,b]$ or $(a,b]_k$. Then $\tau((a,b]_k) = Dh_1 \circ \cdots \circ Dh_k \circ F_k(a)$ where $a = (a_1, \ldots, a_k)$ and $b = (b_1, \ldots, b_k)$ and $h_i = b_i - a_i$ and τ determines a measure by method I.

In order to see that each element $(a,b]_k$ of the covering class is measurable and that $\tau((a,b]_k)$ is its measure, we begin by moving the elements of the space X to $Y = [0,1]^\omega$, the countable cross product of intervals $[0,1]$. To do this, take any strictly increasing continuous function h from $(-\infty,\infty)$ onto

(0,1) and let $h(x) = y$ and each sequence $\{x_n\}$ correspond to the sequence $\{y_n\} = \{h(x_n)\}$. Let $G_k((h(a),h(b)]_k) = F_k((a,b]_k)$. Consistent with this, let

$$\tau'((c,d]_k) = G_k((c,d]_k)$$
$$= Pr_k(c_1<h(x_1)\le d_1, \ldots, c_k<h(x_k)\le d_k).$$

Now the measurability of each $(a,b]_k$ in X is equivalent via h to the measurability of each $(c,d]_k$ in Y as is the equality of $Pr((a,b]_k)$ to $\tau((a,b]_k)$ equivalent to the equality of $Pr((c,d]_k)$ to $\tau'((c,d]_k)$.

In order to make the measure on Y a metric outer measure, we put the standard metric on $[0,1]^\omega$; namely for each $y = \{y_n\}$ and $y' = \{y'_n\}$, with $y_n, y'_n \in [0,1]$, let $d(y,y') = \Sigma|y_n - y'_n|\cdot 2^{-n}$. It is readily checked that this is a metric. Also, under this metric, $[0,1]^\omega$ is compact. Seeing this involves the same reasoning used to show that k-cells are compact. That is, given an open cover of $[0,1]^\omega$ suppose the cover has no finite subcover. But then $Y = [0,1]^\omega = \{y \in Y: 0 \le y_1 \le 1/2\} \cup \{y \in Y: 1/2 \le y_1 \le 1\}$ and one of these must not have a finite subcover. Continuing in this fashion yields a decreasing sequence of sets $\{A_k\}$ of the form

$$\{y \in Y: m_i/2^n \le y_i \le (m_i+1)/2^n, \quad i = 1, 2, \ldots, k\}$$

each of which does not have a finite subcover. But $\cap A_k$ is a single point y_o which is covered by some set G from the original cover. Thus there is a neighborhood $N_\varepsilon(y_o)$ in G and since $\text{diam}(A_k) \le 1/2^k \sum_1^k 2^{-i} + \sum_{k+1}^\infty 2^{-i} < 2^{-k+2}$ and $y_o = \cap A_k$, there is

$A_k \subset G$ contrary to the fact that each A_k had no finite subcover. It follows that $[0,1]^\omega$ is compact.

Now the measurability of each $(c,d]_k$ in Y is guaranteed by considering the measure to be generated by method II, for then the resulting measure is a metric outer measure. Thus it only remains to show that the measure generated by method II gives $Pr((c,d]_k) = \tau'((c,d]_k)$ for each $(c,d]_k$. To see this, note first that the right continuity of each F_k guarantees the right continuity of each G_k. Fix $(c,d]_k$. Given $\varepsilon > 0$ there is $h > 0$ so that $Pr_k([c_h,d]_k) > \tau'((c,d]_k) - \varepsilon$. Now, for any countable cover of $(c,d]_k$ with sets $(c_n,d_n]_{k_n}$ the intervals $(c_n,d_n]_{k_n}$ can be extended to $(c_n,d_{h_n,n})_{k_n}$ with $Pr_{k_n}((c_n,d_{h_n},n]) - \tau'((c_n,d_n]) < \varepsilon/2^n$. Then $\{(c_n,d_{k_n,n})_{k_n}\}$ is an open cover of $[c_h,d]$ and has a finite subcover, say $\{(c_n,d_{h_n,n})_{k_n}: k_n \le K\}$. Then, since Pr_K is a measure

$$\sum_{k_n \le K} \tau'((c,d]_{k_n}) = \sum_{k_n \le K} Pr_K((c,d]_{k_n}) \ge Pr_K((c_h,d]_k) - \varepsilon$$

$$\ge \tau'([c_h,d]_k) - \varepsilon \ge \tau'((c,d]_k) - 2\varepsilon.$$

Since $\varepsilon > 0$ was arbitrary, $Pr((c,d]_k) = \tau'((c,d]_k)$ as required.

We conclude this survey of measures by defining two other useful probability measures. The first is called a <u>Poisson process</u>. Let $\lambda > 0$ and X be the collection of all increasing non-negative sequences of real numbers. (These are thought of as "arrival times" of customers, telephone calls or other events; λ is the "rate of arrival".) One then defines

$$\Pr(\|x_i \in (a,b]\|_c = n) = e^{-\lambda t}(\lambda t)^n/n!$$

there $t = b - a$. For disjoint intervals $(a_j, b_j]$, $j = 1, 2, \ldots, k$, let

$$\Pr(\|x_i \in (a_j,b_j]\| = n_j, \quad j = 1, 2, \ldots, k)$$
$$= \prod_{j=1}^{k} \Pr(\|x_i \in (a_j,b_j]\| = n_j).$$

It, of course, needs to be shown that these sets are measurable and that the value Pr of each set is its measure. There are many variations of this type of measure which represent "arrival times" of a sequence of events.

The final example of a probability measure consists of Brownian motion. On the line, this is a measure defined on the space of all continuous real-valued functions $x(t)$ with $x(0) = 0$ and $t \in [0,\infty)$. With $\sigma > 0$ fixed, one defines for each real number h

$$\Pr(\{\{x(t)\}: x(t+h) - x(t) \le \sigma h\}) = Z_\sigma(h)$$

where Z_σ is the normal distribution with variance σ^2; that is,

$$Z_\sigma(h) = \int_{-\infty}^{h} e^{-\frac{1}{2}t^2/\sigma^2}\, dt \,/\, \sqrt{2\pi\sigma}.$$

Proving measurability of subsets of this space and computing the measure of subsets is not an easy exercise. Variations of this measure for paths in $\mathbb{R}^k$ and other similar measures have been developed. These are studied within the subject which is called stochastic processes.

5.2 Exercises

1. Let C be the Cantor set and $s = (\log 2)/(\log 3)$. Show that $s - m(C) = 1$.
2. Show that Hausdörff one-dimensional measure agrees with length of a curve for all curves in $\mathbb{R}^k$ which are given by a one to one continuous map from $[0,1]$ into $\mathbb{R}^k$.
3. Show that there is a non-empty perfect set $P \subset \mathbb{R}$ so that $P \times P$ is contained in a curve which has finite length.
4. Let $\text{Proj}_y(x,y) = x$. Show that $m^*(\text{Proj}_y(E)) \leq (1 - m)^*(E)$ for each set $E \subset \mathbb{R}^2$.
5. Let F be a non-decreasing function from $[0,1]$ to $[0,1]$. Show that the length of the graph of F is always less than or equal to 2.
6. Show that $(1 - m)(E) = 2$ where E is the graph of the Cantor function.
7. Let $F(I)$ be defined on intervals $I = (a,b]$ in $\mathbb{R}^2$ by $F(I) = 1$ if there is $\varepsilon > 0$ so that $(1+\varepsilon,\ 1+\varepsilon) \in I$ and $F(I) = 0$ otherwise. Show that F is additive and non-negative and that the resulting $F(x)$ is continuous in both variables at $(1,1)$ but $F(x)$ is not continuous on the right at $(1,1)$.
8. (Komolgoroff's extension theorem). Suppose m is defined, non-negative and additive on an algebra $\mathcal{A}$ of subsets of X with $m(\emptyset) = 0$ and $m(X) < \infty$. Suppose that whenever X_i is a decreasing sequence of sets in $\mathcal{A}$ with $\cap X_i = \emptyset$ one has $\lim m(X_i) = 0$. Show that m can be extended to a measure on the smallest σ-algebra containing $\mathcal{A}$.

Chapter Six

6.1 Measurable Functions

This section is concerned with real-valued functions defined on a measure space. Initially, we consider that the space has no other structure; eventually, we will be concerned with spaces in which Borel sets are measurable and finally with the real numbers with Lebesgue measure. Measurable functions are the candidates for the collection of functions for which an integral can be defined. Such functions deserve study in their own right and examining them now is a necessary preliminary to understanding integrals. One could define a measurable function as follows: If $(X,\mathcal{A},m_1)$ is a measure space and f is defined on X and takes its values among the real numbers or extended real numbers, then f is measurable if

$$\{(x,y): x \in X, \quad 0 \le y \le f(x) \quad \text{or} \quad f(x) \le y \le 0\}$$

is a measurable set in $X \times \mathbb{R}$ with respect to the product measure $(m_1 \times m)$. Such a definition would give some idea of forthcoming integration processes. However, we will adopt the traditional and more workable definition of a measurable function. It is from this definition that the properties of measurable functions are more easily deduced.

If f is a real-valued or extended real-valued function defined on a measure space X, f is said to be *measurable* if, for every open set $G \subset \mathbb{R}^+$, $f^{-1}(G)$ is a measurable subset of X. (While we will not be concerned with complex-valued or vector-valued functions, a complex-valued function is called measurable if both its real and imaginary parts are measurable functions and a vector-valued function is called measurable if each of its coordinate functions is measurable. Frequently, results involving such functions can be reduced to consideration of the component real-valued functions.) The following theorem simplifies even further the definition of measurable functions.

(6.1) *Theorem.* *In order that a real-valued or extended real-valued function defined on a measure space* $(X,\mathcal{A},m)$ *be measurable, it is necessary and sufficient that one of the following conditions holds:*

i) *for each* $a \in \mathbb{R}$, $\{x: f(x) > a\} \in \mathcal{A}$,

ii) *for each* $a \in \mathbb{R}$, $\{x: f(x) \ge a\} \in \mathcal{A}$,

iii) *for each* $a \in \mathbb{R}$, $\{x: f(x) < a\} \in \mathcal{A}$,

iv) *for each* $a \in \mathbb{R}$, $\{x: f(x) \le a\} \in \mathcal{A}$.

Proof. We first show that conditions i) - iv) are equivalent. Let $\{r_n\}$ be a dense set of real numbers. Suppose f satisfies i). Since $\{x: f(x) \geq -\infty\} = X \in \mathcal{A}$ and, for $a > -\infty$, $\{x: f(x) \geq a\} = \bigcap_{r_n<a} \{x: f(x) > r_n\} \in \mathcal{A}$, f satisfies ii). Suppose f satisfies ii); then $\{x: f(x) < a\} = X\backslash\{x: f(x) \geq a\} \in \mathcal{A}$, and f satisfies iii). That iii) implies iv) is similar to the proof that i) implies ii). For if f satisfies iii), $\{x: f(x) \leq \infty\} = X \in \mathcal{A}$ and, for $a < \infty$, $\{x: f(x) \leq a\} = \bigcap_{r_n>a} \{x: f(x) < r_n\} \in \mathcal{A}$ and thus iv) holds for f. Finally, if iv) holds for f, then, as in the proof that ii) implies iii), $\{x: f(x) > a\} = X\backslash\{x: f(x) \leq a\} \in \mathcal{A}$. Now, clearly, if $f^{-1}(G) \in \mathcal{A}$ for each open set $G \subset \mathbb{R}^+$, then $\{x: f(x) > a\} \in \mathcal{A}$ and conditions i) - iv) hold. On the other hand, if i) - iv) hold, then for each interval (a,b), $f^{-1}((a,b)) = f^{-1}((-\infty,b)) \cap f^{-1}((a,\infty)) \in \mathcal{A}$. For $G \subset \mathbb{R}$, $G = \cup(a_n,b_n)$, $f^{-1}(G) = \bigcup_n f^{-1}((a_n,b_n)) \in \mathcal{A}$. Since $f^{-1}(\infty) = \bigcap_n \{x: f(x) > r_n\} \in \mathcal{A}$ and $f^{-1}(-\infty) = \bigcap_n \{x: f(x) < r_n\} \in \mathcal{A}$, if G is an open subset of the extended real numbers, it follows that $f^{-1}(G)$ is also measurable.

Note that in Theorem 6.1, the condition that f is measurable is equivalent to the condition that there exist $\{r_n\}$, a dense subset of $\mathbb{R}$, such that each set $\{x: f(x) > r_n\}$ is measurable. In the direction of greater generality, we have: f is measurable if and only if the inverse image of each Borel set under f is a measurable set. This is because, for any sequence of sets $\{A_n\}$ with $A_n \subset \mathbb{R}^+$,

$$f^{-1}(\bigcap_n A_n) = \bigcup_n f^{-1}(A_n), \quad f^{-1}(\bigcap_n A_n) = \bigcap_n f^{-1}(A_n)$$

and for any set $A \subset \mathbb{R}^+$, $f^{-1}(\mathbb{R}^+ \backslash A) = X \backslash f^{-1}(A)$ and from this and the fact that the Borel sets are the smallest σ-algebra containing the open sets, for each Borel set B, $f^{-1}(B)$ is measurable. In particular, if f is a measurable function, then for each x, $f^{-1}(x)$ is a measurable set. However, the inverse image of measurable sets in general need not be measurable. Consider the following:

<u>Example</u> 1. Let P be a nowhere dense perfect subset of $[0,1]$ of positive measure. Then, according to Theorem 3.3, there is an increasing continuous function f which takes P one to one onto the Cantor ternary set. Since f is continuous, the inverse image of open sets are open and hence measurable, and hence f is measurable. However, if $S \subset P$ is not measurable, $f(S) \subset C$ is of measure 0 and hence measurable. But $f^{-1}(f(S)) = S$ is not measurable.

A consequence of this fact is that the composition of two measurable functions of a real variable need not be a measurable function.

<u>Example</u> 2. Let f be as in Example 1, S be a non-measurable subset of P and $g(x) = C_{f(S)}(x)$. Then g is measurable. Indeed, $g^{-1}(G)$ either equals $[0,1]$, $f(S)$, $[0,1] \backslash f(S)$, or $\emptyset$ depending on whether $\{0,1\} \subset G$, $0 \notin G$ and $1 \in G$, $0 \in G$ and $1 \notin G$, or $0 \notin G$ and $1 \notin G$. Thus g is measurable because

$f(S)$ is of measure 0 and hence $f(S)$ and $[0,1]\setminus f(S)$ are measurable. (In fact, we note by the very same argument that $C_E(x)$ is measurable if and only if E is a measurable set.) Now, $g \circ f$ is not a measurable function because $(g \circ f)^{-1}(1) = f^{-1}(g^{-1}(1)) = S$.

Nonetheless, if f is a measurable function and g is continuous or a Baire function of a real variable, then $g \circ f$ is measurable. This is because $(g \circ f)^{-1}(G) = f^{-1}(g^{-1}(G))$ and $g^{-1}(G)$ is open in the event that g is continuous, or a Borel set in the event that g is a Baire function, and hence $f^{-1}(g^{-1}(G))$ is measurable. This being true for each open set G, $g \circ f$ is a measurable function. We will frequently write $f \in \mathcal{M}$ to indicate that f is a measurable function. The context will always clarify whether a particular $\mathcal{M}$ stands for the class of measurable functions or the class of measurable sets.

Measurable functions behave quite well with respect to algebraic and limit operations as is shown by the following theorem. In this theorem and in future theorems, a measurable function is frequently allowed to be undefined on a subset of measure 0 of the measure space. Moreover, two functions are frequently considered to be equivalent if they are defined a.e. and equal at a.e. point in their domain. In fact, it is easily seen that changing the values of a function on a set of measure 0 does not affect the measurability of the function. That is, if $f = g$ a.e., then $f \in \mathcal{M}$ iff $g \in \mathcal{M}$.

(6.2) Theorem. *If* f, g *and each function in a sequence of functions* $\{f_n\}$ *are defined on a measure space* $(X, \mathcal{A}, m)$ *and are measurable, then* $c \cdot f$, $f + g$, $f - g$, $f \cdot g$ *(where* $0 \cdot \infty$ *is understood to equal* 0), $|f|$, $\max(f,g)$, $\min(f,g)$, f/g *(providing* $g(x) \neq 0$), $\sup f_n$, $\inf f_n$, $\overline{\lim}\, f_n$, $\underline{\lim}\, f_n$ *and* $\lim f_n$ *(providing the limit exists* a.e.) *are all measurable functions.*

Proof. Suppose f, g and $\{f_n\}$ are measurable functions defined on X. Since for $c > 0$,

$$\{x: c \cdot f(x) > a\}$$

$= \{x: f(x) > a/c\} \in \mathcal{A}$; and for $c < 0$, $\{x: c \cdot f(x) > a\}$ $= \{x: f(x) < a/c\} \in \mathcal{A}$ for any choice of a, it follows that $c \cdot f \in \mathcal{M}$. Let $\{r_n\}$ be the set of rational numbers. Note that $f(x) + g(x) > a$ iff $f(x) > a - g(x)$ and this holds iff there is a rational number r_n belonging to $(a - g(x), f(x))$; that is, iff $f(x) > r_n$ and $g(x) > a - r_n$ for some rational number r_n. Then

$$\begin{aligned}
&\{x: (f + g)(x) > a\} \\
&\quad = \bigcup_{n=1}^{\infty} \{x: f(x) > r_n \text{ and } g(x) > a - r_n\} \\
&\quad = \bigcup_{n=1}^{\infty} \{x: f(x) > r_n\} \cap \{x: g(x) > a - r_n\}
\end{aligned}$$

belongs to $\mathcal{A}$. Since this is true for each choice of a, $f + g \in \mathcal{M}$. Then $f - g = f + (-1)g \in \mathcal{M}$. If $f \in \mathcal{M}$ and $h(x) = x^2$, then $f^2 = h \circ f \in \mathcal{M}$ by the remark following Example 2 above. Then

$$f \cdot g = [(f + g)^2 - (f - g)^2]/4$$

is a measurable function. If $f \in \mathcal{M}$ and $h(x) = |x|$, then $|f| = h \circ f \in \mathcal{M}$. But then, given $f \in \mathcal{M}$ and $g \in \mathcal{M}$, $\max(f,g) = |f + g|/2 + |f - g|/2 \in \mathcal{M}$. If for each $x \in X$, $g(x) \neq 0$, then for $a > 0$, $\{x: 1/g(x) > a\} = \{x: g(x) < 1/a\} \in \mathcal{A}$ and, for $a < 0$, $\{x: 1/g(x) > a\} = \{x: g(x) > 0\} \cup \{x: g(x) < 1/a\} \in \mathcal{A}$. Thus $1/g \in \mathcal{M}$ and for $f \in \mathcal{M}$, $f/g \in \mathcal{M}$. For $f_n \in \mathcal{M}$,

$$\{x: \sup f_n(x) > a\}$$

$= \bigcup_n \{x: f_n(x) > a\} \in \mathcal{A}$. Since this holds for any choice of a, $\sup f_n \in \mathcal{M}$. Since $\inf f_n = -\sup(-f_n)$, $\inf f_n \in \mathcal{M}$. If $g_k(x) = \sup_n f_{n+k}(x)$, then $\overline{\lim} f_n(x) = \inf_k g_k(x)$ and hence $\overline{\lim} f_n \in \mathcal{M}$. Also $\underline{\lim} f_n(x) = -\overline{\lim}(-f_n(x)) \in \mathcal{M}$. Finally, if $\lim f_n(x)$ exists a.e. it equals the common value of the $\overline{\lim} f_n(x)$ and $\underline{\lim} f_n(x)$ and hence it is measurable.

We come now to an important theorem due to Egoroff which holds for sequences of measurable functions defined on a measure space of finite measure. This theorem, as with those previous to it, assumes no topological or metric character for the measure space.

(6.3) <u>Theorem</u>. (Egoroff's Theorem) <u>If</u> $m(X) < \infty$ <u>and</u> $f_n \in \mathcal{M}$ <u>are defined</u> a.e. <u>on</u> X <u>and approach</u> f a.e. <u>on</u> X, <u>then</u> $f \in \mathcal{M}$ <u>and given</u> $\varepsilon > 0$ <u>there is</u> $E \subset X$ <u>such that</u> E <u>is measurable</u>, $m(X \setminus E) < \varepsilon$ <u>and</u> f_n <u>approaches</u> f <u>uniformly on</u> E.

<u>Proof</u>. Let Z be the set of points where f_n does not converge along with those points where one of the

functions f_n is not defined. Then Z, being a countable union of sets of measure 0, is of measure 0. For each pair of natural numbers N, k let $E_{N,k} = \{x \in X\backslash Z\colon n > N \text{ implies } |f_n(x) - f(x)| < 1/k\}$. Since f_n approaches f a.e., $f \in \mathcal{M}$, $|f_n - f| \in \mathcal{M}$ and each

$$E_{N,k} = \bigcap_{n=N}^{\infty} \{x\colon |f_n(x) - f(x)| < 1/k\}$$

is a measurable set. From the convergence of f_n on $X\backslash Z$, it follows that for every pair M, k of natural numbers, $X\backslash Z = \bigcup_{N=M}^{\infty} E_{N,k}$. Since $m(X\backslash Z) < \infty$, given k and $\varepsilon > 0$ there is a natural number $N_k \geq k$ so that $E_k = \bigcup_{N=N_k}^{\infty} E_{N,k}$ satisfies $m(E_k) > m(X) - \varepsilon/2^k$. Let $E = \bigcap_k E_k$. Then $m(X\backslash E) \leq m(\bigcup_k X\backslash E_k) \leq \sum_k \varepsilon/2^k = \varepsilon$. But now, for every $x \in E$, given any natural number k, the number N_k depends on k but not on x and for $n > N_k$ we have $|f(x) - f_n(x)| < 1/k$. Thus f_n converges uniformly on E.

Note that, in the case that $m(X) = \infty$, a version of Egoroff's Theorem also holds. That is, one can prove that if $m(X) = \infty$ and $f_n \in \mathcal{M}$ with f_n approaching f a.e., then given any number M there is a measurable set $E \subset X$ such that f_n approaches f uniformly on E and $m(E) > M$. That this can not be improved to a set of infinite measure can be easily seen by considering the real functions $f_n(x) = \min(x/n, 1)$ defined on $[0,\infty)$. While all these functions are bounded and approach 0, they do not do so uniformly on a set of infinite Lebesgue measure.

Given a property p which applies to measure spaces, we say that the property "almost p" holds on $(X,\mathcal{A},m)$ if, given any $\varepsilon > 0$, there is a measurable set $E \subset X$ such that $m(X\setminus E) < \varepsilon$ and p holds on E. In this terminology, Egoroff's Theorem asserts that on a space of finite measure pointwise convergence of a sequence of functions is almost uniform convergence. For many measures on metric or topological spaces the measurable functions can be characterized as follows:

(6.4) Theorem. (Lusin's Theorem) *Suppose* $(X,\mathcal{A},m)$ *is a measure on a topological space such that* $m(X) < \infty$, *closed sets are measurable and for any set* $E \in \mathcal{A}$ *and* $\varepsilon > 0$ *there is a closed set* $F \subset E$ *with* $m(E\setminus F) < \varepsilon$. *Then every measurable function on* X *which is a.e. finite is almost continuous.*

Proof. Assume the hypotheses of the theorem hold for X and that f is an a.e. finite measurable function. Given $\varepsilon > 0$, we must find $E \subset X$ so that $m(X\setminus E) < \varepsilon$, E is measurable and f is continuous on E. First, there is a number $M > 0$ so that if $E_1 = \{x: |f(x)| < M\}$ then $m(X\setminus E_1) < \varepsilon/2$. For natural numbers n and integers m with $|m/n| \le M$, let

$$E_{n,m} = \{x \in E_1: m/n \le f(x) < (m+1)/n\}$$

and let $s_n(x) = m/n$ if $x \in E_{n,m}$ and $s_n(x) = 0$ if $x \notin \bigcup_m E_{n,m}$. Since for each n, the sets $E_{n,m}$ are pairwise disjoint, each $s_n(x)$ is defined on E_1. Also, for $x \in E_1$ and natural numbers n,

$$|s_n(x)-f(x)| \le 1/n$$

and thus the sequence of functions $\{s_n\}$ converges to

$f(x)$ uniformly on E_1. Now there are closed sets $F_{n,m} \subset E_{n,m}$ such that $|\bigcup_m E_{n,m} \setminus \bigcup_m F_{n,m}| < \varepsilon/2^{n+1}$. Since for each n, $\bigcup_m E_{n,m}$ is a finite union, the set $\bigcup_m F_{n,m}$, being a finite union of closed sets, is itself closed. Hence s_n is continuous on $\bigcup_m F_{n,m}$. Let $E = \bigcap_n \bigcup_m F_{n,m}$. Since each s_n is continuous on E and s_n converges uniformly to f on E, f is continuous on E. However, $m(X\setminus E) \leq \varepsilon/2 + \Sigma\, \varepsilon/2^{n+1} = \varepsilon$. The existence of such sets E for each $\varepsilon > 0$ shows that f is almost continuous.

While Lusin's Theorem was proved for spaces of finite measure, it also holds on some spaces of infinite measure; in particular, it holds on the line with Lebesgue measure and in Euclidean n-space with n-dimensional measure. In fact, one can observe that the theorem holds in any space $(X,\mathcal{A},m)$ which satisfies the original hypotheses with the hypothesis $m(X) < \infty$ replaced by: For each $\varepsilon > 0$ there is a sequence of measurable sets $\{X_i\}$ of finite measure such that $m(X\setminus \bigcup_{i=1}^{\infty} X_i) < \varepsilon$ and such that for each j, if E_j is the closure of $\bigcup_{i\neq j} X_i$, then $E_j \cap X_j = \emptyset$. Then, whenever f is continuous on closed sets $F_i \subset X_i$, it is continuous on their union. When Euclidean k-space is partitioned into non-overlapping k-cells, such sets X_i can be chosen inside each such cell. From this it follows that Lusin's theorem holds for Lebesgue measure in Euclidean n-space.

We now turn our attention to functions of a real variable. A function f which is continuous on a closed set $F \subset \mathbb{R}$ can easily be extended to a function g which is continuous on the entire line. That is, there is a function g continuous on $\mathbb{R}$ so that $g = f$ at each point of F. In particular, the following function g is continuous: for $x \in F$, $g(x) = f(x)$; if x belongs to an interval (a,b) contiguous to F, the graph of g above this interval consists of the straight line segment connecting $(a,f(a))$ and $(b,f(b))$; if F has a greatest lower bound c, $g(x) = f(c)$ for $x \in (-\infty,c)$; if F has a least upper bound d, $g(x) = f(d)$ for $x \in (d,\infty)$. To see that g is continuous on $\mathbb{R}$, let x be a real number and let $\{x_n\}$ be an increasing sequence whose limit is x. If $x \in (a,b]$ where (a,b) is contiguous to F or $x \in (-\infty,c]$ or if for some $\varepsilon > 0$, $(x-\varepsilon, x]$ is in F, clearly $g(x_n)$ approaches $g(x)$. Otherwise, $x \in F$ and each x_n either belongs to F or to an interval (a_n,b_n) contiguous to F with $b_n \neq x$. Since f is continuous on F, f is uniformly continuous on the compact set $[x_1,x] \cap F$. Thus, given $\varepsilon > 0$ there is $\delta > 0$ so that, if $u,v \in F$ and $|u - v| < \delta$ then $|f(u) - f(v)| < \varepsilon/2$. Then there is a natural number N so that $n > N$ implies $x - x_n < \delta$ and, if $x_n \in (a_n,b_n)$, $x - a_n < \delta$. For each x_n with $n > N$, if $x_n \in F$ then $|g(x_n) - g(x)| = |f(x_n) - f(x)| < \varepsilon$, if $x_n \notin F$ then $x_n \in (a_n,b_n)$ and

$$\begin{aligned} |g(x_n) - g(x)| &\leq |g(x_n) - f(b_n)| + |f(b_n) - f(x)| \\ &\leq |f(a_n) - f(b_n)| + |f(b_n) - f(x)| \\ &\leq \varepsilon/2 + \varepsilon/2 = \varepsilon. \end{aligned}$$

Thus g is continuous on the left at x. Continuity

on the right at each point x is proven in the same way. It follows that the function g is continuous on the entire line.

Now it follows from the proof of Lusin's Theorem that if f is an a.e. finite measurable function on a measurable subset A of $\mathbb{R}$, then there is an $\mathcal{F}_\sigma$ set $E \subset A$ such that $|A\setminus E| = 0$ and there is a sequence of continuous functions f_n with $\lim f_n(x) = f(x)$ at each $x \in E$. Each such function f is thus Baire 1 on E with respect to E. However, measurable functions f, which are finite almost everywhere, do not in general agree with Baire 1 functions on all but a set of measure 0. To see this, we first contruct an $\mathcal{F}_\sigma$ set A contained in $[0,1]$ such that for each $I \subset [0,1]$ both $A \cap I$ and $A^c \cap I$ have positive measure. To do this, let F_1 be a closed nowhere dense subset of $[0,1]$ with 0 and 1 belonging to F_1 and $|F_1| < 1/3$. If F_n has been defined and $I_{n,i}$ are the contiguous intervals to F_n, let $F_{n,i}$ be a closed subset of $I_{n,i}$ of measure greater than 0 and less than $3^{-n}|I_{n,i}|$ containing the midpoint of $I_{n,i}$. Let $F_{n+1} = F_n \cup \cup F_{n,i}$. Note that each F_n is closed and $A = \cup F_n$ has positive measure in each subinterval of $[0,1]$. Moreover, if $I \subset [0,1]$, then there are natural numbers N, i such that $I_{N,i}$ is a contiguous interval to F_N contained in I. But then $|A \cap I_{N,i}| \le \sum_{n=N+1}^{\infty} |I_{N,i} \cap F_n| \le \sum_n 3^{-n}|I_{N,i}| < |I_{N,i}|$ and thus $|A^c \cap I| > 0$. The function $C_A(x)$ does not agree with any Baire 1 function on all but a subset of $[0,1]$ of measure 0. To see this, suppose $g(x) = C_A(x)$ a.e. Then $g(x) = 1$ on a dense subset of $[0,1]$ and $g(x) = 0$ on a dense subset of $[0,1]$. Then every point of $[0,1]$ is a point of discontinuity for g and, since by Theorem 4.7 every Baire 1 function has a residual set of points of continuity, g is not Baire 1.

Nonetheless, the following theorem holds:

(6.5) Theorem. *If f is a measurable function defined on an interval or on the entire line, there is a function h of Baire class 2 such that $f = h$ a.e.*

Proof. First suppose that f is measurable and a.e. finite on an interval or on the entire line. Recall that, if E is a closed set, $C_E(x)$ is a Baire 1 function. By Lusin's Theorem, for each natural number n there is a closed set E_n so that f is continuous on E_n with respect to E_n and $|E_n^c| < 1/n$. By the remarks prior to this theorem there are continuous functions g_n defined on the entire line such that $f(x) = g_n(x)$ at each $x \in A_n = \bigcup_{i=1}^{n} E_i$. Let $h_n = g_n \cdot C_{A_n}$. Then each h_n, being the product of a continuous function and a Baire 1 function, is a Baire 1 function. The sequence of functions $\{h_n\}$ converges everywhere to

$$h(x) = \begin{cases} f(x), & \text{if } x \in \cup E_n \\ 0, & \text{otherwise.} \end{cases}$$

Thus $h(x)$ is a function in Baire class 2. Since, for each natural number n, $|\{x: f(x) \neq h(x)\}| \leq 1/n$, $f(x) = h(x)$ a.e. Now, if g is a measurable function which is not a.e. finite, then E_∞ and $E_{-\infty}$, the sets where g is $+\infty$ and $-\infty$ respectively, are measurable sets. Let $\{A_n\}$ be an increasing sequence of closed subsets of E_∞ so that $|E_\infty \setminus \cup A_n| = 0$ and $\{B_n\}$ be a similar sequence for $E_{-\infty}$. Let $f(x) = 0$ if $x \in E_\infty \cup E_{-\infty}$ and let $f(x) = g(x)$ otherwise. The Baire 1 functions $h_n(x)$ defined above for such a function f converge to $f(x)$ a.e. Thus the Baire 1 functions $h_n + n \cdot C_{A_n} - n \cdot C_{B_n}$ converge to an extended real-valued function which is equal to g a.e.

Clearly, a function which is equal a.e. to a function in Baire class 2 is measurable and hence the above theorem characterizes measurable functions.

We are now going to develop a characterization of Riemann integrable functions in terms of Lebesgue measure. We first note that certain functions which are not Riemann integrable could easily have an integral defined for them. This fact was part of the motivation for the development of the Lebesgue integral.

Recall that for a bounded function f defined on an interval $[a,b]$ and a partition P of $[a,b]$, $U(P,f) = \sum_{i=1}^{n} M_i \Delta x_i$ and $L(P,f) = \sum_{i=1}^{n} m_i \Delta x_i$ where $P = \{x_o, x_1, \ldots, x_n\}$ with $x_{i-1} < x_i$, $x_o = a$ and $x_n = b$, $M_i = \sup\{f(x) : x \in [x_{i-1}, x_i]\}$,

$$m_i = \inf\{f(x) : x \in [x_{i-1}, x_i]\}$$

and $\Delta x_i = x_i - x_{i-1}$. Then, if $\inf_P U(P,f) = \sup_P L(P,f)$, their common value is the Riemann integral of f on $[a,b]$, written $\int_a^b f(x)\, dx$.

Now, if $E = \{1/n\}_{n=1}^{\infty}$, the function $C_E(x)$ is Riemann integrable on $[0,1]$ with its integral equal to 0. However, if $\mathbb{Q}$ is the set of rational numbers, the function $C_{\mathbb{Q}}(x)$ as well as $C_E(x)$ have beneath their graphs and above the x-axis a set consisting of countably many lines of length one. However, $C_{\mathbb{Q}}(x)$ is not Riemann integrable on $[0,1]$ because $\inf U(P,C_{\mathbb{Q}}) = 1$ and $\sup L(P,C_{\mathbb{Q}}) = 0$.

In order to prove the classic theorem which gives necessary and sufficient conditions for a function to be Riemann integrable, the following theorem on the measure of compact sets will be needed.

(6.6) Theorem. *Let E be a closed subset of $[a,b]$. Then, for every $\varepsilon > 0$ there exists $\delta > 0$ so that if P is a partition of $[a,b]$ with $|P| < \delta$ and $I_i = [x_{i-1}, x_i]$ are the intervals determined by P, then*

$$\left|\Sigma'|I_i| - |E|\right| < \varepsilon$$

where Σ' denotes the summation over all I_i with $I_i \cap E \neq \emptyset$ (alternatively, this holds for Σ'' where Σ'' denotes the summation over all I_i which have a point of E in their interior).

Proof. Given E closed, $E \subset [a,b]$, let $G = [a,b]\setminus E$ and $\{J_k\}$ be the set of intervals contiguous to E. Then given $\varepsilon > 0$, there is a natural number K so that $\sum_1^K |J_k| > |G| - \varepsilon/2$. Let $B = \bigcup_1^K \mathring{J}_k$ and note that $E \subset B^c$ and $|E| + \varepsilon/2 > |[a,b]\setminus B|$. If $\delta = \varepsilon/8K$ and $|P| < \delta$ then, if $I_i = [x_{i-1}, x_i]$ with $x_i \in P$,

$\{I_i\colon I_i$ contains an endpoint of some J_k, $k \le K\}$

contains at most $4K$ elements; that is, at most two intervals belong to this set for each of the two endpoints of the J_k. Thus the sum of the lengths of those I_i is less than $4K \cdot \varepsilon/8K = \varepsilon/2$. The other I_i of P which meet E have points of E in their interior. Hence

$$|E| \le \Sigma'|I_i| \le \varepsilon/2 + \sum_{I_i \subset B^c} |I_i| \le \varepsilon/2 + |B^c| \le \varepsilon + |E|.$$

Thus $\left|\Sigma'|I_i| - |E|\right| < \varepsilon$ where Σ' denotes summation over all I_i with $I_i \cap E \ne \emptyset$. Alternately, if Σ'' denotes summation over all I_i from P which have a point of E in their interior, using the same estimates, it follows that

$$\Sigma''|I_i| - |E| \le \Sigma'|I_i| - |E| < \varepsilon$$

and $\left|\Sigma''|I_i| - |E|\right| < \varepsilon$.

We now show that a function f is Riemann integrable on an interval if and only if f is bounded and continuous a.e. on the interval.

(6.7) Theorem. *In order that a function f be Riemann integrable on $[a,b]$, it is necessary and sufficient that f be bounded on $[a,b]$ and that the set of points of discontinuity of f on $[a,b]$ be of measure 0.*

Proof. Clearly, if f is not bounded above on $[a,b]$, then $U(P,f) = \infty$ for each partition P of $[a,b]$ and if f is not bounded below, then $L(P,f) = -\infty$ for each partition P. Suppose f is bounded and that

$$E = \{x \in [a,b]: f \text{ is discontinuous at } x\}$$

is not of measure 0. Then $E = \cup E_n$ where $E_n = \{x \in [a,b]: \omega_f(x) \geq 1/n\}$. Recall that each set E_n is closed and hence measurable. There is a natural number N so that $|E_N| > 0$. Thus, given $\varepsilon > 0$ with $\varepsilon < |E_N|$, by Theorem 6.6 there is δ so that $|P| < \delta$ implies $\Sigma''\Delta x_i \geq |E| - \varepsilon > 0$. But then, since $\omega_f(x) \geq 1/N$ at some interior point of each interval involved with Σ'', it follows that $M_i - m_i \geq 1/N$ at each such interval and $U(P,f) - L(P,f) \geq \Sigma''(M_i - m_i)\Delta x_i > (1/N)(|E| - \varepsilon) > 0$. Then $\inf U(P,f) - \sup L(P,f) > 0$ and f is not Riemann integrable. Thus it is not possible that the set of points of discontinuity be of positive measure.

Before proceeding to the proof of sufficiency note that: If on a closed interval $I = [c,d]$ there is an $\varepsilon > 0$ and at each $x \in I$ $\omega_f(x) < \varepsilon$, then I can be partitioned into smaller intervals I_i such that $M_i - m_i < \varepsilon$ on each I_i. This is because

$$\{J: J = (x-\delta, x+\delta) \text{ and } \sup_{t\in J} f(t) - \inf_{t\in J} f(t) < \varepsilon\}$$

is an open cover of I and has a finite subcover. The intervals from the finite subcover which are not contained in a union of others is a cover of $[c,d]$ of the form $(x_1', x_1''), (x_2', x_2''), \ldots, (x_n', x_n'')$ where $x_i'' \in (x_{i+1}', x_{i+1}'')$ $i = 1, \ldots, n-1$, and $x_1' \in (x_{i-1}', x_{i-1}'')$ $i = 2, \ldots, n$. Then $x_o = c$, $x_1 \in (x_2', x_1''), \ldots, x_{n-1} \in (x_n', x_{n-1}'')$, $x_n = d$ is the required partition of I.

Now, to complete the proof of Theorem 6.7, suppose f is bounded and the set of points of discontinuity of f in $[a,b]$ is of measure 0. Given $\varepsilon > 0$, there is a natural number N so that $1/N < \varepsilon$. Let M be an upper bound for $|f(x)|$.

Choose $\delta > 0$ so that, if $|P| < \delta$, then $\Sigma'|I_i| < \varepsilon$ where Σ' is over all I_i with $I_i \cap E_N \neq \emptyset$. This is possible due to Theorem 6.6 and the fact that E_N is a closed set of measure 0. Starting with a partition Q with $|Q| < \delta$, the intervals in Q which do not meet E_N satisfy the conditions noted above and they can be partitioned into smaller intervals for which $M_i - m_i < 1/N < \varepsilon$. The resulting partition P satisfies $|P| < \delta$ and if $I_i \cap E_N = \emptyset$ where $I_i = [x_{i-1}, x_i]$ with $x_{i-1}, x_i \in P$, then $M_i - m_i < \varepsilon$. Then

$$U(P,f) - L(P,f) = \Sigma'(M_i - m_i)\Delta x_i + \Sigma''(M_i - m_i)\Delta x_i$$

where Σ'' is over all i with $I_i \cap E_N = \emptyset$. Thus

$$\begin{aligned} U(P,f) - L(P,f) &\leq 2M\cdot\Sigma'\Delta x_i + \varepsilon\cdot\Sigma''\Delta x_i \\ &\leq 2M\cdot\varepsilon + \varepsilon(b - a). \end{aligned}$$

Since $\varepsilon > 0$ is arbitrary, the upper and lower Riemann sums can be made arbitrarily close to each other; that is, the Cauchy criterion for Riemann integrability holds. Thus f is Riemann integrable.

This section is concluded with an example on [0,1] of a differentiable function F with a bounded derivative f which is not Riemann integrable. This is another example of a function for which an integral could be defined. From the above theorem it is sifficient to produce a differentiable function F with bounded derivative f so that the points of discontinuity of f form a set of positive measure.

Let P be a nowhere dense perfect subset of [0,1] of positive measure with $0,1 \in P$. Let $\{(a_n,b_n)\}$ be the sequence of intervals contiguous to P. Let (c_n,d_n) have the same midpoint as (a_n,b_n) and satisfy $d_n - c_n = (1/n)(b_n - a_n)$. Let $F(x) = 0$

if $x \notin \bigcup_{n=1}^{\infty} (c_n, d_n)$. If $x \in (c_n, d_n)$, let

$$F(x) = (x - c_n)^2 (d_n - x)^2 / (d_n - c_n)^3.$$

Then $F(x)$ is defined on $[0,1]$. If $x \in P$ and $x+h \notin \cup (c_n, d_n)$, then $F(x+h) = F(x) = 0$. If $x \in P$ and $x+h \in (c_n, d_n)$,

$$|F(x+h)| \le \left(\frac{d_n - c_n}{2} \right)^4 \Big/ (d_n - c_n)^3 = (d_n - c_n)/16.$$

Thus

$$\left| \frac{F(x+h) - F(x)}{h} \right| \le \frac{d_n - c_n}{16(c_n - a_n)}$$

$$\le \left(\frac{1}{n}\right) \frac{b_n - a_n}{16(b_n - a_n)(n-1)/2n} = \frac{1}{8(n-1)}$$

For $x \in P$, $\frac{F(x+h) - F(x)}{h}$ approaches 0 as h approaches 0 and $F'(x) = 0$ for each $x \in P$. Since $F'(c_n) = F'(d_n) = 0$, $F'(x) = 0$ for each $x \notin \bigcup_{n=1}^{\infty} (c_n, d_n)$. If $x \in (c_n, d_n)$,

$$F'(x) = 2 \frac{(x - c_n)(d_n - x)^2 - (x - c_n)^2 (d_n - x)}{(d_n - c_n)^3}$$

and $F'\left(\frac{3}{4}(c_n) + \frac{1}{4}(d_n)\right) = 2\left(\frac{1}{4}\right)\left(\frac{3}{4}\right)^2 - 2\left(\frac{1}{4}\right)^2\left(\frac{3}{4}\right) = 3/16$. Since each point $x \in P$ is the limit of a sequence of points of the form $\frac{3}{4}(c_n) + \frac{1}{4}(d_n)$, $F'(x)$ is not continuous at any point of P. However, on (c_n, d_n) an upper bound can be computed for $|F'(x)|$ by replacing the x's in the formula for $F'(x)$ with c_n and d_n. That is

$$|F'(x)| \le 2 \frac{(d_n - c_n)(d_n - c_n)^2 + (d_n - c_n)^2 (d_n - c_n)}{(d_n - c_n)^3}$$

$= 4$. Thus F is a differentiable function with a bounded derivative on $[0,1]$ and $F'(x)$ is not Riemann integrable.

6.1 Exercises

1. Two measurable functions f and g are said to be equimeasurable on $[a,b]$ if for each $[c,d]$, $m(f^{-1}([c,d])) = m(g^{-1}([c,d]))$. Show that for each measurable function f defined on $[a,b]$ there is a non-decreasing function g defined on $[a,b]$ so that f and g are equimeasurable.
2. For $(x,y) \in [0,1] \times [0,1]$, if x and y have decimal expansion $.x_1x_2 \ldots$ and $.y_1y_2 \ldots$ let $f(x,y) = .x_1y_1x_2y_2 \ldots$. Show that f is defined a.e. and is a measurable function.
3. Suppose each f_n and f are measurable and defined a.e. on $[0,1]$. Prove that $f_n \to f$ a.e. iff for each $\varepsilon > 0$,
$$\lim_{N\to\infty} m(\{x: |f_n(x)-f(x)| < \varepsilon \text{ for every } n>N\}) = 0.$$
4. Give an example of a collection of functions $f_t(x)$ for $t \in (0,\infty)$ each measurable but satisfying $\overline{\lim_{t\to\infty}} f_t(x)$ is not measurable.
5. Given any function f, let $g(x) = f(x)$ if f is continuous at x and $g(x) = 0$ otherwise. Show that g is a measurable function.
6. Let $f \in \mathcal{M}$ and $g(x) = 1/q$ when $f(x) = p/q$ and $g(x) = 0$ otherwise. Prove that g is also measurable.
7. Let $f \in \mathcal{M}$ and $g(x) = n$ if the nth digit in the decimal expansion of $f(x)$ is a 0 but no previous digit equals 0; if $f(x)$ has no digit

0 in its expansion, let $g(x) = 0$. Prove $g \in \mathcal{M}$.

8. If $f \in \mathcal{M}$ on $[0,1]$, show that the planar measure of the graph of f is 0.
9. If $F(x)$ is continuous on $[0,1]$ and $S(y) = \|f^{-1}(y)\|_C$, where $S(y) = \infty$ if $\|f^{-1}(y)\|_C$ is infinite, show that $S(y)$ is a measurable function.

6.2 Density Topology and Approximate Continuity

Let f be a real-valued function of a real variable. A number L (possibly $+\infty$ or $-\infty$) is the <u>approximate</u> <u>limit</u> of f as t approaches x, written $\lim_{t \to x} \text{ap } f(t) = L$, provided that x is a point of density of a measurable set E which is contained in the domain of f and the limit of the values $f(t)$ as t approaches x with $t \in E$ is L. Since two measurable sets, each having x as a point of density, have points in common in every neighborhood of x, the approximate limit, when it exists, is unique. A function f is said to be <u>approximately</u> <u>continuous</u> at x if $f(x)$ is finite and $\lim_{t \to x} \text{ap } f(t) = f(x)$. One-sided approximate limits and continuity are defined at x by requiring only that the measurable set E have one-sided density at x of 1; then it is easily seen that $\lim_{t \to x} \text{ap } f(t) = L$ iff $\lim_{t \to x^+} \text{ap } f(t) = L$ and $\lim_{t \to x^-} \text{ap } f(t) = L$.

Clearly, for a function f defined on an interval I with $x \in I$ and $\lim_{t \to x} f(t) = L$, one has

$\lim\limits_{t\to x} \text{ap } f(t) = L$. A simple example shows that the approximate limit is more general than the ordinary limit. Let $A = \bigcup\limits_{n=1}^{\infty} (1/(n+1), 1/n - 1/2^{n+1})$ and $B = A \cup (-\infty,0]$. Then B has 0 as a point of density. To see this, let $\varepsilon > 0$ be given. If $I = [a,b]$ is an interval which contains 0 and $b - a < (N+1)/2^N < \varepsilon$, then, if $n > N$ and $b \in (1/(n+1), 1/n]$, $m((0,b] \cap A) \geq b - \sum\limits_{i=n+1}^{\infty} 2^{-i} = b - 2^{-n} > b(1 - \varepsilon)$ and $m([a,b] \cap B) \geq (b - a)(1 - \varepsilon)$. Since $\varepsilon > 0$ was arbitrary, B has 0 as a point of density. Now, let f be continuous at 0 and g be any function with $g(x) = f(x)$ at each $x \in A \cup (-\infty,0)$, then $\lim\limits_{t\to 0} \text{ap } g(t) = f(0)$.

Note that, if a function f has an approximate limit at x and a function g is equal to f a.e., then g has the same approximate limit at x. Also, if $\lim\limits_{t\to x} \text{ap } f(t) = L$ and c is a real number, $\lim\limits_{t\to x} \text{ap } c\cdot f(t) = c\cdot L$. Furthermore, if $\lim\limits_{t\to x} \text{ap } g(t) = M$, then $\lim\limits_{t\to x} \text{ap } (f + g)(t) = L + M$, $\lim\limits_{t\to x} \text{ap } (f\cdot g)(t) = L\cdot M$ and $\lim\limits_{t\to x} \text{ap } (f/g)(t) = L/M$ provided $M \neq 0$ (equality also holds in the extended sense for the other cases provided $L + M$ and $L\cdot M$ are defined). Seeing this requires only the fact that the intersection of two measurable sets each of which has x as a point of density also has x as a point of density; this is proven below as part of Theorem 6.8.

From the notion of the approximate limit, one obtains that of the approximate derivative. The <u>approximate derivative</u> of a function f at x, written $f'_{ap}(x)$, is $\lim_{t \to x} \text{ap}\,[f(t) - f(x)]/(t - x)$ provided this limit exists. The usual formulas for the derivative of a constant multiple of a function and the derivative of the sum, product, and quotient of two functions will clearly hold for approximate derivatives. In addition, if f has a finite approximate derivative at x, then f is approximately continuous at x. This follows from the equation

$$\begin{aligned}\lim_{t \to x} \text{ap}\,[f(t) - f(x)] &= \lim_{t \to x} \text{ap}\,[f(t) - f(x)]/(t - x) \cdot \lim_{t \to x} \text{ap}\,(t - x) \\ &= 0\end{aligned}$$

and thus when $f'_{ap}(x)$ is finite, $\lim_{t \to x} \text{ap}\, f(t) = f(x)$.

Again, a simple example can illustrate the difference between approximate differentiability and ordinary differentiability. Let $f(x)$ be defined in a neighborhood of 0 and be differentiable at 0. Let $g(x) = f(x)$ for x in the set B described above and let $g(x)$ be defined in anyway whatsoever on B^c; then $g'_{ap}(0) = f'(0)$.

An alternate approach to defining the approximate limit consists of defining the upper and lower approximate limits as

$$\overline{\lim_{t \to x}} \text{ap}\, f(t) = \inf\{y: f^{-1}((y,\infty)) \text{ has } x \text{ as a point of dispersion}\}$$

and

$\underline{\lim}_{t\to x}$ ap $f(t) =$
sup{y: $f^{-1}((-\infty,y))$ has x as a point of dispersion}.

(Recall that x is a point of dispersion of E if $\overline{\lim}_{I\to x} m^*(E\cap I)/|I| = 0$.) Note that, when the outer density of the domain of f is positive at x, one always has

$$\underline{\lim}_{t\to x} f(t) \le \underline{\lim}_{t\to x} \text{ap } f(t) \le \overline{\lim}_{t\to x} \text{ap } f(t) \le \overline{\lim}_{t\to x} f(t).$$

The inequalities involving the limit supremum and limit infimum always hold because, if $y = \overline{\lim}_{t\to x} f(t)$, then x is an isolated point of each set $f^{-1}((y + \varepsilon, \infty))$, for any $\varepsilon > 0$, and hence x is a point of dispersion of each of these sets; similarly, x is an isolated point of each set

$$f^{-1}((-\infty, \underline{\lim}_{t\to x} f(t) - \varepsilon)).$$

If $a = \underline{\lim}_{t\to x}$ ap $f(t)$ were greater than $b = \overline{\lim}_{t\to x}$ ap $f(t)$, then x would be a point of dispersion of the entire domain of f and a would equal ∞, b would equal $-\infty$. Thus, if x is a point of positive outer density of the domain of f, then $\underline{\lim}_{t\to x}$ ap $f(t) \le \overline{\lim}_{t\to x}$ ap $f(t)$.

Note that, whenever $\lim_{t\to x}$ ap $f(t) = L$, one has $\overline{\lim}_{t\to x}$ ap $f(t) = L = \underline{\lim}_{t\to x}$ ap $f(t)$. This is because each set $\{t\colon f(t) \in [L-\varepsilon, L+\varepsilon]\}$, for any $\varepsilon > 0$, contains a neighborhood of x intersected with the measurable set E which has x as a point of density and for which the limit of $f(t)$ as $t \to x$ with $t \in E$ is

L. Thus, each $\{t: f(t) \notin [L-\varepsilon, L+\varepsilon]\}$ has x as a point of dispersion; hence $\overline{\lim_{t\to x}} \text{ ap } f(t) = L$ and $\underline{\lim_{t\to x}} \text{ ap } f(t) = L$.

The converse also holds; namely, if the domain of f contains a measurable set E which has x as a point of density and if $\overline{\lim_{t\to x}} \text{ ap } f(t) = L = \underline{\lim_{t\to x}} \text{ ap } f(t)$, then $\lim_{t\to x} \text{ ap } f(t) = L$. To see this, let

$$E_n = f^{-1}([L - (1/n), L + (1/n)])$$

and note that each set E_n contains a measurable set E'_n and each E'_n has x as a point of density. To simplify the calculations showing that $\lim_{t\to x} \text{ ap } f(t) = L$, we suppose without loss of generality that $x = 0$ and construct a measurable set E which has density 1 for the right side of 0 so that $\lim_{t\to 0^+} f(t) = L$ when t approaches 0 through points of E. Given E'_n as above, choose a sequence $\{\delta_n\}_{n=1}^{\infty}$ decreasing to 0 so that $|E'_n \cap (0,\delta)| > (1 - 2^{-n-1})\delta$ when $\delta < \delta_n$. Let $A_n = E'_n \cap (\delta_{n+1}, \delta_n]$ and $E = \bigcup_{n=1}^{\infty} A_n$. Then if $I = [0,\delta]$ and $\delta_{n+1} < \delta \le \delta_n$, then

$$\begin{aligned}
|E \cap I| &= \sum_{i=n+1}^{\infty} |A_i| + |A_n \cap [\delta_n,\delta]| \\
&= \sum_{i=n+1}^{\infty} |E'_i \cap (0,\delta_i)| - \sum_{i=n+1}^{\infty} |E'_i \cap (0,\delta_{i+1})| \\
&\quad + |A_n \cap (0,\delta)| - |A_n \cap (0,\delta_{n+1})|
\end{aligned}$$

$$\geq \sum_{i=n+1}^{\infty} (1 - 2^{-i-1})\delta_i - \sum_{i=n+1}^{\infty} \delta_{i+1}$$
$$+ (1 - 2^{-n})\delta - \delta_{n+1}$$
$$\geq (1 - 2^{-n})\delta - \sum_{i=n+1}^{\infty} 2^{-i-1}\delta_i$$
$$\geq (1 - 2^{-n} - 2^{-n-1})\delta$$
$$= (1 - 2^{-n} - 2^{-n-1})|I|.$$

It then follows that E has 0 as a point of density on the right and that, if $t \in E$ and $t < \delta_n$, then $f(t) \in [L-(1/n), L+(1/n)]$. Consequently, $\lim_{t \to 0^+} \text{ap } f(t) = L$.

The approximate limit and approximate continuity were introduced by Denjoy. It turns out that the approximate limit is actually a limit with respect to a topology on the domain which is stronger (has more open sets) than the usual topology. Recall that according to the Lebesgue density theorem, almost every point of a Lebesgue measurable subset of the reals is a point of density of the set. A measurable set $E \subset \mathbb{R}$, satisfying the condition that every point of E is a point of density of E, is called a d-<u>open</u> <u>set</u>. The usual open sets are all examples of d-open sets. Sets which are complements of sets of measure 0 are also d-open. If A is the set described in the example above of an approximately continuous function, then $A \cup \{0\} \cup -A$ is a d-open set. Given any measurable set E, let E^d be the set of all points of E which are points of density of E. Since almost every point of E belongs to E^d and $E^d \subset E$, each interval I satisfies

$|E^d \cap I| = |E \cap I|$ and thus each point of E^d is a point of density of E^d. Consequently, whenever $E \in \mathcal{M}$, E^d is a d-open set.

(6.8) *Theorem*. *The collection of all of the d-open subsets of the line forms a topology; that is, this collection is closed under arbitrary unions, finite intersections and contains* $\emptyset$ *and* $\mathbb{R}$.

Proof. Suppose $\{E_\alpha\}$ is a collection of d-open sets and $E = \cup E_\alpha$. If $x \in E$, there is α_0 so that $x \in E_{\alpha_0}$ and since x is a point of density of E_{α_0}, $1 \geq \underline{\lim}_{I \to x}\, m^*(E \cap I)/|I| > \underline{\lim}_{I \to x}\, m(E_{\alpha_0} \cap I)/|I| = 1$. So each $x \in E$ is a point of density of E. To show that E is d-open, it remains to show that E is measurable. But for each $x \in E$ there is an α so that $x \in E_\alpha$ and there is $\varepsilon_{x,\alpha}$ so that, if $x \in I$ and $|I| < \varepsilon_{x,\alpha}$, then $|E_\alpha \cap I|/|I| > 1/2$. Then for each such I, there is a closed set $F_{\alpha,I}$ so that $F_{\alpha,I} \subset E_\alpha \cap I$ and $m(F_{\alpha,I}) > (1/2)|I|$. The collection of all such $F_{\alpha,I}$ covers E in the sense of Vitali. Thus, there is a sequence of these sets $\{F_n\}$ which are pairwise disjoint such that $m(E \backslash \cup F_n) = 0$. Since each $F_n \subset E$ and each $F_n \in \mathcal{M}$, it follows that E is a countable union of closed sets along with a set of measure 0 and that E is measurable. Thus the arbitrary union of d-open sets is d-open. To see that the finite intersection of d-open sets is d-open, consider two such sets E_1 and E_2. Clearly, $E_1 \cap E_2$ is measurable. To show that $E_1 \cap E_2$ is d-open, it remains to show that each point $x \in E_1 \cap E_2$ is a point of density of $E_1 \cap E_2$. Given $x \in E_1 \cap E_2$ and $\varepsilon > 0$ choose $\delta_i > 0$, $i = 1, 2$, so that if $x \in I$ and $|I| < \delta_i$, then $|E_i \cap I|/|I| \geq 1 - \varepsilon/2$. Then, if $\delta = \min(\delta_1,\delta_2)$ and $|I| < \delta$ with $x \in I$, $|E_i^c \cap I| \leq (\varepsilon/2)|I|$ and thus $|(E_1^c \cup E_2^c) \cap I| \leq \varepsilon|I|$; therefore, $|E_1 \cap E_2 \cap I|/|I| \geq 1 - \varepsilon$. Since $\varepsilon > 0$ was arbitrary, x is a point of density of $E_1 \cap E_2$; that is, each point of $E_1 \cap E_2$ is a point of density of $E_1 \cap E_2$. It follows by induction that the finite intersection of d-open sets is d-open. Since $\emptyset$ and $\mathbb{R}$ are clearly d-open sets, the collection of d-open sets forms a topology.

The resulting topology, consisting of the d-open sets, is called the density topology. When the density topology is put on the domain of a real-valued function f defined on an interval, or on a d-open set, and the usual topology is put on the range, the points of continuity of f with respect to these topologies are exactly the points of approximate continuity of the function f. To see this, note that at each point x of approximate continuity of f there is a measurable set E in the domain of f with density 1 at x so that $\lim f(t) = f(x)$ when t is restricted to E. Since E^d is a d-open set, and $\lim_{\substack{t \to x \\ t \in E^d}} f(t) = f(x)$, it follows that f is continuous at x with the density topology on the domain and the usual topology on the range. Conversely, if f is continuous at x with respect to these two topologies, then for each $\varepsilon > 0$, $f^{-1}((f(x)-\varepsilon,\ f(x)+\varepsilon))$ contains a d-open set E_ε with

$x \in E_\varepsilon$. Then the sets $f^{-1}((f(x)+\varepsilon, \infty))$ and $f^{-1}((-\infty, f(x)-\varepsilon))$ are contained in E_ε^c and have x as a point of dispersion. But then $\overline{\lim_{t\to x}} \text{ ap } f(t) = \underline{\lim_{t\to x}} \text{ ap } f(t) = f(x)$ and hence $\lim_{t\to x} \text{ ap } f(t) = f(x)$ and f is approximately continuous at x.

The concept of approximate continuity can be used to provide a characterization of measurable functions.

(6.9) Theorem. *A real-valued function f defined on a measurable set E is measurable on E if and only if f is approximately continuous a.e. on E.*

Proof. Let f be a measurable function defined on $E \in \mathcal{M}$. By Lusin's theorem there is a sequence of closed sets $\{F_n\}$ such that f is continuous on each F_n and

$$m((E \cap [-n,n])\setminus F_n) < 1/n$$

and hence $m(E\setminus \cup F_n) = 0$. For fixed n and $x \in F_n^d$, $\lim_{\substack{t\to x \\ t\in F_n}} f(t) = f(x)$ and thus each such x is a point of approximate continuity of f. But almost every point of E belongs to some F_n and almost every point of each F_n is a point of density of F_n and thus a point of approximate continuity of f. It follows that f is approximately continuous at almost every point of E. For the converse, let f be approximately continuous at a.e. point of $E \in \mathcal{M}$. For each real number y, let $A_y = \{x \in E: f(x) > y\}$. It remains to show that each set A_y is measurable. Fix $y = a$.

For each $x \in A_a$ select a set $E_x \in \mathcal{M}$ with x a point of density of E_x and such that $\lim_{\substack{t \to x \\ t \in E_x}} f(t) = f(x)$. Then by the density of E_x at x and the continuity of f at x with respect to E_x there is $\delta_x > 0$ so that $|E_x \cap [x-\delta, x+\delta]| > \delta$ when $\delta < \delta_x$ and each $t \in E_x \cap [x-\delta_x, x+\delta_x]$ satisfies $f(t) > a$. Now for each $x \in A_a$ and $0 < \delta < \delta_x$ there is a closed set $F_{x,\delta}$ with $x \in F_{x,\delta} \subset E_x \cap [x-\delta, x+\delta]$ and $|F_{x,\delta}| > \delta$. Then the collection of all such $F_{x,\delta}$ covers A_a in the sense of Vitali. Thus there is a sequence of sets $\{F_n\}$ such that $m(A_a \setminus \cup F_n) = 0$. Since each $F_{x,\delta} \subset A_a$, it follows that A_a is a measurable set. Finally, for each $a \in \mathbb{R}$, $A_a \in \mathcal{M}$ and f is a measurable function on E.

We now turn to an examination of some properties of the density topology. Because the open sets of the usual topology on the line are also d-open, the density topology is Hausdörff. If E is any subset of $\mathbb{R}$, the d-<u>interior</u> of E, denoted by E^d, is the interior of E with respect to the density topology. Clearly, for $E \in \mathcal{M}$, this definition agrees with the previous use of the term E^d. In general, E^d is the set of points of a set E for which

$$\underline{D}_E(x) = \varliminf_{I \to x} m_*(E \cap I)/|I| = 1;$$

thus E^d is the union of all d-open sets contained in E and hence E^d is always a measurable set. The d-<u>closure</u> of E, written $d\text{-cl}(E)$, is the complement of $(E^c)^d$ and $d\text{-cl}(E)$ is E along with the set of all points x for which $\overline{D}_E^+(x) = \overline{\lim_{I \to x}} m^*(E \cap I)/|I| > 0$. Then the d-<u>boundary</u> of E, d-bd(E), is d-cl(E)$\setminus E^d$. The d-closure and d-boundary of a set E are clearly also measurable sets. If $E \in \mathcal{M}$, then d-bd(E) is a set of measure 0 because it consists of $E \setminus E^d$ and $E^c \setminus (E^c)^d$ both of which are of measure 0.

Each interval on the line is connected in the density topology. To see this consider any interval I and suppose I is not connected in the density topology. Then $I = A \cup B$ where $x_1 \in A$, $x_2 \in B$, $x_1 < x_2$, and d-cl(A) $\cap B = A \cap$ d-cl(B) $= \emptyset$. Then $A \cap [x_1,x_2]$ and $B \cap [x_1, x_2]$ are d-open subsets of $[x_1, x_2]$. Let $F(x) = m(A \cap [x_1,x])$. That $F'(x) = D_A(x)$ is an easy computation using derivatives. This makes F a differentiable function whose derivative takes on only the values 0 and 1 with $F'(x_1) = 1$ and $F'(x_2) = 0$ which, by the mean value theorem for derivatives, cannot happen. It follows that each I is connected in the density topology.

Note that each set of measure 0 is d-closed since its complement is clearly d-open. Moreover, each set of measure 0 clearly consists solely of d-isolated points; conversely, if all the points of a set Z are d-isolated points, then $m(Z) = 0$. If the converse were not true, and a set Z with $m^*(Z) > 0$ existed, since $\overline{D}_Z^+(x) = 1$ at every point of Z except for a set of measure 0, there would be points of Z where $\overline{D}_Z^+(x) = 1$ and these points would not be d-isolated points.

That the density topology is regular is a consequence of the following theorem.

(6.10) Theorem. (Lusin-Menchoff) *If E is a measurable set and F is a closed subset of E^d, then there is a perfect set P with $F \subset P \subset E$ so that each point of F is a point of density of P.*

Proof. Without loss of generality, suppose $E \subset [0,1]$. By hypothesis, $E \in \mathcal{M}$ and F is a closed subset of E^d. For each natural number n, let

$$E_n = \{x\colon d(x,bd(F)) \in (1/(n+1),\ 1/n)\}$$

where, as usual, $bd(F) = \overline{F}\backslash\mathring{F}$ and

$$d(x,A) = \inf\{d(x,y)\colon y \in A\}.$$

Then each set E_n consists of a finite collection of intervals; this is because E_n is open and each component interval of E_n is adjacent to an interval in a finite cover of $bd(F)$ consisting of the intervals which make up the open set

$$\{x\colon d(x,bd(F)) < 1/(n+1)\}.$$

For each n choose a perfect set $P_n \subset E_n \cap E$ so that $m(P_n) > m(E_n \cap E) - 2^{-n}$; here $P_n = \emptyset$ when $m(E_n \cap E) = 0$. Since every point of $bd(F)$ is a limit point of $\cup P_n$ due to the density of such points in E and since $P = \bigcup_{n=1}^{\infty} P_n \cup F$ is closed, P is a perfect set containing F. It remains to show that $F \subset P^d$. Consider any $x \in F$. If $x \in \mathring{F}$, then x is a point of density of F and hence a point of density of P. So suppose $x \in bd(F)$. Let $\{I_i\}_{i=1}^{\infty}$ be a sequence of intervals with $x \in I_i$ and $\lim \text{diam}(I_i) = 0$. Because x is a point of density of E and $bd(F)$ is nowhere dense, for each i there is a least natural number n_i so that $I_i \cap P_{n_i} \neq \emptyset$. Hence

$$m(P \cap I_i) \geq m(bd(F) \cap I_i) + \sum_{n>n_i} m(P_n \cap I_i)$$
$$\geq m(bd(F) \cap I_i) + \sum_{n>n_i} m(E_n \cap I_i) - 2^{-n}$$
$$\geq m(E \cap I_i) - 1/2^{n_i-1}$$
$$\geq m(E \cap I_i) - |I_i|(n_i+1)/2^{n_i-1}.$$

Here the last inequality follows from the fact that $|I_i| > 1/(n_i+1)$. Then

$$\frac{m(P \cap I_i)}{|I_i|} \geq \frac{m(E \cap I_i)}{|I_i|} + (n_i+1)/2^{n_i-1}.$$

Now, as i approaches ∞, n_i approaches ∞ and since x is a point of density of E, $\lim_{i\to\infty} \frac{m(P \cap I_i)}{|I_i|} = 1$ and hence $D_p(x) = 1$ and x is a point of density of P. Thus each point of F is a point of density of P and $F \subset P \subset E$.

The Lusin-Menchoff theorem actually shows that, given any d-closed set $A = (E^d)^c$ and any closed set F disjoint from A, there is a d-open set $H = P^d$ containing F and an open set $G = P^c$ containing A so that $H \cap G = \emptyset$. Thus, as a special case of the theorem, one has that the density topology is regular; this follows from the useful fact that, given $F = \{x\}$ with $x \in E^d = A^c$, there is a perfect set $P \subset E^d$ such that x is a point of density of P; then P^d and P^c are disjoint d-open sets containing respectively x and A.

Now consider $A = \{x_n\}$ and $B = \{y_n\}$, two disjoint sequences of real numbers. It is then possible to find two disjoint d-open sets U and V so that $A \subset U$ and $B \subset V$. To do this, let P_1 be a perfect set which misses B and has $x_1 \in P_1^d$ as guaranteed by the Lusin-Menchoff theorem. Let Q_1 be a perfect set which misses the d-closed set $A \cup P_1$ such that $y_1 \in Q_1^d$. Let P_2 be a perfect set which misses the d-closed set $Q_1 \cup B$ with $\{x_1, x_2\} \subset P_2^d$. Continuing alternately in this fashion yields perfect sets P_n and Q_n so that each P_{n+1} misses $\bigcup_{i=1}^{n} Q_i \cup B$ and $\{x_1, \ldots, x_{n+1}\} \subset P_{n+1}^d$ and also each Q_{n+1} misses $\bigcup_{i=1}^{n} P_i \cup A$ and $\{y_1, \ldots, y_{n+1}\} \subset Q_{n+1}^d$. Then $U = \bigcup_{n=1}^{\infty} P_n^d$ and $V = \bigcup_{n=1}^{\infty} Q_n^d$ are disjoint d-open sets with $A \subset U$ and $B \subset V$. Moreover, the above argument could be applied equally well to sets C and D where $C \cap D = \emptyset$, both C and D are $\mathcal{F}_\sigma$ sets and both C and D are d-closed (in particular, if $C, D \in \mathcal{F}_\sigma$ and $m(C) = m(D) = 0$).

Nonetheless, the density topology is not normal. To see this, first let $X = \{x_i\}$ and $Y = \{y_i\}$ be disjoint dense sets. Then, for any two d-open sets U and V, with $X \subset U$, $Y \subset V$ and $U \cap V = \emptyset$, as guaranteed by the above, we first show that U and V cannot have disjoint d-closures. Note that, since X and Y are dense, given such U and V, one can select a sequence $\{I_n\}$ of closed intervals with $|I_n| < 1/n$ so that $m(I_1 \cap U) > |I_1| \cdot 1/2$, $I_2 \subset I_1$ and

$m(I_2 \cap V) > |I_2| \cdot 2/3$ and, in general, $I_n \subset I_{n-1}$ and, if n is even, $m(I_n \cap U) > |I_n| \cdot n/(n+1)$ and, if n is odd, $m(I_n \cap V) > |I_n| \cdot n/(n+1)$. Then $\bigcap_{n=1}^{\infty} I_n$ contains a single point x_o and, clearly, $\bar{D}_U(x_o) = \bar{D}_V(x_o) = 1$. Hence x_o belongs to both d-cl(U) and to d-cl(V). It now follows that if X and Y are countable dense sets and U and V are d-open sets with $X \subset U$, $Y \subset V$, $U \cap V = \emptyset$, then V^c and Y are disjoint d-closed sets. We now show that given U, V as above, the d-closed sets V^c and Y are not contained in disjoint d-open sets. To see this suppose $V^c \subset U_1$ and $Y \subset V_1$ with U_1, V_1 d-open. Then $X \subset U \subset \text{d-cl}(U) \subset V^c$ and $Y \subset V_1 \subset \text{d-cl}(V_1) \subset V_1^c \subset V$ would imply that U and V_1 contain X and Y respectively, and have disjoint d-closures, which cannot happen. Thus the density topology is not normal. An immediate consequence is that it is not possible to put a metric on the line so that the open subsets of the metric space are the d-open sets.

A space X is called <u>first countable</u> if for each $x \in X$ there is a sequence $\{G_{n,x}\}$ of neighborhoods of x so that any neighborhood of x contains one of the $G_{n,x}$. That metric spaces are first countable follows by considering $\{N_{1/n}(x)\}$. That the density topology is not metrizable also follows from the fact that it is not first countable. To see this, let x be any point and $\{E_n\}_{n=1}^{\infty}$ be a sequence of d-open sets which contain x. For each natural number n, let $y_n \in E_n$ with $y_n \neq x$. Then $E_1 \setminus \{y_n\}_{n=1}^{\infty}$ is a d-open set which does not contain any of the sets E_n. Thus the density topology is not first countable.

Considerable study has been given to the class of those functions defined on a fixed interval and approximately continuous at every point of the interval. Attention has also been focused on the class of functions which are approximate derivatives. Both the class of approximately continuous functions and that of approximate derivatives form vector spaces. Every such function is of Baire class one, is Darboux and satisfies a property called the Denjoy-Clarkson property; namely, satisfies "the inverse image of each open set is either empty or of positive measure". Both classes are closed under uniform limits. Moreover, every bounded approximately continuous function is a derivative and every Baire 2 function is the limit of a sequence of approximately continuous functions. The above facts have been stated without proof contrary to the spirit of this text; proving them would lead us too far astray.

In $\mathbb{R}^k$, a slight modification of the Lebesgue density theorem yields a topology called the ordinary density topology. Here one puts the restriction on sequences of intervals $\{I_n\}$ approaching a point x that each interval in the sequence must satisfy the same parameter of regularity. That is, for $E \subset \mathbb{R}^k$, the upper outer density of E at x is

$$\overline{D}_E^+(x) = \sup \lim_{I_n \to x} m^*(E \cap I_n)/|I_n|$$

where the supremum is over all sequences I_n of k-cells where each k-cell in a given sequence satisfies the same parameter of regularity. The upper inner density, lower outer density, and lower inner density are defined analogously. Again x is a point of density of E if $\underline{D}_E^+(x) = 1$ and a point of dispersion of E if $\overline{D}_E^+(x) = 0$. The same proof as that given for the Lebesgue density theorem using parameters of regularity shows that almost every point of a measurable set E is a point of density of E and a point of dispersion of E^c. Also all of the results stated for the density topology on the line hold for the ordinary density topology in $\mathbb{R}^k$. Most of the proofs involve slight modification (connectedness of intervals, however, requires a more complicated proof).

There is also a strong density topology in $\mathbb{R}^k$ which is obtained by dropping the restriction of a parameter of regularity on the intervals I_n approaching a point x. With this restriction dropped, it can be shown that almost every point of a measurable set is a point of strong density of the set.

6.2 Exercises

1. Show that $\lim_{t\to 0} \text{ap}\, f(t) = L$ implies that for almost every x, $\lim_{n\to\infty} f(x/2^n) = L$.
2. Does the converse to the statement of Exercise 1 hold?
3. Prove that the d-boundary of a set E is of measure 0 iff $E \in \mathcal{M}$.
4. Give an example of a continuous function defined on $[0,1]$ which is everywhere approximately derivable but is not differentiable on a set of positive measure.
5. Show that if $E \subset \mathbb{R}$ is not measurable, then

$$A = \{x \in E: x \text{ is a point of density of } E^c\}$$

is not measurable; moreover, show that this set contains no measurable subset of positive measure and that $E \setminus A$ is measurable.

6. Suppose P is a non-empty perfect subset of $\mathbb{R}$ and that every point of P, except an at most countable set of points, is a point of density of P. Show that P must contain an interval.
7. Show that the sets of first category with respect to the density topology consist exactly of the sets which are of Lebesgue measure 0.

Chapter Seven

7.1 The Lebesgue Integral

Several approaches to the development of the Lebesgue integral have been taken by different authors. The basic geometrical idea is that the integral of a real-valued non-negative function of a real variable should be the area measure of the region between the graph of the function and the x-axis. If a function takes on negative and positive values, the integral ought to be the measure of the area between the part of the graph which is above the x-axis and the axis less the area of the part of the graph which is below the x-axis and the axis. In an abstract setting, given a non-negative real-valued measurable function f

defined on a measure space X, the integral can be defined as the product measure of

$$\{(x,y): 0 \le y \le f(x), \quad x \in X\}.$$

The appropriate subtraction can be used when f takes on positive and negative values. An infinite value for the positive part or for the negative part can be allowed, but clearly not for both. This is one approach to defining the integral.

The approach we are going to begin with gives rise to the most general integral of the Lebesgue type and is close to the approach used by Lebesgue. Specifically, the range of the function rather than the domain is partitioned. A third approach involves starting with <u>simple functions</u>; that is, with measurable functions whose range consists of a finite number of values. The integral is defined naturally for such functions and then extended to more general measurable ones. This approach has benefits and will be discussed after the integral is developed using partitions of the range.

We proceed in steps to the general definition of the Lebesgue integral for measurable functions defined on a measure space. This will be done by first considering the integral of a bounded measurable function defined on a space of finite measure, then the integral of a non-negative measurable function on a space of finite measure, then the integral of a non-negative measurable function and, finally, the integral of a general measurable function when it exists. At each stage we will establish that the following properties hold for the integral on each measurable set E in the space and each f integrable on E. Here f is said to be <u>integrable</u> on E if $\int_E f$ denotes a finite number. For such f and E, it will be established that the following hold: for each f and g integrable and c a real number,

i) $\int_E c \cdot f = c \cdot \int_E f$

ii) $\int_E f + \int_E g = \int_E (f + g)$

iii) $f \geq g$ implies $\int_E f \geq \int_E g$ and, in particular, $\int_E |f| \geq |\int_E f|$ (since $|f| \geq f$ and $|f| \geq -f$).

iv) The bounded convergence theorem (Lebesgue). If $\{f_n\}$ is a sequence of measurable functions all bounded by a number M and defined on a measurable set E of finite measure and $f_n \to f$ a.e., then $\lim \int_E f_n = \int_E f$.

v) Fatou's Lemma. If $\{f_n\}$ is a sequence of non-negative measurable functions, $\int_E \underline{\lim}\, f_n \leq \underline{\lim} \int_E f_n$.

vi) The monotone convergence theorem (Lebesgue). If $\{f_n\}$ is a non-decreasing sequence of non-negative measurable functions, $\lim \int_E f_n = \int_E \lim f_n$.

vii) Countable additivity of the integral. If $E = \cup E_n$ where $\{E_n\}$ is a sequence of pairwise disjoint measurable sets, then, if f is integrable on E, $\int_E f = \sum_n \int_{E_n} f$.

viii) The dominated convergence theorem (Lebesgue). If $\{f_n\}$ is a sequence of measurable functions and $f_n \to f$ a.e. and there is an integrable function g such that for each n, $|f_n| \le g$, then $\lim_n \int_E f_n = \int_E f$.

Establishing these properties at each stage of the definition results in their having easy proofs which illuminate the process of integration. In some stages the properties will be obvious or will follow from their having been established previously. When we are finished, we will have these eight properties for the general Lebesgue integral. Each stage will be shown to be a generalization of the previous one. That is, the value of the integral of a function on a set at one stage agrees with the value obtained in the next stage. For real-valued functions f defined on an interval $[a,b]$, it will be shown that, whenever the Riemann integral of f exists, so does the Lebesgue integral and the two integrals have the same value. It will be clear as we proceed that the Lebesgue integral of a non-negative function f on a set E is the product measure of the set of points beneath the graph of f; that is of

$$\{(x,y): 0 \le y \le f(x), \quad x \in E\}.$$

We begin by assuming that the functions under consideration are bounded and measurable with respect to a measure m and defined on a measurable set E of finite measure. If $|f| < M$, we denote a partition of $[-M,M]$ by P where $P = \{y_i\}_{i=0}^{n}$ and $-M = y_o < y_1 < \dots < y_n = M$. As with the Riemann integral which is developed by means of upper and lower sums, $U(P,f)$ and $L(P,f)$, we will utilize Lebesgue sums and make use of some similar temporary notation. Namely, if A_i

$= E \cap f^{-1}([y_{i-1}, y_i))$, let $U(f,P) = \sum_{i=1}^{n} y_i \cdot m(A_i)$ and $L(f,P) = \sum_{i=1}^{n} y_{i-1} \cdot m(A_i)$. Some observations on Lebesgue sums analogous to those for Riemann sums are needed.

(7.1) Theorem. *If $P' \supset P$, then $U(f,P') \leq U(f,P)$, $L(f,P') \geq L(f,P)$ and each lower sum is less than or equal to each upper sum.*

Proof. Suppose that P' contains one additional point y which is not in P and that $y \in (y_{i-1}, y_i)$. Then $U(f,P) - U(f,P')$ consists of three terms. Specifically, if $A = f^{-1}([y_{i-1}, y))$ and $B = f^{-1}([y, y_i))$, then $A_i = A \cup B$ and

$$\begin{aligned} U(f,P) - U(f,P') &= y_i \cdot m(A_i) - y \cdot m(A) - y_i \cdot m(B) \\ &\geq y_i \cdot m(A_i) - y_i \cdot m(A) - y_i \cdot m(B) \\ &\geq 0. \end{aligned}$$

Similarly,

$$\begin{aligned} L(f,P) - L(f,P') &= y_{i-1} \cdot m(A_i) - y_{i-1} \cdot m(A) - y \cdot m(B) \\ &\leq y_{i-1} \cdot m(A_i) - y_{i-1} \cdot m(A) - y_{i-1} \cdot m(B) \\ &\leq 0. \end{aligned}$$

So $U(f,P) \geq U(f,P')$ and $L(f,P) \leq L(f,P')$. Then, if P' is any partition which contains P, the inequalities follow by successively adding points to P until P' is obtained. To see that the rest of the observation holds, let P and Q be two partitions and let $P' = P \cup Q$. In exact analogy to the case for Riemann sums, it follows that $L(f,P) \leq L(f,P') \leq U(f,P') \leq U(f,Q)$.

The Lebesgue integral of a bounded measurable

function f on a measurable set E of finite measure is then defined to be $\sup L(f,P)$ and is written $\int_E f\,dm$ or simply $\int_E f$. Note that if $A \subset E$ is measurable, then $L(C_A,P) = m(A)$ if $1 \in P$. Thus $\int_E C_A = m(A \cap E)$. Two other observations are given by the next two theorems.

(7.2) __Theorem__. _If f is bounded and measurable on E with $m(E) < \infty$, f is integrable on E and $\int_E f = \sup L(f,P) = \inf U(f,P)$._ (If either f or E or both were not measurable, upper and lower integrals could be defined using m^* for the upper sums and m_* for the lower ones. It is an easy exercise to see that if f is not measurable on E, $\sup L(f,P)$ would be less than $\inf U(f,P)$.).

__Proof__. Given $\varepsilon > 0$, if P satisfies $\max\{y_i - y_{i-1}\} < \varepsilon$, then $U(f,P) - L(f,P) = \Sigma(y_i - y_{i-1})\cdot m(A_i) < \varepsilon\cdot m(E)$. Since $\varepsilon > 0$ is arbitrary,

$$\inf U(f,P) = \sup L(f,P).$$

(7.3) __Theorem__. _If the intervals $[y_{i-1},y_i)$ in the Lebesgue sum were replaced by $(y_{i-1},y_i]$, the same value would be obtained for the integral of a bounded measurable f on a measurable set E with $m(E) < \infty$._

__Proof__. Given $\varepsilon > 0$ and P satisfying $\max\{y_i-y_{i-1}\} < \varepsilon$, let $A_i' = f^{-1}((y_{i-1},y_i))$ and $B_i' = f^{-1}(y_{i-1})$. Then, if $|f| < M$, $B_o' = B_n' = \emptyset$ and we readily determine that the number $\sum_{i=1}^{n} y_{i-1}\cdot m(B_{i+1}' \cup A_i)$ is the lower sum determined by $(y_{i-1}, y_i]$ and that

$$L(f,P) = \sum_{i=1}^{n} y_{i-1} \cdot m(A_i' \cup B_i').$$

The difference between these two sums does not exceed

$$\left| \sum_{i=1}^{n} y_{i-1} \cdot m(B_{i+1}') - \sum_{i=1}^{n} y_{i-1} \cdot m(B_i') \right|$$

$$= \sum_{i=1}^{n} (y_i - y_{i-1}) \cdot m(B_i') \leq \varepsilon |E|.$$

Since $\varepsilon > 0$ is arbitrary, lower sums using $(y_{i-1}, y_i]$ have the same supremum as the sums $L(f,P)$. Likewise upper sums using $(y_{i-1}, y_i]$ have the same infimum as $\inf U(f,P)$.

We now proceed to show:

(7.4) Theorem. _The integral of measurable functions satisfy i) through viii) provided that the functions are bounded by a common bound and defined on a measurable set E of finite measure._

Proof.

i) For $c > 0$ both cf and f are integrable on E and, since $L(cf, cP) = c \cdot L(f,P)$ where $cP = \{cy_i : y_i \in P\}$; it follows that $\int_E cf = c \cdot \int_E f$. For $c = 0$, clearly $\int_E cf = c \cdot \int_E f = 0$. For $c = -1$, Theorem 7.3 implies that the collection of all numbers $L(-f,P)$ has the same supremum as the collection of all numbers $-U(f,P)$. But

$$\sup\{-U(f,P)\} = -\inf\{U(f,P)\} = -\sup\{L(f,P)\} = -\int_E f.$$

Thus $\int_E f = -\int_E -f$ and then, for $c < 0$,

$$\int_E cf = |c| \int_E -f = c \cdot \int_E f.$$

ii) Let $h = f + g$ where f and g are bounded and measurable on E with $m(E) < \infty$. Given partitions $P = \{y_i\}_{i=1}^n$ and $Q = \{v_j\}_{j=1}^m$ containing the range of f and g respectively, let

$$P + Q = \{z_k\colon z_k = y_i + v_j, \quad y_i \in P, \quad v_j \in Q\}$$

Let $A_i = f^{-1}([y_{i-1}, y_i)) \cap E$, $B_j = g^{-1}([v_{j-1}, v_j)) \cap E$ and $C_k = h^{-1}([z_{k-1}, z_k)) \cap E$. Since the A_i, the B_j and the C_k are respectively pairwise disjoint, so are the sets $A_i \cap B_j \cap C_k$ all pairwise disjoint. Now

$$L(h, P+Q) = \sum_k z_{k-1} \cdot m(C_k) = \sum_{i,j} \sum_k z_{k-1} \cdot m(A_i \cap B_j \cap C_k).$$

For fixed i and j either $A_i \cap B_j \cap C_k = \emptyset$ or $[z_{k-1}, z_k) \subset [y_{i-1}+v_{j-1}, y_i+v_j)$. Thus for fixed i and j

$$\begin{aligned} \sum_k z_{k-1} \cdot m(A_i \cap B_j \cap C_k) &\geq \sum_k (y_{i-1} + v_{j-1}) \cdot m(A_i \cap B_j \cap C_k) \\ &\geq (y_{i-1} + v_{j-1}) \cdot m(A_i \cap B_j) \end{aligned}$$

because $A_i \cap B_j = \bigcup_k (A_i \cap B_j \cap C_k)$ and these sets are pairwise disjoint. Thus

$$\begin{aligned} L(h, P+Q) &\geq \sum_{i,j} (y_{i-1} + v_{j-1}) \cdot (m(A_i \cap B_j) \\ &\geq \sum_{i,j} y_{i-1} \cdot m(A_i \cap B_j) + \sum_{i,j} v_{j-1} \cdot m(A_i \cap B_j) \\ &\geq \sum_i y_{i-1} \cdot m(A_i) + \sum_j v_{j-1} \cdot m(B_j) \\ &\geq L(f, P) + L(g, Q). \end{aligned}$$

Proceeding in the same fashion, it follows that

$$U(h, P+Q) \leq U(f, P) + U(g, Q).$$

Thus $L(f,P) + L(g,Q) \leq \int_E h \leq U(f,P) + U(g,Q)$ and, since this is true for all partitions P and Q containing the range of f and g,

$$\int_E f + \int_E g \leq \int_E h \leq \int_E f + \int_E g$$

and

$$\int_E h = \int_E f + \int_E g.$$

iii) Note that if $m \le f \le M$ on E and P is given with $y_o < m$ and $y_n > M$, it follows that $y_o \cdot m(E) < L(f,P)$ and $U(f,P) < y_n \cdot m(E)$. From this observation it follows that $m \cdot m(E) \le \int_E f \le M \cdot m(E)$. In particular, if $h \ge 0$ on E, $\int_E h \ge 0$. It is then immediate that, if $g \le f$ on E, then $h = f - g \ge 0$ on E and $0 \le \int_E h = \int_E (f - g) = \int_E f - \int_E g$. Thus $\int_E g \le \int_E f$.

iv) Suppose that $\{f_n\}$ is a sequence of measurable functions with each f_n satisfying $|f_n| \le M$ on E with $m(E) < \infty$ and that $f_n \to f$ a.e. on E. Assume $m(E) > 0$, for otherwise there is nothing to prove because all the integrals have the value 0. Given $\varepsilon > 0$, Egoroff's theorem implies that there is a set $E' \subset E$ such that $f_n \to f$ uniformly on E', $m(E') > 0$ and $m(E \backslash E') < \varepsilon/4M$. Then there exists a natural number N so that $n > N$ and $x \in E'$ implies $|f_n(x) - f(x)| < \varepsilon/2|E'|$. Then

$$\begin{aligned} &|\int_E (f_n(x) - f(x))| \\ &\quad = |\int_{E'} (f_n(x) - f(x)) + \int_{E\backslash E'} (f_n(x) - f(x))| \\ &\quad \le \int_{E'} |f_n(x) - f(x)| + \int_{E\backslash E'} |f_n(x) - f(x)| \\ &\quad \le (\varepsilon/2|E'|)\cdot|E'| + (\varepsilon/4M)2M = \varepsilon. \end{aligned}$$

Therefore $\int_E (f_n(x) - f(x))$ approaches 0 as n approaches ∞ and $\lim \int_E f_n = \int_E f$.

v) If f_n is a sequence of non-negative functions each bounded by M on E with $m(E) < \infty$, let $g_n = \inf\{f_n, f_{n+1}, \ldots\}$. Then g_n is a non-decreasing sequence of functions and $\lim g_n = \underline{\lim}\, f_n$ and, for $k \geq n$, $f_k \geq g_n$ at each $x \in E$. Thus, for $k \geq n$, $\int_E g_n \leq \int_E f_k$ and taking the lim inf over k yields $\int_E g_n \leq \underline{\lim} \int_E f_n$. Since the g_n are bounded by M, by iv) above, $\int_E \underline{\lim}\, f_n = \int_E \lim g_n \leq \underline{\lim} \int_E f_n$. Note that in the case where the f_n are bounded by some number M on E of finite measure, $\int \underline{\lim}\, f_n \leq \underline{\lim} \int_E f_n$ holds true even if the f_n are not non-negative.

vi) If f_n is a bounded increasing sequence of measurable functions defined on E with $m(E) < \infty$, the limit of the sequence exists and iv) implies $\int_E \lim f_n = \lim \int_E f_n$.

vii) Suppose $E = \cup E_n$ where the E_n are pairwise disjoint and measurable. Given a bounded measurable f, let $A_n = \bigcup_1^n E_i$ and $f_n = f \cdot C_{A_n}$. Then by iv), $\lim f_n = f$ and hence $\lim \int_E f_n = \int_E f$. But then

$$\int_E f_n = \int_E f \cdot C_{A_n} = \int_E \sum_{i=1}^{n} f \cdot C_{E_i}$$

$$= \sum_{i=1}^{n} \int_E f \cdot C_{E_i} = \sum_{i=1}^{n} \int_{E_i} f$$

and as n approaches ∞ the limit of these terms exists and equals $\int_E f$. That is $\int_E f = \sum_{n=1}^{\infty} \int_{E_n} f$.

viii) This is the same as iv) in this context.

Before proceeding to the next stage in the definition of the Lebesgue integral, we are in a position to prove:

(7.5) <u>Theorem</u>. <u>If</u> $f(x)$ <u>is Riemann integrable on</u> $[a,b]$, <u>then</u> $f(x)$ <u>is Lebesgue integrable and the integrals are equal</u>.

<u>Proof</u>. Since $f(x)$ is Riemann integrable on $[a,b]$, it is bounded and continuous a.e. on $[a,b]$. Functions which are continuous a.e. are easily seen to be measurable. Let $P = \{x_o, x_1, \ldots, x_n\}$ be a partition of $[a,b]$. Let M be a bound for $f(x)$ and let Q be a partition of $[-M,M]$ such that if $M_i = \sup\{f(x): x \in [x_{i-1},x_i]\}$ then $M_i \in Q$. Then, if $Q = \{y_o, y_1, \ldots, y_m\}$, $A_j = f^{-1}([y_{j-1}, y_j))$

$$\begin{aligned} U(f,Q) &= \sum_j y_j \cdot m(A_j) = \sum_{i,j} y_j \cdot m(A_j \cap [x_{i-1},x_i]) \\ &\le \sum_i \sum_j M_i \cdot m(A_j \cap [x_{i-1},x_i]) \\ &\le \sum_i M_i \cdot m([x_{i-1},x_i]) = U(P,f). \end{aligned}$$

Similarly, if Q contains each

$$m_i = \inf\{f(x): x \in [x_{i-1},x_i]\},$$

then $L(f,Q) \ge L(P,f)$. Since the Lebesgue sums are between the upper and lower Riemann sums (and this holds in general whether or not f is Riemann integrable), it follows that, if f is Riemann integrable, its Riemann integral agrees with its Lebesgue integral.

We now consider unbounded non-negative measurable functions defined on a set of finite measure. For such an f, given a natural number N, let f^N denote the function $\min(f,N)$. Then $\int_E f$ is defined to be $\lim_N \int_E f^N$. This limit may be $+\infty$ but it does exist because $\int_E f^N$ is a non-decreasing sequence of numbers. Also the value of the integral clearly agrees with that of the integral of a bounded measurable function defined on a set of finite measure when the function is non-negative.

(7.6) Theorem. The integrals of non-negative measurable functions defined on a set E of finite measure satisfy i) through viii) where in i) it is assumed that $c \geq 0$.

Proof.

i) When $f \geq 0$ and $c \geq 0$, $\int_E cf = \lim \int_E cf^N = c \cdot \lim \int_E f^N = c \cdot \int_E f$.

ii) If $h = f + g$, $h^N \leq f^N + g^N \leq h^{2N}$. It follows readily that $\int_E (f + g) = \int_E f + \int_E g$.

iii) Suppose $f \geq g$. Then for $M \geq N$, $f^M \geq g^N$ and hence $\int_E f^M \geq \int_E g^N$. Letting M approach ∞, we have $\int_E f \geq \int_E g^N$ and letting N approach ∞ yields $\int_E f \geq \int_E g$.

iv) The bounded convergence theorem holds in this context due to the agreement of the integral with the integral of bounded functions.

v) Suppose $\{f_n\}$ is a sequence of non-negative measurable functions defined on E, a measurable set of finite measure. Again let $g_n = \inf\{f_n, f_{n+1}, \ldots\}$. Then $g = \lim g_n = \underline{\lim}\, f_n$ and for each N, $g^N = \lim_n g_n^N = \underline{\lim}_n\, f_n^N$. Thus, for each N, $\int_E g^N =$

$\int_E \lim g_n^N = \int_E \underline{\lim}_n F_n^N \le \underline{\lim} \int_E f_n^N \le \underline{\lim} \int_E f_n$. Letting N approach ∞, we have $\int_E g \le \underline{\lim} \int_E f_n$; that is, $\int_E \underline{\lim}\, f_n \le \underline{\lim} \int_E f_n$.

vi) If f_n is a non-decreasing sequence of non-negative measurable functions defined on E, let $f = \lim f_n$. Then $\int_E f = \int_E \underline{\lim}\, f_n \le \underline{\lim} \int_E f_n = \lim \int_E f_n$. If $\int_E f = \infty$ then $\lim \int_E f_n = \infty$. Otherwise, $\int_E f_n \le \int_E f$, $\lim \int_E f_n \le \int_E f$ and $\int_E f = \lim \int_E f_n$.

vii) Let f be a non-negative measurable function and $E = \cup E_n$ with all these sets measurable, $m(E) < \infty$ and the sets E_n pairwise disjoint. Then, as before, if $A_n = \overset{n}{\underset{1}{\cup}} E_i$ and $f_n = f \cdot C_{A_n}$, f_n satisfies vi) and hence $\lim \int_E f_n = \int_E f$. Since $\int_E f_n = \sum_{i=1}^{n} \int_{E_i} f$, it follows that $\int_E f = \sum_{n=1}^{\infty} \int_{E_n} f$.

viii) If $\{f_n\}$ are non-negative measurable functions satisfying $f_n \le g$ where $\int_E g < \infty$, then $\int_E \underline{\lim}\, f_n \le \underline{\lim} \int_E f_n \le \int_E g$ by v) above. Also $\int_E \underline{\lim}\,(g - f_n) \le \underline{\lim} \int_E (g - f_n)$. But $\int_E \underline{\lim}\,(g - f_n) = \int_E g - \int_E \overline{\lim}\, f_n$ and $\underline{\lim} \int_E (g - f_n) \ge \int_E g - \overline{\lim} \int_E f_n$, by v). It follows that $\overline{\lim} \int_E f_n \le \int_E \overline{\lim}\, f_n$. Then, if f_n approaches f, $\overline{\lim} \int_E f_n \le \int_E \lim f_n \le \underline{\lim} \int_E f_n$ and thus $\int_E f = \lim \int_E f_n$.

We now consider the case where $m(E)$ is possibly infinite and the functions under consideration are measurable and non-negative. Given such an $f \geq 0$ on E, if there is an $\varepsilon > 0$ such that $m\{x \in E: f(x) > \varepsilon\} = \infty$, define $\int_E f = \infty$. Otherwise we let $\int_E f = \sup \int_A f$ where the supremum is over all measurable $A \subset E$ with $m(A) < \infty$. Note that, if there does not exist $\varepsilon > 0$ so that $m\{x \in E: f(x) > \varepsilon\} = \infty$, then $\{x \in E: f(x) > 0\} = \cup_n \{x \in E: f(x) > 1/n\}$ and, each of these sets being of finite measure, $\{x \in E: f(x) > 0\}$ has σ-finite measure. This is a necessary condition for the integral of f to be finite on E and represents no real restriction on the domain of f, even though there are many useful measures which are not σ-finite. Note that, as before, for each measurable set $A \subset E$, $\int_E C_A = m(A)$ is implied by the above definition. Now we prove:

(7.7) Theorem. *The integrals of non-negative measurable functions satisfy* i) *through* viii) *where in* i) *it is again assumed that* $c \geq 0$.

Proof. All the sets and functions used in the proof will be assumed to be measurable.

i) For $f \geq 0$ and $c \geq 0$ and $A \subset B \subset E$ with $m(B) < \infty$, if $\int_E f < \infty$, $\int_A cf = c \cdot \int_A f \leq c \cdot \int_B f$ and $c \cdot \int_A f = \int_A cf \leq \int_B cf$. Taking the supremum over $B \subset E$ and then over $A \subset E$ yields $\int_E cf \leq c \cdot \int_E f \leq \int_E cf$. Thus $\int_E cf = c \cdot \int_E f$. Note that if $\int_E f = \infty$, where E has σ-finite measure, the formula still holds provided $0 \cdot \infty$ is understood to be 0.

ii) If $h = f + g$ and $A \subset B \subset C \subset E$ with $m(C) < \infty$, then

$$\int_A h = \int_A f + \int_A g \leq \int_B f + \int_C g.$$

Taking the supremum over C first, then B, then A, yields $\int_E h \leq \int_E f + \int_E g$. Since $\int_E h \geq \int_A f + \int_B g$ taking supremums over B and then A yields the other inequality and thus $\int_E (f + g) = \int_E f + \int_E g$.

iii) If $f \geq g$ and $A \subset B \subset E$ with $m(B) < \infty$, then

$$\int_B f \geq \int_B g \geq \int_A g.$$

Taking the supremum over B and then A yields $\int_E f \geq \int_E g$.

iv) This needs no proof except the observation that the integral at this stage agrees with the previous integral. For if f equals 0 except on a set $B \subset E$ with $m(B) < \infty$, $\sup_{A \subset E} \int_A f = \int_B f$. This is because, if $A \supset B$ and $m(A) < \infty$, $\int_A f = \int_B f + \int_{A \setminus B} f = \int_B f$, since $\int_{A \setminus B} f = 0$.

v) Suppose f_n are non-negative on E and $A \subset E$ with $m(A) < \infty$. Since $\int_A f_n \leq \int_E f_n$, it follows that $\int_A \underline{\lim}\, f_n \leq \underline{\lim} \int_A f_n \leq \underline{\lim} \int_E f_n$. Taking the supremum over A yields $\int_E \underline{\lim}\, f_n \leq \underline{\lim} \int_E f_n$.

vi) Suppose f_n are non-negative and $\{f_n\}$ is non-decreasing on E. By v) $\int_E \underline{\lim}\, f_n = \int_E \lim f_n \leq \underline{\lim} \int_E f_n = \lim \int_E f_n$. But $f_n \leq f = \lim f_n$ and so $\int_E f_n \leq \int_E f$ and $\lim \int_E f_n \leq \int_E f$. Thus, $\int_E f = \lim \int_E f_n$. The formula holds whether or not $\int_E f$ is finite.

vii) Suppose $E = \cup E_n$ with the E_n measurable and pairwise disjoint. Again for $f \geq 0$ on E, f

measurable, $\int_E f = \sum_{n=1}^{\infty} \int_{E_n} f$ follows from vi) with the exact same argument as in the previous stages.

viii) If $|f_n| \le g$ and $\int_E g < \infty$, the exact same proof as in the previous stage yields that, if $\lim f_n = f$, then $\lim \int_E f_n = \int_E f$.

We are now ready for the definition of the integral of a measurable function (when the integral exists) on any measurable set E. Given such an f, let $f+ = \max(f,0)$ and $f- = -\min(f,0)$. Since $f+$ and $f-$ are non-negative $\int_E f+$ and $\int_E f-$ are defined, though possibly infinite. Providing that at most one of them is infinite, the integral of f on E is defined by $\int_E f = \int_E f+ - \int_E f-$. If $\int_E f$ is finite f, as before, is said to be Lebesgue integrable on E and we write $f \in \mathscr{L}(E)$ and, if E is the whole measure space, $f \in \mathscr{L}$.

(7.8) Theorem. The Lebesgue integral satisfies properties i) through viii).

Proof.

i) For $c \ge 0$ since $(cf)+ = c \cdot f+$ and $(cf)- = cf-$ we have $\int_E cf = \int_E (cf)+ - \int_E (cf)- = c\int_E f+ - c\int_E f- = c\int_E f$. For $c < 0$, $(cf)+ = |c|f-$ and $(cf)- = |c|f+$ so that $\int_E cf = \int_E |c|f- - \int_E |c|f+ = -|c|\int_E f = c\int_E f$.

ii) Note that if $E = A \cup B$, $A \cap B = \emptyset$ and $\int_E f$ is defined, then $\int_E f = \int_E f+ - \int_E f- = \int_A f+ + \int_B f+ - \int_A f- - \int_B f- = \int_A f + \int_B f$. Now let

$h = f + g$. These are the six possibilities for the sign of $f(x)$, $g(x)$, and $h(x)$:

1. On $E_{+++} = \{x \in E: f(x) \geq 0, g(x) \geq 0\}$, we have $\int h = \int f + \int g$.
2. On $E_{---} = \{x \in E: f(x) < 0, g(x) < 0\}$, we have $\int h = \int f + \int g$.
3. On $E_{+-+} = \{x \in E: f(x) \geq 0, g(x) < 0, h(x) \geq 0\}$, $f = |g| + h$ which implies $\int f = -\int g + \int h$.
4. On $E_{+--} = \{x \in E: f(x) \geq 0, g(x) < 0, h(x) < 0\}$, $|g| = f + |h|$ implies $-\int g = \int f - \int h$.
5. On $E_{-++} = \{x \in E: f(x) < 0, g(x) \geq 0, h(x) \geq 0\}$, the situation is analogous to E_{+-+}.
6. On $E_{-+-} = \{x \in E: f(x) < 0, g(x) \geq 0, h(x) < 0\}$, the situation is analogous to E_{+--}.

In each of these six sets, $\int f + \int g = \int (f + g)$. It follows that this is also true on E.

iii) If $f \geq g$ and $\int_E g = -\infty$, $\int_E f \geq \int_E g$. Otherwise $f - g \geq 0$ and $\int_E (f - g) \geq 0$ holds from the previous stage. Now by ii) above $\int_E (f - g) = \int_E f - \int_E g \geq 0$. Hence $\int_E f \geq \int_E g$.

iv), v), and vi) follow from earlier arguments.

vii) If $E = \cup E_n$, where the E_n are pairwise disjoint measurable sets and $\int_E f$ is defined, then

$$\int_E f+ = \sum_{n=1}^{\infty} \int_{E_n} f+ \quad \text{and} \quad \int_E f- = \sum_{n=1}^{\infty} \int_{E_n} f- \quad \text{implies that}$$

$$\int_E f = \sum_{n=1}^{\infty} \int_{E_n} f+ - \sum_{n=1}^{\infty} \int_{E_n} f- = \sum_{n=1}^{\infty} \int_{E_n} f.$$

viii) If $|f_n| \leq g$ and $\int_E g < \infty$ and $f_n \to f$ on E, then $0 \leq f_n + g \leq 2g$ and by the previous stage,

$\lim \int_E (f_n + g) = \int_E (f + g)$. But $\lim \int_E (f_n + g) = \int_E g + \lim \int_E f_n$ and then $\lim \int_E f_n = \int_E f$.

The Lebesgue integral has now been developed and properties i) - iii) and the convergence theorems and countable additivity of the integral iv) - viii) have been established. It is standard in the statement of each of i) - viii) to allow for a set or sets of measure 0 on which the property does not hold or convergence does not happen or at which the functions are not defined at all or are infinite. If i) - viii) were to be restated in this fashion, all of the properties, convergence theorem and countable additivity would still be valid. It is, in fact, traditional to disregard sets of measure 0 in the definition of the Lebesgue integral because a function which is not defined on a set of measure 0 would have the same integral no matter how it were defined on that set. This is because changing the value of a function on a set of measure 0 would not change the value of the Lebesgue sums. Thus, i) through viii) could be restated taking this into account wherever possible. For example, iii) and v) would read:

iii) If $f(x)$ and $g(x)$ are defined a.e. on E and $f(x) \geq g(x)$ a.e. on E, then $\int_E f \geq \int_E g$.

v) If $\{f_n\}$ is a sequence of measurable functions each defined except on a set of measure 0 of E, each non-negative at almost every x in E, then $\int_E \underline{\lim}\, f_n \leq \underline{\lim} \int_E f_n$. (Here $\underline{\lim}\, f_n$ exists a.e. on E.)

The alternate development of the Lebesgue integral using simple functions provides a useful way of viewing the subject. A [non-negative] <u>simple function</u> is a function of the form $s(x) = \sum_{i=1}^{n} a_i C_{E_i}$ where the E_i are measurable sets [and $a_i \geq 0$]. The integral of $s(x)$ over the measure space X is $\int_X s(x) = \Sigma a_i m(E_i)$. If this alternate development is taken, it is first necessary to show that, if a simple function is written in two different ways, the integral so obtained is unique. This, of course follows from $\int C_A = m(A)$, i) and ii). That is, if

$$s(x) = \sum_{i=1}^{n} a_i C_{A_i} = \sum_{j=1}^{m} b_j C_{B_j},$$

then

$$\int_X s(x) = \sum_i a_i m(A_i) = \sum_j b_j m(B_j).$$

Once the integral is uniquely defined for simple functions, for non-negative measurable functions f, the integral of f can be defined as

$$\int_X f = \sup\{\int_X s(x): s(x) \leq f(x)\}.$$

For general measurable functions f, the integral of f is $\int_X f = \int_X f+ - \int_X f-$ providing one of $\int_X f+$ or $\int_X f-$ is finite. Again a function is said to be integrable if it has a finite integral.

It is not difficult to see that this approach gives rise to the same integral as was obtained by partitioning the range. Suppose $f(x)$ is a non-negative measurable function bounded above by M on E with $m(E) < \infty$. Let $\{y_i\}_{i=0}^{n}$ be a partition of $[0,M]$ and $A_i = E \cap f^{-1}([y_{i-1}, y_i))$. Then $s(x) = \sum_{i=1}^{n} y_{i-1} C_{A_i}$ is a simple function satisfying $s(x) \leq f(x)$ and $\int_X s(x) \geq \int_E f(x) - \max\{y_i - y_{i-1}\} \cdot m(E)$.

Thus the integral defined using simple functions agrees with that defined by partitioning the range for bounded measurable functions defined on measurable sets of finite measure. It also agrees with the Lebesgue integral for non-negative measurable functions. In fact, every non-negative measurable function is the limit of a non-decreasing sequence of simple functions. Specifically, if

$$A_{i,n} = f^{-1}([\frac{i-1}{2^n}, \frac{i}{2^n}]) \quad i = 1, 2, \ldots, n \cdot 2^n,$$

then $s_n(x) = \sum_{i=1}^{n \cdot 2^n} \frac{i-1}{2^n} C_{A_{i,n}}$ is a monotone non-decreasing sequence of functions whose limit is f. By the monotone convergence theorem $\lim \int_X s_n = \int_X f$. Properties i) through viii) have been proved directly by other authors using the simple function approach. Some further theorems using this approach have easier proofs and we will in what follows utilize whichever approach is most suitable. For example, in the following theorem, the concept of partitioning the range is appropriate:

(7.9) Theorem. *If f is a finite valued measurable function defined on a space X which has σ-finite measure m_1, then the graph of f is of measure 0 in the product space $X \times \mathbb{R}$.*

Proof. It is sufficient to proceed with $m_1(X) < \infty$. Let f be a measurable function. Fix $\varepsilon > 0$ and for each integer n, let

$$A_n = \{x \in X: f(x) \in [(n-1)\varepsilon, n\varepsilon)\}.$$

Then the graph of f, which is $\{(x,y): y = f(x)\}$, is

contained in $B = \bigcup_n (A_n \times [(n-1)\varepsilon, n\varepsilon))$. Since $(m_1 \times m)(B) \le \sum_n m_1(A_n)\cdot\varepsilon \le \varepsilon\cdot m_1(X)$, it follows that the product measure of the graph of f is 0.

Using the properties of the Lebesgue integral, we can now provide a simple proof of the assertion made earlier about the product measure of the product of two measurable sets.

(7.10) <u>Theorem</u>. *If m_1 is a measure on X, m_2 a measure on Y, and A and B are two measurable subsets of X and Y respectively, then*

$$(m_1 \times m_2)(A \times B) = m_1(A)\cdot m_2(B).$$

<u>Proof</u>. By definition, $(m_1 \times m_2)(A \times B)$ equals

$$\inf\{ \sum m_1(A_i)\cdot m_2(B_i) : A \times B \subset \bigcup_i A_i \times B_i \},$$

where the A_i and B_i are measurable sets in X and Y. Let A_i, B_i be measurable sets with $A \times B \subset \bigcup_i A_i \times B_i$. Note that at each $x \in X$, the function $\sum_{i=1}^{\infty} m_2(B_i)\cdot C_{A_i}(x)$ equals $\Sigma'_x\, m_2(B_i)$ where Σ'_x is over all sets B_i with $x \in A_i$. But the union of all sets B_i for which $x \in A_i$ contains B and thus, for each $x \in A$, $\sum_{i=1}^{\infty} m_2(B_i)\cdot C_{A_i}(x) \ge m_2(B)$. Using this fact and the monotone convergence theorem, we have:

$$\sum_{i=1}^{\infty} m_1(A_i)\cdot m_2(B_i) = \sum_{i=1}^{\infty} \int m_2(B_i)\cdot C_{A_i}\, dm_1$$

$$= \lim_{n\to\infty} \sum_{i=1}^{n} \int m_2(B_i)\cdot C_{A_i}\, dm_1$$

$$= \lim_{n\to\infty} \int \sum_{i=1}^{n} \left(m_2(B_i)\cdot C_{A_i}\right) dm_1$$

$$= \int \sum_{i=1}^{\infty} m_2(B_i)\cdot C_{A_i}\, dm_1$$

$$\geq \int m_2(B)\cdot C_A\, dm_1$$

$$\geq m_2(B)\cdot \int C_A\, dm_1$$

$$\geq m_1(A)\cdot m_2(B).$$

Thus $(m_1 \times m_2)(A \times B) \geq m_1(A)\cdot m_2(B)$ and since the opposite inequality follows from the definition, $(m_1 \times m_2)(A \times B) = m_1(A)\cdot m_2(B)$.

At this point it is natural to complete the discussion of the meaning of the Lebesgue integral by showing that for non-negative functions f, the integral of f over a set X is the product measure of $\{(x,y)\colon 0 \leq y \leq f(x),\ x \in X\}$. That this can be taken as the definition of the integral of a non-negative measurable function f is a consequence of the following theorem. The setting is restricted to that of a σ-finite measure as is natural in considering the process of integration.

(7.11) <u>Theorem</u>. <u>Given a non-negative function</u> f <u>defined on a measure space</u> X <u>having σ-finite measure</u> m_1, <u>for each</u> $E \in \mathcal{M}$,

$$\int_E f = (m_1 \times m)\left(\{(x,y)\colon 0 \leq y \leq f(x),\ x \in E\}\right).$$

<u>Proof</u>. It is clearly sufficient to consider $m_1(X) < \infty$ and that $E = X$. Given $f \in \mathcal{M}$, let $f_n(x) =$

$\min(f(x),n)$ and $B_n = \{(x,y): 0 \le f_n(x) \le y\}$. Fix $\varepsilon > 0$ and a partition $\{y_i\}$ of an interval containing $[0,n]$ with $\max(y_i - y_{i-1}) < \varepsilon$. Note that, if $A_i = f_n^{-1}([y_{i-1}, y_i))$, then $\bigcup_i A_i \times [0,y_i]$ contains B_n which in turn contains $\bigcup_i A_i \times [0,y_{i-1}]$. Since the A_i are pairwise disjoint,

$$\begin{aligned}\int_X f_n &\ge \Sigma\, y_{i-1}\cdot m_1(A_i)\\ &= \Sigma\, (m_1 \times m)(A_i \times [0,y_{i-1}])\\ &\ge (m_1 \times m)(B_n).\end{aligned}$$

Letting $n \to \infty$, it follows that $\int_X f \ge (m_1 \times m)(B)$, where $B = \{(x,y): 0 \le y \le f(x)\} = \lim B_n$.

On the other hand,

$$\begin{aligned}\int_X f_n &\le \sum_i y_i\cdot m(A_i) \le \sum_i y_{i-1}\cdot m(A_i) + \varepsilon\cdot\sum_i m(A_i)\\ &\le (m_1 \times m)(B_n) + \varepsilon\cdot m(X).\end{aligned}$$

Letting $n \to \infty$, we have $\int_X f \le (m_1 \times m)(B) + \varepsilon\cdot m(X)$. Since $\varepsilon > 0$ was arbitrary, both inequalities hold and $\int_X f\, dm_1 = (m_1 \times m)(B)$.

We proceed now to the theorem which allows equating an integral on a product space with an iterated integral over the spaces forming the product, Fubini's theorem.

(7.12) Theorem. (Fubini). *If X and Y are measure spaces with σ-finite measures m_1 and m_2 respectively, and if $f(x,y)$ is a measurable [integrable] function with respect to $m_1 \times m_2$, then for m_1 a.e. x, $f(x,y)$ is a measurable [integrable] function of y and for m_2 a.e. y, $f(x,y)$ is a measurable [integrable] function of x. If $f(x,y)$ is integrable [of constant sign] on $X \times Y$, then the integrals in*

<u>the following equations exist</u> [<u>possibly infinite</u>] <u>and the equations themselves hold true</u>:

$$\int_{X\times Y} f(x,y)\, d(m_1 \times m_2) = \int_X\left(\int_Y f(x,y)\, dm_2\right)dm_1 = \int_Y\left(\int_X f(x,y)\, dm_1\right)dm_2.$$

<u>Proof</u>. Because of the countable additivity of the integral, we may assume that $X \times Y$ is of finite $m_1 \times m_2$ measure. It is clear that the finite sum of functions which satisfy this theorem and constant multiples of functions which satisfy the theorem also satisfy it. The rest of the proof of the theorem proceeds in a series of steps.

i) The theorem is true if $f(x,y) = C_{A\times B}$ where A and B are sets measurable m_1 and m_2 respectively.

ii) If the theorem is true for $f_n(x,y) \geq 0$ and $\{f_n\}$ is a non-decreasing sequence of non-negative measurable functions, then it is true for $f(x,y) = \lim f_n(x,y)$.

iii) If the theorem is true for $f_n(x,y) \geq 0$ where $\{f_n\}$ is a non-increasing sequence of non-negative measurable functions, then it is true for $\lim f_n(x,y)$ where the integral equation holds providing $\int f_1\, d(m_1 \times m_2)$ is finite.

iv) If Z is a set of $m_1 \times m_2$ measure 0, the theorem is true for C_Z.

v) If E is a set which is $m_1 \times m_2$ measurable, the theorem is true for C_E.

vi) The theorem is true for non-negative simple functions.

vii) The theorem is true for non-negative $m_1 \times m_2$ measurable functions.

viii) The theorem is true in general for $m_1 \times m_2$ measurable functions.

Because of the symmetry in the statement of the theorem, we clearly need to prove only half of the measurability [integrability] conditions.

Proof.

i). If A and B are respectively m_1 and m_2 measurable sets, then

$$\int_{X\times Y} C_{A\times B}\, d(m_1 \times m_2) = (m \times m_2)(A \times B) = m_1(A)\cdot m_2(B)$$

by Theorem 7.10. But for each $x \in X$, if $x \in A$, then $\int_Y C_{A\times B}\, dm_2 = m_2(B)$ and if $x \notin A$, then $\int_Y C_{A\times B}\, dm_2 = 0$. Thus, for $x \in X$, $\int_Y C_{A\times B}\, dm_2 = m_2(B)\cdot C_A(x)$. But then

$$\int_X (\int_Y C_{A\times B}\, dm_2) dm_1 = \int_X m_2(B)\cdot C_A(x)\, dm_1 = m_2(B)\cdot m_1(A)$$

and i) is proved.

ii). Suppose $f_n(x,y)$ are non-negative and measurable, satisfy the conclusions of the theorem and that $\{f_n\}$ is a non-decreasing sequence of functions. Then for almost every x, $f_n(x,y)$ is m_2 measurable and thus, if Z_n is the set of x for which $f_n(x,y)$ is not measurable as a function of y, $f(x,y) = \lim f_n(x,y)$ is a measurable function of y for each x not in $\cup Z_n$; that is, for a.e. $x \in E$. If in addition each $f_n(x,y)$ is Lebesgue integrable, since each $f_n(x,y)$ satisfies the conclusion of the theorem, it follows that, for almost every y, $f_n(x,y)$ is integrable with respect to x and for each n

$$\int_{X\times Y} f_n(x,y)\, d(m_1 \times m_2) = \int_Y(\int_X f_n(x,y)\, dm_1)dm_2.$$

Now $\int_X f_n(x,y)\, dm_1$ is a monotone non-decreasing sequence of non-negative measurable functions defined for m_2 a.e. y. Applying the Lebesgue monotone convergence theorem to both sides of the equation yields:

$$\begin{aligned}\int_{X\times Y} f(x,y)\, d(m_1\times m_2) &= \lim_{n\to\infty} \int_{X\times Y} f_n(x,y)\, d(m_1\times m_2) \\ &= \lim_{n\to\infty} \int_Y\left(\int_X f_n(x,y)\, dm_1\right)dm_2 \\ &= \int_Y\left(\lim_{n\to\infty} \int_X f_n(x,y)\, dm_1\right)dm_2 \\ &= \int_Y\left(\int_X f(x,y)\, dm_1\right)dm_2.\end{aligned}$$

By symmetry, the integral equals the other iterated one and ii) is proved.

iii). If $\{f_n\}$ is a non-increasing sequence of non-negative $m_1 \times m_2$ measurable functions satisfying the conclusions of the theorem, let $g_n = f_1 - f_n$. Then $\{g_n\}$ is a non-decreasing sequence of $m_1 \times m_2$ measurable functions. Clearly, since f_1 and $-f_n$ satisfy the conclusions of the theorem their sums, g_n, satisfy the conclusions. By ii) with $f = \lim f_n$,

$$\int_{X\times Y}(f_1 - f)\, d(m_1\times m_2) = \int_Y\left(\int_X(f_1 - f)dm_1\right)dm_2.$$

But

$$\begin{aligned}&\int_{X\times Y}(f_1 - f)\, d(m_1 \times m_2) \\ &\qquad = \int_{X\times Y} f_1\, d(m_1 \times m_2) - \int_{X\times Y} f\, d(m_1 \times m_2)\end{aligned}$$

and

$$\begin{aligned}&\int_Y\left(\int_X(f_1 - f)\, dm_1\right)dm_2 \\ &\qquad = \int_Y\left(\int_X f_1\, dm_1 - \int_X f\, dm_1\right)dm_2 \\ &\qquad = \int_Y\left(\int_X f_1\, dm_1\right)dm_2 - \int_Y\left(\int_X f\, dm_1\right)dm_2.\end{aligned}$$

It follows that

$$\int_{X\times Y} f(x,y)\, d(m_1 \times m_2) = \int_Y\left(\int_X f(x,y)\, dm_1\right)dm_2$$

and by symmetry the other equation holds and iii) is proved.

Note. Let E be a $m_1 \times m_2$ measurable set. By the definition of the $m_1 \times m_2$ measure, there are sets $A_{i,n} \subset X$, $B_{i,n} \subset Y$ with $A_{i,n}$ and $B_{i,n}$ respectively m_1 and m_2 measurable such that

$$E \subset E_n = \bigcup_{i=1}^{\infty} A_{i,n} \times B_{i,n}$$

and $(m_1 \times m_2)(E_n \setminus E) < 1/n$. Moreover, the $A_{i,n}$ and $B_{i,n}$ can be chosen so that for any given n the sets $A_{i,n} \times B_{i,n}$ are pairwise disjoint and so that E_n is a non-increasing sequence of sets. To see how this last assertion can be accomplished consider the following: If R is a measurable rectangle and C is a finite union of pairwise disjoint measurable rectangles, then both $R \cap C$ and $R \setminus C$ can be written as a finite union of pairwise disjoint measurable rectangles. In fact, if $C = \bigcup_{i=1}^{n} A_i \times B_i$ and $R = A \times B$, then $R \cap C = \bigcup_{i=1}^{n} (A_i \cap A) \times (B_i \cap B)$. Also $R \setminus A_1 \times B_1 = R \cap (A_1 \times B_1^c) \cup R \cap (A_1^c \times B_1) \cup R \cap (A_1^c \times B_1^c)$ is a finite union of pairwise disjoint measurable rectangles. By successively removing $A_i \times B_i$, it can be seen that $R \setminus C$ can be written as a finite union of pairwise disjoint measurable rectangles. Now suppose E_1 is a countable union of measurable rectangles such that $E \subset E_1$ and $(m_1 \times m_2)(E \setminus E_1) < 1$. Let $E_1 = \bigcup_{i=1}^{\infty} (A_{i,1} \times B_{i,1})$. By writing each $(A_{i,1} \times B_{i,1}) \setminus \bigcup_{j<i} (A_{j,1} \times B_{j,1})$ as a finite union of disjoint measurable rectangles, E_1 has been written as the at most countable union of all the resulting

measurable rectangles. Suppose E_n has been obtained as a disjoint union of a sequence of measurable rectangles and $E \subset E_n$ and $(m_1 \times m_2)(E_n \backslash E) < 1/n$. Let E'_{n+1} be the union of any sequences of measurable rectangles such that $E \subset E'_{n+1}$ and $(m_1 \times m_2)(E'_{n+1}) < 1/(n+1)$. The collection of all intersections of each of those rectangles with the rectangles forming E_n is an at most countable collection of rectangles whose union is $E_{n+1} = E_n \cap E'_{n+1}$. The set E_{n+1} can now be written as a countable union of pairwise disjoint measurable rectangles in the same way that E_1 was. Proceeding in this fashion yields a non-increasing sequence of sets E_n each of which is an at most countable union of pairwise disjoint measurable rectangles. Moreover, for each n, $E \subset E_n$ and $(m_1 \times m_2)(E \backslash E_n) < 1/n$.

iv). Let Z be a set of measure 0 and let E_n be the sets described in the above note so that $Z \subset E_n = \bigcup_i A_{i,n} \times B_{i,n}$ with $(m_1 \times m_2)(E_n) < 1/n$. Since for each n, $A_{i,n} \times B_{i,n}$ are pairwise disjoint if $E_{j,n} = \bigcup_{i=1}^{j} A_{i,n} \times B_{i,n}$, the conclusion of the theorem holds for $C_{E_{j,n}}$ because $C_{E_{j,n}} = \sum_{i=1}^{j} C_{A_{i,n} \times B_{i,n}}$. It follows from ii) that the conclusion of the theorem holds for $C_{E_n} = \lim_{j \to \infty} C_{E_{j,n}}$ because $C_{E_{j,n}}$ is a non-decreasing sequence of non-negative measurable functions for which the conclusions hold. If $E = \bigcup_{i=1}^{\infty} E_n$ then since the E_n are non-increasing, $\{C_{E_n}\}$

is a non-increasing sequence of functions and hence the conclusion of the theorem holds for $C_E = \lim_n C_{E_n}$. Now it follows that the integral of C_Z with respect to $m_1 \times m_2$ and the iterated integrals all equal zero. Hence $\int_Y C_E \, dm_2 = 0$ for a.e. x and thus for m_1 a.e. x, $\{y: (x,y) \in E\}$ has m_2 measure 0. Likewise, for m_2 a.e. y, $\{x: (x,y) \in E\}$ has m_1 measure 0. Since $Z \subset E$, it follows that for m_1 a.e. x, $\{y: (x,y) \in Z\}$ has m_2 measure 0 and for m_2 a.e. y, $\{x: (x,y) \in Z\}$ has m_1 measure 0. Thus

$$\int_{X \times Y} C_Z \, d(m_1 \times m_2) = 0 = \int_Y (\int_X C_Z \, dm_1) dm_2$$
$$= \int_X (\int_Y C_Z \, dm_2) dm_1$$

and the theorem holds for C_Z.

v). If E is an $m_1 \times m_2$ measurable set, using the note prior to iv) we can enclose E in $\bigcap_n E_n$ where $E_n = \bigcup_i A_{i,n} \times B_{i,n}$ and $(m_1 \times m_2)(E_n \setminus E) < 1/n$. It follows from ii) that C_{E_n}, which is an increasing limit of functions for which the conclusions hold, is also a function for which the conclusions hold. Also $C_{\cap E_n}$ is a decreasing limit of such functions and thus the conclusions hold for $C_{\cap E_n}$. But if $Z = \cap E_n \setminus E$, Z is a set of $m_1 \times m_2$ measure 0 and the conclusions hold for C_Z. Since $C_E = C_{\cap E_n} - C_Z$, the conclusions hold for C_E.

vi). This follows from the observation that, whenever the conclusions hold for a finite collection of functions, they hold for the sum of constant multiples of these functions; therefore they hold for simple functions.

vii). If f is a non-negative measurable [integrable] function, then $f = \lim s_n$ where the s_n are simple functions and $\{s_n\}$ is a non-decreasing sequence of functions. By ii) it follows that the conclusions hold for such f.

viii). This follows from the initial observation that whenever the theorem holds for two functions (in this case $f+$ and $-f-$) it holds for their sum.

Using Fubini's theorem, we can make an improvement on Theorem 7.11 by adding to it the following theorem.

(7.13) <u>Theorem</u>. <u>A real-valued or extended real-valued function</u> f <u>defined on a σ-finite measure space</u> X <u>with measure</u> m_1 <u>is measurable iff</u>

$$\{(x,y)\colon 0 \le y \le f(x)\} \quad \underline{\text{and}} \quad \{(x,y)\colon f(x) \le y \le 0\}$$

<u>are measurable in</u> $X \times \mathbb{R}$ <u>with respect to</u> $m_1 \times m$.

<u>Proof</u>. Without any loss, it can be assumed that $m_1(X) < \infty$ and that f is non-negative. Suppose first that f is a measurable function on X. Then if

$$A_{i,n} = f^{-1}\left(\left[\frac{i - 1}{2^n}, \frac{i}{2^n}\right)\right),$$

each $A_{i,n}$ is m_1 measurable and $A_{i,n} \times [0, (i-1)/2^n)$ is $m_1 \times m_2$ measurable. But $\{(x,y)\colon 0 \le y \le f(x)\}$ is the union of $\bigcup_n \bigcup_i A_{i,n}$ with $f^{-1}(\infty) \times \mathbb{R}$ and with the graph of f, all three of which are measurable sets, the graph of f being of measure 0 by Theorem

7.9. Conversely, suppose $E = \{(x,y): 0 \le y \le f(x)\}$ is an $m_1 \times m$ measurable subset of $X \times \mathbb{R}$ (or $X \times \mathbb{R}^+$). Then for $r \ge 0$ let $g_r(x) = \min(f(x), r)$. Then

$$A_r = \{(x,y): 0 \le g_r(x) \le y\}$$

is $E \cap (X \times [0,r])$ and is $m_1 \times m$ measurable. Since A_r is a set of finite $m_1 \times m$ measure, C_{A_r} is integrable in $X \times \mathbb{R}$ and

$$\int_{X \times Y} C_{A_r} d(m_1 \times m_2)$$

equals $\int_X (\int_R C_{A_r} dm) dm_1 = \int_X g_r(x)\ dm_1$. It follows that $g_r(x)$ is integrable and hence measurable and, since $f(x) \ge y$ iff $g_y(x) \ge y$, it follows that f is measurable.

The following corollary to Fubini's theorem is worth consideration for its description of sets which are measurable in a product measure.

(7.14) <u>Corollary</u>. <u>If</u> E <u>is an</u> $m_1 \times m_2$ <u>measurable subset of</u> $X \times Y$, <u>then for</u> m_1 <u>almost every</u> x, $E_x = \{y: (x,y) \in E\}$ <u>is</u> m_2 <u>measurable</u>; <u>and for</u> m_2 a.e. y, $E_y = \{x: (x,y) \in E\}$ <u>is</u> m_1 <u>measurable and</u>

$$(m_1 \times m_2)(E) = \int_Y m(E_y)\ dm_2 = \int_X m(E_x)\ dm_1.$$

This is simply the statement of the theorem for C_E.

The statement of Fubini's theorem suggests that it holds for all measurable $f(x,y)$. The measurability part of the theorem does. However, the equality of the two iterated integrals requires in the theorem the existence of the integral of $f(x,y)$. A test for this can be obtained frequently by determining whether $|f(x,y)|$ is integrable or whether $g(x,y)$ is integrable where $g(x,y)$ is larger than $|f(x,y)|$. If $f(x,y)$ is not Lebesgue integrable, the iterated integrals need not be equal even if both exist. For example, consider the double sequence of numbers $\{a_{ij}\}$, where $a_{ij} = 0$ if $j < i$, $a_{ii} = 1$, $a_{ij} = -2^{i-j}$ if $j > i$. Then

$$\sum_i \sum_j a_{ij} = \sum_i \left(1 - \sum_{n=1}^{\infty} 2^{-n}\right) = \sum 0 = 0$$

and

$$\sum_j \sum_i a_{ij} = \sum_j \left(1 - \sum_{n=1}^{j-1} 2^{-n}\right) = \sum_{j=1}^{\infty} 2^{1-j} = 2.$$

This behavior for double sequences can occur in iterated integrals. Thus, for example, on $[0,1] \times [0,1]$, let $r_1 = 1$ and $r_n \to 0$ and

$$R_{ij} = (r_{i+1} - r_i)(r_{j+1} - r_j)$$

be the area of the rectangle $(r_{i+1}, r_i] \times (r_{j+1}, r_j]$ and let $f(x,y) = a_{ij}R_{ij}^{-1}$ if $(x,y) \in (r_{i+1}, r_i] \times (r_{j+1}, r_j]$. Then

$$\int_0^1 \left(\int_0^1 f(x,y)\, dy\right)dx = \sum_i \int_{r_{i+1}}^{r_i} \left(\int_0^1 f(x,y)\, dy\right)dx$$

$$= \sum_i \sum_j a_{ij} = 2$$

and

$$\int_0^1 \int_0^1 f(x,y)\, dx\, dy = \sum_i \sum_j a_{ij} = 0.$$

All of the iterated integrals are Lebesgue integrals and are not "improper" integrals. However, $|f(x,y)|$ is not integrable on $[0,1] \times [0,1]$ and this accounts for the possibility of unequal iterated integrals.

We conclude with an example of a function which is "improper" Lebesgue integrable but not Lebesgue integrable. For $n = 1, 2, \ldots,$ let

$$f(x) = \begin{cases} 4n+2, & \text{if } x \in [1/(2n+1), 1/2n] \\ -(4n-2), & \text{if } x \in (1/2n, 1/(2n-1)] \end{cases}$$

Then $\int_{I_n} f(x)\, dx = 1/n$ if $I_n = (1/(2n+1), 1/2n]$ and $\int_{I_n} f(x)\, dx = -1/n$ if $I_n = (1/2n, 1/(2n-1)]$. Thus $|\int_h^1 f(x)\, dx| \le 1/n$ if $h < 1/(2n+1)$ and hence $\lim_{h \downarrow 0} \int_h^1 f(x)\, dx = 0$. However, $f(x)$ is not Lebesgue integrable on $[0,1]$ because $\int_0^1 |f(x)|dx = \Sigma \frac{2}{n} = \infty$. More general integrals which include improper Lebesgue integrals are the subject of Chapter 9.

7.1 Exercises

1. Let $f \in \mathcal{L}$ on $[0,1]$ and $\varepsilon > 0$. Show that there is a function $g(x) = \sum_{i=1}^{n} a_i \cdot C_{I_i}(x)$ where $\{I_i\}_{i=1}^{n}$ is a finite sequence of intervals so that $\int_0^1 |f(x) - g(x)|dx < \varepsilon$.
2. Show that if $f \in \mathcal{L}$ on $[0,1]$ and $g \in \mathcal{M}$ is bounded on $[0,1]$, then $f \cdot g \in \mathcal{L}$ on $[0,1]$.
3. Show that for any g which satisfies on $[0,1]$, $f \cdot g \in \mathcal{L}$ whenever $f \in \mathcal{L}$, there is a set E with $m([0,1] \setminus E) = 0$ so that f is bounded on E.
4. If $0 \le f_n(x) \le 1$ for each $x \in (0,1]$ and $\lim \int_0^1 f_n(x)dx = 0$, show that $\lim f_n(x) = 0$ a.e.
5. Show that there is a finite valued measurable function defined on $\mathbb{R}$ which is not Lebesgue integrable on any interval.

6. Show that for every $E \in \mathcal{M}$,
$$\int_0^1 \int_0^1 \sin(\pi t \cdot C_E(x))dx\, dt = 2 \cdot m(E)/\pi.$$

7. Is $\int_0^1 \underline{\lim}\, e^{1+\cos(nt)} dt = e^2$?

8. Show that $\lim_{n\to\infty} \int_0^1 (x^n + 1)/(x^n + 2)\, dx = 1/2.$

9. Let $f*g(x) = \int_{-\infty}^{\infty} f(t)g(x - t)\, dt.$ If f, g and h are non-negative measurable functions, show $f*g = g*f$ and $(f*g)*h = f*(g*h)$.

10. If $f \in \mathcal{M}$ is bounded on (a,b) and there is a set E in (a,b) with x a point of density of E and $\lim_{\substack{t\to x \\ t\in E}} f(t) = f(x)$, then show $F'(x) = f(x)$ where $F(x) = \int_a^x f(t)\, dt.$

7.2 Introduction to $\mathcal{L}^p$ Spaces

If f and g are Lebesgue integrable functions on a space $(X,\mathcal{A},m)$, it is not true, in general, that $f \cdot g$ is Lebesgue integrable, even if $m(X) < \infty$. For example, $f(x) = g(x) = x^{-1/2}$ on $(0,1]$ are Lebesgue integrable but $f(x) \cdot g(x) = x^{-1}$ is not. A partial answer to the question, "When is the product of two integrable functions integrable?", is obtained as follows: Let $\phi(x)$ be a strictly increasing continuous function on $[0,\infty)$ with $\phi(0) = 0$. Let $\psi(x) = \phi^{-1}(x)$. Consider any positive number a and let $b = \phi(a)$. Then the rectangle $[0,a] \times [0,b]$ is contained in

$$\{(x,y)\colon 0 \le y \le \phi(x),\quad 0 \le x \le a\}$$
$$\cup \{(x,y)\colon 0 \le x \le \psi(y),\quad 0 \le y \le b\}.$$

The measure of $[0,a] \times [0,b]$ is $a\cdot b$ and, if $\Phi(x) = \int_0^x \phi(t)\, dt$ and $\Psi(y) = \int_0^y \psi(t)\, dt$, then

$$(*) \quad a\cdot b \le \int_0^a \phi(x)\, dx + \int_0^b \psi(y)\, dy = \Phi(a) + \Psi(b).$$

Now, given f, g, Φ and Ψ as above so that $\Phi \circ |f|$ and $\Psi \circ |g|$ are Lebesgue integrable on X, it follows that $f\cdot g$ is Lebesgue integrable on X. This is simply a consequence of the fact that $|f(x)\cdot g(x)| \le \Phi(|f(x)|) + \Psi(|g(x)|)$.

The functions Φ and Ψ are said to be _conjugate functions_ and a function f, for which $\Phi(|f|)$ is Lebesgue integrable on X, is said to belong to $\mathcal{L}_X^\Phi$ or simply to $\mathcal{L}^\Phi$. Of particular interest is the case where, for $p > 1$, $\phi(x) = x^{p-1}$. Then $\Phi(x) = x^p/p$. Since $y = \phi(x) = x^{p-1}$ and $\psi = \phi^{-1}$, $\psi(y) = y^{1/(p-1)}$ and $\Psi(y) = y^q/q$ where q is defined by $q = p/(p-1)$. Such pairs of numbers p and q with $p,q > 1$ and $q = p/(p-1)$ are said to be _conjugates_ and it is helpful for later calculations to note that p and q could alternatively be defined by $\frac{1}{p} + \frac{1}{q} = 1$, $p\cdot q = p + q$ or $p = q/(q-1)$. For $0 < p < \infty$, the functions f which satisfy

$$\int_X |f(x)|^p\, dm < \infty$$

are said to belong to _the space_ $\mathcal{L}_X^p$ or simply to $\mathcal{L}^p$. The following theorem implies that each $\mathcal{L}^p$ is a linear space.

(7.15) Theorem. If $0 < p < \infty$ and $f \in \mathcal{L}^p$, $g \in \mathcal{L}^p$ and c is a real number, then $f + g \in \mathcal{L}^p$ and $c \cdot f \in \mathcal{L}^p$. If $p, q > 1$, $1/p + 1/q = 1$, $f \in \mathcal{L}^p$ and $h \in \mathcal{L}^q$, then $f \cdot h$ is Lebesgue integrable. Furthermore, if $m(X) < \infty$ and $0 < p' < p < \infty$, then $f \in \mathcal{L}^p$ implies $f \in \mathcal{L}^{p'}$.

Proof. Suppose f and g belong to $\mathcal{L}^p$ on X. Then

$$\begin{aligned} \int_X |f + g|^p \, dm &\le \int_X (2 \cdot \max(|f|, |g|))^p \, dm \\ &\le \int_X ((2|f|)^p + (2|g|)^p) dm \\ &\le 2^p \cdot \int_X |f|^p \, dm + 2^p \cdot \int_X |g|^p \, dm. \end{aligned}$$

Thus $f + g \in \mathcal{L}^p$. Clearly $\int_X |cf|^p \, dm = |c|^p \cdot \int_X |f|^p \, dm$ and hence, for any real number c, $cf \in \mathcal{L}^p$ whenever $f \in \mathcal{L}^p$. If $p, q > 1$, $1/p + 1/q = 1$ and $f \in \mathcal{L}^p$, $h \in \mathcal{L}^q$, it follows that $f \in \mathcal{L}^\Phi$ and $h \in \mathcal{L}^\Psi$ where $\Phi(x) = x^p/p$ and $\Psi(x) = x^q/q$. Hence $f \cdot g$ is Lebesgue integrable. Finally, suppose $m(X) < \infty$, $0 < p' < p < \infty$ and $f \in \mathcal{L}^p$. Let $X_o = \{x: |f(x)| < 1\}$ and

$$X_1 = \{x: |f(x)| \ge 1\}.$$

Then

$$\begin{aligned} \int_X |f|^{p'} dm &= \int_{X_o} |f(x)|^{p'} dm + \int_{X_1} |f(x)|^{p'} dm \\ &\le \int_{X_o} 1 \, dm + \int_{X_1} |f(x)|^p \, dm \\ &\le m(X) + \int_X |f|^p \, dm < \infty \end{aligned}$$

and $f \in \mathcal{L}^{p'}$ on X.

Note that, for $m(X) = \infty$ and $0 < p' < p < \infty$, it does not follow that $f \in \mathcal{L}^p$ implies $f \in \mathcal{L}^{p'}$. For example, $f(x) = x^{-1}$ satisfies $f \in \mathcal{L}^2_{[1,\infty)}$ but $f \notin$

$\mathcal{L}^1_{[1,\infty)}$ since $\int_1^\infty x^{-1}dx = \infty$.

The collection of all measurable functions f which are equal almost everywhere to a bounded function make up <u>the space</u> $\mathcal{L}^\infty$. A <u>Banach space</u> is defined to be a complete normed linear space and the spaces $\mathcal{L}^p$, $1 \le p \le \infty$ are some of the classical examples of Banach spaces. The norm on $\mathcal{L}^p$ for $1 \le p < \infty$ is $\|f\|_p = (\int_X |f|^p\, dm)^{1/p}$ and the norm for f in $\mathcal{L}^\infty$ is defined to be $\|f\|_\infty = \inf\{M: |f(x)| \le M \text{ a.e.}\}$. Points in these spaces consist of equivalence classes of functions, each class consisting of those functions which differ only on a set of measure 0. Thus, if $f = g$ a.e., we will treat f and g for this purpose as the same function. We, of course, need to show that $\| \|_p$ and $\| \|_\infty$ are norms and that the resulting spaces are complete. To see that $\| \|_1$ is a norm, note first that $\int_X |f|\, dm = 0$ iff $f(x) = 0$ a.e. Secondly, $\int_X |cf|\, dm = |c|\cdot\int_X |f|\, dm$. Finally, $\int_X |f + g|\, dm \le \int_X |f|\, dm + \int_X |g|\, dm$ follows from the fact that at each x in X we have $|f(x) + g(x)| \le |f(x)| + |g(x)|$. Thus, for all f and g in $\mathcal{L}^1$ and real numbers c,

$$\|f\|_1 = 0 \text{ iff } f = 0,$$

$$\|cf\|_1 = |c|\|f\|_1$$

and

$$\|f + g\|_1 \le \|f\|_1 + \|g\|_1$$

and $\| \|_1$ is a norm on $\mathcal{L}^1$. To see that $\| \|_\infty$ is a norm on $\mathcal{L}^\infty$ note first that $\|f\|_\infty = 0$ if $f = 0$ a.e. Secondly, $\inf\{M: |cf(x)| \le M \text{ a.e.}\} =$

$$|c|\cdot\inf\{M: |f(x)| \le M \text{ a.e.}\}$$

is obvious if $c = 0$ and for $c \neq 0$ follows from the fact that $m(\{x: |cf(x)| > M\}) = 0$ iff

$$m(\{x: |f(x)| > M/|c|\}) = 0.$$

Thus $\|cf\|_\infty = |c|\|f\|_\infty$. Finally, if $\|f\|_\infty = M_1$ and $\|g\|_\infty = M_2$ then

$$m(\{x: |f(x)| > M_1\}) = 0 = m(\{x: |g(x)| > M_2\})$$

and, since $m(\{x: |f(x) + g(x)| \geq M_1 + M_2\})$ must also be 0 because the set in question is contained in the previous two sets, we have $\|f + g\|_\infty \leq \|f\|_\infty + \|g\|_\infty$ and $\| \ \|_\infty$ is a norm.

For the spaces $\mathcal{L}^p$, $1 < p < \infty$, the first two defining properties of a norm are readily obtained. For, if $\|f\|_p = (\int_X |f|^p \, dm)^{1/p} = 0$, then $\int_X |f|^p \, dm = 0$ and $f = 0$ a.e. Also

$$\begin{aligned}(\textstyle\int_X |cf|^p \, dm)^{1/p} &= (\textstyle\int_X |c|^p |f|^p \, dm)^{1/p} \\ &= (|c|^p \cdot \textstyle\int_X |f|^p \, dm)^{1/p} \\ &= |c|(\textstyle\int_X |f|^p \, dm)^{1/p}\end{aligned}$$

and hence $\|cf\|_p = |c|\|f\|_p$. The triangle inequality for $\| \ \|_p$, $1 < p < \infty$ follows from the second of two famous inequalities which are stated here as theorems.

(7.16) Theorem. (Hölder's Inequality) *If f and g are measurable functions and $1/p + 1/q = 1$, then*

$$\int_X |f \cdot g| dm \leq (\int_X |f|^p \, dm)^{1/p} \cdot (\int_X |g|^q \, dm)^{1/q}.$$

Proof. If $\int_X |f|^p \, dm$ equals 0 or if $\int_X |g|^q \, dm$ equals 0, then the inequality follows from the fact that either f or g equals 0 a.e. and thus $f \cdot g$ also equals 0 a.e. Also, if either of these two integrals is infinite, the inequality is obvious. So suppose both integrals are finite and non-zero, let $F(x) = f(x)/\|f\|_p$ and $G(x) = g(x)/\|g\|_q$. Putting $\Phi(x) = x^p/p$ and $\Psi(x) = x^q/q$ in the formula (*) which was derived at the beginning of this section, we have for all $a,b > 0$, $a \cdot b \le \frac{a^p}{p} + \frac{b^q}{q}$. Thus, for each x in X, $|F(x) \cdot G(x)| \le \frac{|F(x)|^p}{p} + \frac{|G(x)|^q}{q}$. That is,

$$\frac{|f(x)|}{\|f\|_p} \cdot \frac{|g(x)|}{\|g\|_q} \le \frac{1}{p}\frac{|f(x)|^p}{\|f\|_p^p} + \frac{1}{q}\frac{|g(x)|^q}{\|g\|_q^q}.$$

Integrating both sides of this inequality yields

$$\frac{\int_X |f(x) \cdot g(x)|\, dm}{\|f\|_p \cdot \|g\|_q} \le \frac{1}{p} + \frac{1}{q} = 1$$

and thus $\int_X |f(x) \cdot g(x)|\, dm \le \|f\|_p \cdot \|g\|_q$ which is the statement of Hölder's inequality.

(7.17) <u>Theorem</u>. (Minkowski's Inequality) <u>If</u> f <u>and</u> g <u>belong to</u> $\mathcal{L}^p$ <u>on</u> X <u>where</u> $1 \le p < \infty$, <u>then</u>

$$\left(\int_X |f + g|^p\, dm\right)^{1/p} \le \left(\int_X |f|^p\, dm\right)^{1/p} + \left(\int_X |g|^p\, dm\right)^{1/p}.$$

<u>Proof</u>. Suppose f and g belong to $\mathcal{L}^p$, $1 < p < \infty$. Then

$$\begin{aligned}\int_X |f + g|^p\, dm &= \int_X |f + g||f + g|^{p-1}\, dm \\ &\le \int_X |f||f + g|^{p-1}\, dm \\ &\qquad + \int_X |g||f + g|^{p-1}\, dm.\end{aligned}$$

Applying Hölder's inequality to each of the integrals on the right yields

$$\int_X |f + g|^p dm$$
$$\leq ((\int|f|^p dm)^{1/p} + (\int|g|^p dm)^{1/p}) (\int|f + g|^{(p-1)\cdot q} dm)^{1/q}.$$

Since $(p-1) q = p$ we have, upon dividing both sides by $(\int|f + g|^p \, dm)^{1/q}$, that

$$(\int|f + g|^p \, dm)^{1-(1/q)} \leq (\int|f|^p \, dm)^{1/p} + (\int|g|^p \, dm)^{1/p}.$$

Since $1 - (1/q) = 1/p$, Minkowski's inequality is proven.

Before considering the completeness of the spaces $\mathcal{L}^p$, we will examine several ways in which a sequence of functions can approach a given function. A function f is said to be the <u>limit in mean</u> of a sequence of functions $\{f_n\}$ if $\lim_{n\to\infty} \|f_n - f\|_1 = 0$; this is written $\lim_{n\to\infty} f_n = f$ [mean] or $f_n \to f$ [mean]. Other types of convergence are similarly abbreviated. For example, $f_n \to f$ [a.e.] and $f_n \to f$ [unif.] refer to convergence almost everywhere and uniform convergence. Recall that Egoroff's theorem asserts that on a space of finite measure $f_n \to f$ [a.e.] implies that f_n approaches f almost uniformly; the latter type of convergence is abbreviated $f_n \to f$ [a. un.]; this refers to the fact that for every $\varepsilon > 0$ there is a set E_ε such that $|E_\varepsilon^c| < \varepsilon$ and for $x \in E_\varepsilon$, f_n approaches f uniformly. Given $p > 0$, if

$$\lim_{n\to\infty} \int|f_n - f|^p \, dm = 0,$$

f_n is said to approach f [mean p].

In order to show that the $\mathcal{L}^p$ spaces $1 \leq p \leq \infty$ are complete, we will utilize an additional mode of convergence. A sequence of functions $\{f_n\}$ is said to

<u>converge</u> <u>in</u> <u>measure</u> to a function f if, for each $\varepsilon > 0$,

$$\lim_{n\to\infty} m^*(\{x: |f_n(x) - f(x)| > \varepsilon\}) = 0.$$

This is written $f_n \to f$ [meas.].

If a sequence of functions converges to a function [a. un.] the sequence clearly also converges [a.e.] and [meas.]. Note, however, that even on a space of finite measure, a sequence of measurable functions can converge in measure to a function without converging at a single point. For example, if the collection of intervals of the form $[i/2^j, (i+1)/2^j]$, $0 \le i < 2^j$ is enumerated in a sequence, say $\{I_n\}$, the sequence of functions $f_n(x) = C_{I_n}(x)$ converges in measure to 0. This is because, for each $\varepsilon > 0$, there is a natural number N so that all intervals I_k with $|I_k| > \varepsilon$ occur in the sequence $\{I_n\}$ with indices $k < N$. Thus, for $k \ge N$, $|I_k| < \varepsilon$ and

$$m(\{x: |C_{I_k}(x) - 0| > \varepsilon\}) < \varepsilon.$$

However, each $x \in [0,1]$ belongs to infinitely many I_n and also belongs to infinitely many I_n^c. Thus, for each $x \in [0,1]$, $\overline{\lim}\, f_n(x) = 1$ and $\underline{\lim}\, f_n(x) = 0$ and $\{f_n\}$ does not converge at a single point. It is easily seen that f_n approaches 0 [mean] and [mean p], and hence, these modes of convergence do not imply convergence at a single point. Moreover, $g_n(x) = 2^n \cdot f_n(x)$ converges to 0 [meas.] but does not converge [mean] or [mean p].

Notwithstanding these examples, convergence in measure behaves predictably, as is shown by the following theorem:

(7.18) <u>Theorem</u>. <u>Suppose</u> $\{f_n\}$ <u>and</u> $\{g_n\}$ <u>are sequences of measurable functions defined on a measure space</u> X. <u>Suppose</u> $f_n \to f$ [meas.] <u>and</u> $g_n \to g$ [meas.]. <u>Then</u>

i) <u>if</u> $f_n \to h$ [meas.], <u>then</u> $f = h$ a.e.

ii) <u>there is a subsequence</u> $\{f_{n_k}\}$ <u>of</u> $\{f_n\}$ <u>such that</u> $f_{n_k} \to f$ [a. un.] <u>and hence</u> $f_{n_k} \to f$ [a.e.]

iii) f <u>is a measurable function</u>

iv) $f_n + g_n \to f + g$ [meas.]; <u>also, if</u> c <u>is any real number</u>, $c \cdot f_n \to cf$ [meas.].

<u>Proof</u>.

i) Suppose $f_n \to f$ [meas.] and $f_n \to h$ [meas.]. Then, given $\varepsilon > 0$ there is a natural number N_1 so that $n > N_1$ implies $m(\{x: |f_n(x) - f(x)| > \varepsilon/2\}) < \varepsilon/2$. There is also N_2 so that $n > N_2$ implies $m(\{x: |f_n(x) - h(x)| > \varepsilon/2\}) < \varepsilon/2$. Since

$\{x: |f(x) - h(x)| > \varepsilon\} \subset$

$\{x: |f_n(x) - f(x)| > \varepsilon/2\} \cup \{x: |f_n(x) - h(x)| > \varepsilon/2\}$,

it follows by considering n larger than N_1 and N_2 that $m(\{x: |f(x) - h(x)| > \varepsilon\}) < \varepsilon$. Since this is true for each $\varepsilon > 0$, $f = h$ a.e.

ii) Suppose $f_n \to f$ [meas.]. Let $\varepsilon_k = 1/2^k$. For each k choose n_k so that for each $n \geq n_k$, $m(\{x: |f_n(x) - f(x)| > \varepsilon_k\}) < \varepsilon_k$. Let

$$A_k = \{x: |f_{n_k}(x) - f(x)| > \varepsilon_k\}.$$

To see that $f_{n_k} \to f$ [a. un.] let $\varepsilon > 0$ be given. Choose K with $2^{-K} < \varepsilon/2$ and let $E_\varepsilon = \bigcup_{k=K}^{\infty} A_k$. Then

$m(E_\varepsilon) \le \sum_{k=K}^{\infty} m(A_k) \le \sum_{k=K}^{\infty} 2^{-k} = 2^{-K+1} < \varepsilon$. Moreover, if $k > N > K$ and $x \in E_\varepsilon^C$, $|f_{n_k}(x) - f(x)| \le \varepsilon_N$. Then f_{n_k} approaches f uniformly on E_ε^C and, since $\varepsilon > 0$ was arbitrary, $f_{n_k} \to f$ [a. un.].

iii) Simply note that, if $f_n \to f$ [meas.], the above sequence $\{f_{n_k}\}$ approaches f a.e. and the fact that each f_{n_k} is measurable implies that f is measurable.

iv) Let $f_n \to f$ [meas.], $g_n \to g$ [meas.]. Given $\varepsilon > 0$ there is a natural number N_1 so that if $n > N_1$, $m(\{x: |f_n(x) - f(x)| > \varepsilon/2\}) < \varepsilon/2$. For this ε, there is N_2 so that $m(\{x: |g_n(x) - g(x)| > \varepsilon/2\}) < \varepsilon/2$. But $\{x: |f_n(x) + g_n(x) - f(x) - g(x)| > \varepsilon\}$ is contained in

$$\{x: |f_n(x) - f(x)| > \varepsilon/2\} \cup \{x: |g_n(x) - g(x)| > \varepsilon/2\}.$$

Hence, if $n > N = \max(N_1, N_2)$,

$$m(\{x: |f_n(x) + g_n(x) - f(x) - g(x)| > \varepsilon\}) < \varepsilon.$$

Consequently, $f_n + g_n \to f + g$ [meas.]. For real numbers $c \ne 0$,

$$\{x: |c\cdot f_n(x) - c\cdot f(x)| > \varepsilon\} = \{x: |f_n(x) - f(x)| > \varepsilon/|c|\}.$$

Thus, if $f_n \to f$ [meas.] and $c \ne 0$, $c\cdot f_n \to cf$ [meas.]. Since it is clear that for $c = 0$, $c\cdot f_n \to cf$ in measure because $c\cdot f_n = 0 = cf$, it follows that for real numbers c, whenever $f_n \to f$ [meas.], $c\cdot f_n \to cf$ [meas.].

The relationship between convergence in measure

and convergence in mean or convergence mean p is made explicit in the following theorem.

(7.19) Theorem. *If $\{f_n\}$ is a sequence of measurable functions and $f_n \to f$ [mean p], then $f_n \to f$ [meas.]. If the sequence $\{f_n\}$ of measurable functions is uniformly bounded on X with $m(X) < \infty$, then $f_n \to f$ [meas.] implies that for each $p > 0$, $f_n \to f$ [mean p].*

Proof. If $p > 0$ and $f_n \to f$ [mean p], then $(\int|f_n - f|^p\,dm)^{1/p} \to 0$ and $\int|f_n - f|^p\,dm \to 0$. Given $\varepsilon, \delta > 0$, there is a natural number N so that $n > N$ implies $\int|f_n - f|^p\,dm < \varepsilon^p \cdot \delta$. If

$$A_{n,\varepsilon} = \{x\colon |f_n(x) - f(x)| > \varepsilon\},$$

then $\varepsilon^p \cdot \delta > \int_X|f_n - f|^p\,dm \geq \int_{A_{n,\varepsilon}}|f_n - f|^p\,dm \geq \varepsilon^p|A_{n,\varepsilon}|$. Hence, for $n > N$, $|A_{n,\varepsilon}| < \delta$ and $\lim_{n\to\infty} |A_{n,\varepsilon}| = 0$. Since this is true for each $\varepsilon > 0$, $f_n \to f$ [meas.]. Now suppose that $m(X) < \infty$ and $\{f_n\}$ is a sequence of functions which are uniformly bounded on X and that $f_n \to f$ [meas.]. Again let $A_{n,\varepsilon} = \{x\colon |f_n(x) - f(x)| > \varepsilon\}$. Then

$$\int_X|f_n - f|^p\,dm = \int_{A_{n,\varepsilon}}|f_n - f|^p\,dm + \int_{A^c_{n,\varepsilon}}|f_n - f|^p\,dm$$

$$\leq |2M|^p \cdot m(A_{n,\varepsilon}) + \varepsilon^p \cdot m(X)$$

where M, a bound for $\{f_n(x)\colon x \in X,\ n = 1, 2, \ldots\}$, is also a bound for $f(x)$ at a.e. point x by ii) of the previous theorem. Since $\varepsilon > 0$ is arbitrary and $m(A_{n,\varepsilon}) \to 0$ as $n \to \infty$, it follows that $\int_X |f_n - f|^p \, dm \to 0$ as $n \to \infty$; that is, $f_n \to f$ [mean p].

Convergence [mean p] is convergence with respect to the metric on $\mathcal{L}^p$ space. The other forms of convergence [meas.], [a.e.], and [a. un.] occur without there being any natural metric on the functions involved. Nonetheless, there are Cauchy criteria for each of these modes of convergence. Specifically:

A sequence $\{f_n\}$ is said to be <u>Cauchy</u> <u>[a.e.]</u> if there is a set Z with $m(Z) = 0$ and given any $\varepsilon > 0$ there is N so that $|f_n(x) - f_m(x)| < \varepsilon$ when $x \notin Z$ and $n,m > N$.

A sequence $\{f_n\}$ is said to be <u>Cauchy</u> <u>[a. un.]</u> if given any $\varepsilon > 0$ there is N and a set E_ε with $|E_\varepsilon| < \varepsilon$ and such that $|f_n(x) - f_m(x)| < \varepsilon$ for each $x \notin E_\varepsilon$ and $n,m > N$.

A sequence $\{f_n\}$ is said to be <u>Cauchy</u> <u>[meas.]</u> if given $\varepsilon > 0$ there is N so that

$$m(\{x: |f_n(x) - f_m(x)| > \varepsilon\}) < \varepsilon$$

whenever $n,m > N$.

Clearly, $\{f_n\}$ is Cauchy [a.e.] if and only if $\{f_n\}$ converges a.e. Also, if $\{f_n\}$ is Cauchy [a. un.] then f_n converges [a. un.] to a function f. This is because, given $\varepsilon > 0$, for each natural number k there are sets $E_{\varepsilon/2^k}$ of measure less than $\varepsilon/2^k$ and natural numbers N_k so that for $x \notin E_{\varepsilon/2^k}$ and $n,m > N_k$, $|f_n(x) - f_m(x)| < \varepsilon/2^k$. If $E_\varepsilon = \cup_k E_{\varepsilon/2^k}$, then $\{f_n\}$ is uniformly Cauchy on E_ε^c, $m(E_\varepsilon) < \varepsilon$ and hence $\{f_n\}$ converges uniformly on E_ε^c. It follows that there is a function f defined a.e. and $f_n \to f$ [a. un.]. Since convergence of $\{f_n\}$ almost uniformly clearly implies Cauchy [a. un.], a sequence converges [a. un.] if and only if it is Cauchy [a. un.]. That Cauchy [mean p] implies Cauchy [meas.] which in turn implies convergence in measure is the content of the following theorem.

(7.20) Theorem. If $\{f_n\}$ is a sequence of measurable functions which are Cauchy [meas.], then there is a function f so that $f_n \to f$ [meas.]. If $\{f_n\}$ is a sequence of functions which is Cauchy [mean p], then $\{f_n\}$ is Cauchy [meas.].

Proof. If $\{f_n\}$ is Cauchy [meas.], then there is an increasing sequence of natural numbers $\{N_k\}$ so that $m(\{x: |f_n(x) - f_m(x)| > 1/2^k\}) < 1/2^k$ whenever $n,m > N_k$. Let $E_K = \bigcup_{k=K}^{\infty} \{x: |f_{N_k}(x) - f_{N_{k+1}}(x)| > 1/2^k\}$. Then $|E_K| < 1/2^{K-1}$. Since $\{f_{N_k}\}$ is Cauchy on each E_K^c, f_{N_k} converges [a. un.] to some measurable function f. But then, given $\varepsilon > 0$,

$$A_{n,\varepsilon} = \{x: |f_n(x) - f(x)| > \varepsilon\}$$

is contained in

$$\{x: |f_n(x) - f_{N_k}(x)| > \varepsilon\} \cup \{x: |f_{N_k}(x) - f(x)| > \varepsilon\}$$

and, since $\{f_n\}$ is Cauchy [meas.] and $f_{N_k} \to f$ [a. un.], for each $\varepsilon > 0$ the measure of $A_{n,\varepsilon}$ approaches 0 as n approaches ∞; hence $f_n \to f$ [meas.]. Now suppose $\{f_n\}$ is a sequence of functions which is Cauchy [mean p]. Then for every $\varepsilon > 0$ there is N so that, for $n,m > N$, $\int_X |f_n - f_m|^p \, dm < \varepsilon^{p+1}$. But then for $n,m > N$ it follows that

$$m(\{x: |f_n(x) - f_m(x)| > \varepsilon\}) < \varepsilon;$$

for otherwise, $\int_X |f_n - f_m|^p \, dm$ would be larger than $\varepsilon^p \cdot \varepsilon$. Hence, if $\{f_n\}$ is Cauchy [mean p], $\{f_n\}$ is Cauchy [meas.].

We are now ready to show the completeness of the $\mathcal{L}^p$ spaces $1 \le p \le \infty$.

(7.21) Theorem. (Riesz-Fisher) *The spaces $\mathcal{L}^p$, $1 \le p \le \infty$, are complete normed linear spaces.*

Proof. First consider p with $1 \le p < \infty$. Let $\{f_n\}$ be a sequence of measurable functions which is Cauchy [mean p]. By the previous theorem, $\{f_n\}$ is Cauchy [meas.]. Moreover, the subsequence $\{f_{N_k}\}$, described in the proof of the theorem, converges to a function f [a.e.]. Since $\{f_n\}$ is Cauchy [mean p], given $\varepsilon > 0$

there is a natural number N so that $\int |f_n - f_m|^p \, dm < (\varepsilon/2)^p$ whenever $n, m > N$. Fix k so that $N_k > N$ and note that $|f_{N_k} - f_{N_i}|^p$ approaches $|f_{N_k} - f|^p$ a.e. By Fatou's lemma,

$$\int |f_{N_k} - f|^p \, dm \le \varliminf_i \int |f_{N_k} - f_{N_i}|^p \, dm \le (\varepsilon/2)^p.$$

Then, by Minkowski's inequality,

$$\begin{aligned}(\int |f|^p \, dm)^{1/p} & \\ &\le (\int |f_{N_k} - f|^p \, dm)^{1/p} + (\int |f_{N_k}|^p \, dm)^{1/p} \\ &< \infty\end{aligned}$$

and it follows that $f \in \mathcal{L}^p$. A second application of Minkowski's inequality yields

$$\begin{aligned}(\int |f - f_n|^p \, dm)^{1/p} & \\ &\le (\int |f_{N_k} - f|^p \, dm)^{1/p} + (\int |f_n - f_{N_k}|^p \, dm)^{1/p} \\ &< \varepsilon\end{aligned}$$

whenever $n > N$. Thus $f_n \to f$ [mean p]. Now consider a Cauchy sequence $\{f_n\}$ in the space $\mathcal{L}^\infty$. Given $\varepsilon > 0$, there is N so that $n, m \ge N$ implies $|f_n(x) - f_m(x)| < \varepsilon$ except on a set of measure 0. That is, $\{f_n\}$ is Cauchy [a.e.] and hence converges a.e. to a function f. If $|f_N| \le M$ a.e. then $|f(x)| \le |f_N(x) - f(x)| + |f_N(x)| \le M + \varepsilon$ almost everywhere. Thus $f \in \mathcal{L}^\infty$. But

$$\begin{aligned}|f(x) - f_n(x)| &\le |f_m(x) - f(x)| + |f_m(x) - f_n(x)| \\ &\le |f_m(x) - f(x)| + \varepsilon\end{aligned}$$

except on a set of measure 0. Letting $m \to \infty$, it follows that, for each $n > N$, $|f(x) - f_n(x)| \le \varepsilon$ except on a set of measure 0. Since $\varepsilon > 0$ was arbitrary, $f_n \to f$ [mean ∞]. It has been noted previously that the spaces $\mathcal{L}^p$, $1 \le p \le \infty$, are normed linear spaces; we now have that they are complete normed linear spaces.

A function f is in $\mathcal{L}_X^\Phi$ if and only if $\Phi \circ f$ has a finite integral over the space X. Compositions of the form $\Phi \circ f$, where f is a real-valued function defined on a measure space and Φ is a real-valued function of a real variable, occur naturally in the setting where X is a probability space. We conclude this section by considering such a situation and the formula used for computing integrals in that setting.

If X(s) is a real-valued measurable function defined on a probability space S, X(s) is called a <u>random variable</u>. The idea behind calling such functions random variables is that such an X(s) assigns to each of the elements of the probability space a numerical value. If the elements of the probability space are viewed as possible outcomes of an experiment, the value of the variable is considered to be taken on "at random". The requirement that X(s) be measurable is equivalent to the assertion that, given any interval I, $\{s: X(s) \in I\}$ is measurable and thus the probability that X(s) belongs to I is defined. While random variables can be allowed to take on values in a vector space, we are only concerned here with real-valued functions.

If m denotes the measure on S, it is standard to define the <u>mean</u> of X(s) or the <u>expected value</u> of X, written E(X) to be $\int_S X(s)\, dm$ provided the integral exists. The <u>nth moment</u> of X(s) is defined to be

$$E(X^n) = \int_S (X(s))^n\, dm.$$

Clearly, the nth moment exists if and only if $X \in \mathcal{L}_S^n$. If $g(x)$ is a continuous real-valued function of a real variable, or merely a Baire function, then $g(X(s))$ is a measurable function and hence a random variable. Then $E(g(X))$ by definition equals $\int_S g(X(s))\, dm$, provided the integral exists. This integral, however, can be evaluated indirectly without specifically using the measure on S; this is accomplished by integrating the function g over the real line with respect to a measure which is produced from the measure on S and the measurable function $X(s)$.

Given a measurable function $X(s)$ defined on S, the function $F(x) = m(\{s: X(s) \le x\})$ is called the <u>distribution function</u> for $X(s)$. Clearly, $F(x)$ is non-decreasing, continuous on the right at each x and satisfies $\lim_{x\to-\infty} F(x) = 0$ and $\lim_{x\to\infty} F(x) = 1$. The integral of a function $g(x)$ with respect to the measure generated by a non-decreasing F is called the <u>Lebesgue-Stieltjes integral</u> and is written $\mathcal{LS}\int g(x)\, dF$ or $\mathcal{LS}\int_{-\infty}^{\infty} g(x)\, dF$ and this integral is used to evaluate $E(g(x))$. Here, $\mathcal{LS}\int g(x)\, dF$ is the integral with respect to the measure generated by

$$m_F^*(E) = \inf\{\Sigma F(I_n): E \subset \cup \mathring{I}_n\}.$$

Then if $F(x)$ is a non-decreasing function of a real variable, $\mathcal{LS}\int_a^b 1\, dF = F(b+) - F(a-) = m_F([a,b])$. However, this does not in general agree with the value of the Riemann-Stieltjes integral because $\mathcal{RS}\int_a^b 1\, dF = F(b) - F(a)$. Also, for $c \in (a,b)$,

$$\mathcal{LS}\int_a^c 1\ dF + \mathcal{LS}\int_c^b 1\ dF = \mathcal{LS}\int_a^b 1\ dF + F(c+) - F(c-)$$

because $m_F(\{c\})$ is accounted for both in $[a,c]$ and again in $[c,b]$. That the Lebesgue-Stieltjes integral is not in general additive on adjacent closed intervals can be remedied by defining what is called the Stieltjes integral of g with respect to F by

$$\mathcal{S}\int_a^b g(x)\ dF = \mathcal{LS}\int_a^b g(x)\ dF - g(b)\big(F(b+) - F(b)\big) - g(a)\big(F(a) - F(a-)\big).$$

Then $\mathcal{S}\int_a^b g\ dF$ is additive on adjacent closed intervals and does not take into account the values of F outside of $[a,b]$. We will need the fact that a bounded function g measurable with respect to m_F and Riemann-Stieltjes integrable on $[a,b]$ is also Stieltjes integrable on $[a,b]$ and the two integrals agree. The proof of this involves the same argument as the proof that on $[a,b]$, Riemann integrable functions are Lebesgue integrable and the integrals agree (Theorem 7.5).

We are now ready to show that $\mathcal{LS}\int_{-\infty}^{\infty} g(x)\ dF$ gives the value of $E(g(X))$ when F is the distribution function for $X(s)$.

(7.22) Theorem. *If S is a probability space, $X(s)$ a random variable defined on S, $F(x)$ the distribution function for X, and $g(x)$ is a Baire function, then $E(g(X))$ exists (finite or infinite) if and only if*

$$\mathcal{LS}\int_{-\infty}^{\infty} g(x)\ dF$$

exists and, when both do exist, $E(g(X)) = \mathcal{LS}\int_{-\infty}^{\infty} g(x)\ dF$.

Proof. Let $X(s)$ be a random variable defined on a probability space S and let $F(x)$ be the distribution function of $X(s)$. First suppose that $g(x)$ is continuous and non-negative. Then g is Riemann-Stieltjes integrable with respect to F on each interval $[a,b]$. Since g is bounded and is measurable with respect to m_F, g is also Stieltjes integrable on each $[a,b]$ and $\mathcal{S}\int_a^b g\ dF = \mathcal{RS}\int_a^b g\ dF$. Given a closed interval $[a,b]$ and $\varepsilon > 0$, let $a = x_o < x_1 < \dots < x_n = b$ be a partition of $[a,b]$ such that

(*) $$\left|\mathcal{RS}\int_a^b g(x)\ dF - \Sigma g(x_i)(F(x_i) - F(x_{i-1}))\right| < \varepsilon$$

and

(**) the oscillation of g on each $[x_{i-1}, x_i]$ is less than ε.

Let $S_i = \{s\colon x_{i-1} < X(s) \le x_i\}$ and

$$S_{a,b} = \{s\colon a < X(s) \le b\}.$$

Since $S_i = \{s\colon X(s) \le x_i\} \setminus \{s\colon X(s) \le x_{i-1}\}$, it follows that

(***) $$m_F(S_i) = F(x_i) - F(x_{i-1}) \qquad i = 1, 2, \dots, n.$$

Because of (**), $|g(x_i)\cdot m_F(S_i) - \int_{S_i} g(X(s))dm| \le$

$\varepsilon \cdot m(S_i)$ and thus $|\Sigma g(x_i) \cdot m_F(S_i) - \int_{S_{a,b}} g(X(s))dm| \le \varepsilon \cdot \Sigma m(S_i) \le \varepsilon$. Combining this inequality with (***) and (*), since F is continuous on the right at a, we have $|\mathcal{RS}\int_a^b g\,dF - \int_{S_{a,b}} g(X(s))dm| \le 2\varepsilon$. Since $\varepsilon > 0$ is arbitrary, $\mathcal{RS}\int_a^b g\,dF = \int_{S_{a,b}} g(X(s))dm$. Letting $b \to \infty$ and then $a \to -\infty$, we have

$$\mathcal{LS}\int_{-\infty}^{\infty} g\,dF = \int_S g(X(s))dm.$$

Here, the integrals are either both finite and equal or both equal $+\infty$. Now, if g is allowed to be any continuous function, $g = g+ - g-$ and both g+ and g- are continuous. Then either both of the integrals do not exist because the integrals of positive and negative parts of g are both infinite or, if one integral does exist (finite or infinite), so does the other and they are equal. Now suppose that g is a Baire function in Baire class α and that, for each $\beta < \alpha$ and each function h in Baire class β, $\int_{-\infty}^{\infty} h(x)dF$ exists if and only if $\int_S h(X(s))ds$ does and when these integrals exist they are equal. Since g is in Baire class α, $g = \lim h_n$ where each h_n belongs to class β_n with $\beta_n < \alpha$. Let $g+(x) = \max(g(x), 0)$ and $g_N^+(x) = \min(g+(x), N)$; also let $h_n^+(x) = \max(h_n(x), 0)$ and $h_{n,N}^+(x) = \min(h_n^+(x), N)$. Then g+ and each g_N^+ belong to Baire class α; each h_n^+ and $h_{n,N}^+$ belong to Baire class β_n. Since m_F assigns measure 1 to the real line and $m(S) = 1$, applying the bounded convergence theorem we have

$$\begin{aligned}\int_S g_N^+(X(s))dm &= \lim_n \int_S h_{n,N}^+(X(s))ds \\ &= \lim_n \mathcal{LS}\int_{-\infty}^{\infty} h_{n,N}^+(x)\, dF \\ &= \mathcal{LS}\int_{-\infty}^{\infty} g_N^+(x)\, dF\end{aligned}$$

and these integrals are finite. Now, by taking limits on N, the monotone convergence theorem yields

$$\begin{aligned}\int_S g^+(X(s))dm &= \lim_N \int_S g_N^+(X(s))dm \\ &= \lim_N \mathcal{LS}\int_{-\infty}^{\infty} g_N^+(x)\, dF \\ &= \mathcal{LS}\int_{-\infty}^{\infty} g^+(x)\, dF.\end{aligned}$$

Similar considerations with respect to g^- yield that $\int_S g^-(X(s))dm = \mathcal{LS}\int_{-\infty}^{\infty} g^-(x)\, dF$. Again $\int_S g(X(s))dm$ exists (finite or infinite) if and only if $\mathcal{LS}\int_{-\infty}^{\infty} g(x)\, dF$ does and when they do exist they are equal. It now follows by induction that this is true for all Baire functions defined on the line.

7.2 Exercises

1. A <u>step function</u> f is a function for which there is a finite collection of intervals (possibly degenerate, open, closed or half-open) on each of which f is constant. Prove that the step functions are dense in $\mathcal{L}^1_{[0,1]}$.
2. Suppose $m(X) < \infty$ and functions which are equal a.e. are considered equivalent. Show that

$$d(f,g) = \inf\{\varepsilon: m(\{x: |f(x) - g(x)| > \varepsilon\}) < \varepsilon\}$$

is a metric on the measurable functions defined and finite a.e. on X.

3. Show that neither the hypothesis that $m(X) < \infty$ nor the hypothesis that the functions are finite a.e. can be dropped in the statement of Exercise 2.
4. If $d(f,g)$ is as in Exercise 2 and $X = [0,1]$ with Lebesgue measure for the measure, show that in general
$$d(cf,0) \neq |c|d(f,0) \quad \text{but} \quad d(f,g) = d\big((f,g),0\big).$$
5. If $m(X) < \infty$ and $d(f,g)$ is defined as in Exercise 2, show that $f_n \to f$ [meas.] iff $d(f_n,f) \to 0$; also show $\{f_n\}$ is Cauchy [meas.] if $\{f_n\}$ is Cauchy with respect to d.
6. The space ℓ^p is the collection of all sequences $\{x_n\}$ such that $\Sigma|x_n|^p < \infty$. Show that $\Sigma|x_n|^p$ is a norm on ℓ^p.
7. Prove that if $\{x_n\} \in \ell^p$ and $\{y_n\} \in \ell^q$ with $1/p + 1/q = 1$, then $\Sigma|x_n y_n| < \infty$.
8. Suppose $f \in \mathcal{L}^p$ and $g \in \mathcal{L}^q$ with $1/p + 1/q = 1/r$. Show that $f\cdot g \in \mathcal{L}^r$.
9. Suppose $f,g \in \mathcal{L}^p$, $f \neq g$, and $\|f\|_p = \|g\|_p = 1$. Show that for each $t\in(0,1)$, $\|tf + (1 - t)g\|_p < 1$.
10. Show that there are functions f defined on $[0,1]$ so that for each $p \geq 1$, $f \in \mathcal{L}^p$ but $f \notin \mathcal{L}^\infty$.
11. If $f \in \mathcal{L}^\infty_{[0,1]}$, show that $\|f\|_\infty = \lim\limits_{p\to\infty} \|f\|_p$.
12. Show that $f\cdot g = \int_0^1 f(x)\cdot g(x)\,dx$ is an inner product on $\mathcal{L}^2_{[0,1]}$.

Chapter Eight

8.1 Differentiation of the Integral

The derivative of a function of a real variable has several interpretations and the subject is naturally related to that of integration. In fact, the two are frequently thought of as inverse operations. It is useful to think of this as developing out of the simplest situation; namely, that multiplication and division are inverse operations. Thus, when the velocity of an object is constant, distance traversed equals velocity multiplied by time; that is, $s = v \cdot t$. Knowing v and t gives one the distance traversed and, conversely, knowing s and t gives $v = s/t$. When s varies with t, the definition of $v(t)$ as the derivative of $s(t)$ is a natural one and equally natural is the following: Suppose $v(t)$ is plotted between $t = a$ and $t = b$ where $v(t)$ is continuous and non-negative. Then a Riemann partition of the area under the graph indicates almost immediately that this area equals the distance traversed between time a and time b. For the velocity between time t_{i-1} and t_i is approximately $v(t_i)$ and $\Sigma\, v(t_i)\Delta t_i$ is then an approximation to both the area under the graph and to the distance traveled. It is expected that the approximations are more exact as the norm of the partition approaches 0. The integral is then an antiderivative of the function f providing f is continuous. The extent to which this relationship continues to hold for the Lebesgue integral is the subject matter of this section.

In considering differentiation, it is worthwhile to consider both real-valued functions of a point $F(x)$ and real-valued functions of an interval $F(I)$. If $F(x)$ is any finite function of a point, there is associated with $F(x)$ the natural interval function $F(I)$ where, if $I = [a,b]$, $F(I) = F(b) - F(a)$. For any $F(x)$, $F(I)$ is an additive function of an interval; that is, if $I = I_1 \cup I_2$ with I_1 and I_2 non-overlapping, then $F(I) = F(I_1) + F(I_2)$. Continuity of $F(x)$ at x_o is then equivalent to $\lim_{I\to x_o} F(I) = 0$, where $\lim_{I\to x_o} F(I)$ means that the lim sup and lim inf are equal where

$$\limsup_{I\to x_o} F(I) = \lim_{\varepsilon\to 0^+} \sup\{F(I):\ x_o \in I,\ \ |I| < \varepsilon\}$$

and

$$\liminf_{I \to x_o} F(I) = \lim_{\varepsilon \to 0^+} \inf\{F(I)\colon x_o \in I, \ |I| < \varepsilon\}.$$

Thus, if $\lim_{x \to x_o} F(x) = F(x_o)$ then $\lim_{x \to x_o} [F(x) - F(x_o)] = 0$ and

$$\lim_{I \to x_o} F(I) = \lim_{I \to x_o} [F(b) - F(a)]$$

$$= \lim_{I \to x_o} [F(b) - F(x_o) + F(x_o) - F(a)] = 0.$$

Thus the continuity of $F(x)$ is equivalent to $\lim_{I \to x} F(I) = 0$. The oscillation of F on I is $\theta(F;I) = \sup\{|F(J)|\colon J \subset I\}$ and F is continuous iff $\lim_{I \to x} \theta(F;I) = 0$. Differentiability of a function of an interval F at a point x is defined by $\lim_{I \to x} \frac{F(I)}{|I|}$. Similarily, $F'(x_o) = m$ iff $\lim_{I \to x_o} \frac{F(I)}{|I|} = m$. For if

$$\lim_{x \to x_o} \frac{F(x) - F(x_o)}{x - x_o} = m, \quad \text{then} \quad \lim_{I \to x_o} \frac{F(b) - F(a)}{b - a} =$$

$$= \lim_{I \to x_o} \left(\frac{F(b) - F(x_o)}{b - x_o} \cdot \frac{b - x_o}{b - a} + \frac{F(x_o) - F(a)}{x_o - a} \cdot \frac{x_o - a}{b - a} \right)$$

for $x_o \in (a,b)$

$$= \lim_{I \to x_o} \left((F'(x_o) + \varepsilon_1) \cdot \frac{b - x_o}{b - a} + (F'(x_o) + \varepsilon_2) \cdot \frac{x_o - a}{b - a} \right)$$

where $\varepsilon_1, \varepsilon_2 \to 0$

$$= F'(x_o) + \lim_{I \to x_o} \varepsilon_1 \left(\frac{b - x_o}{b - a} \right) + \varepsilon_2 \left(\frac{x_o - a}{b - a} \right) = m.$$

Conversely, if $\lim\limits_{I \to x_o} \frac{F(I)}{|I|} = m$, by choosing x_o to be an endpoint of I, it is clear that $F'(x_o) = m$. The interval notation for the derivative, as well as for other concepts is sometimes more convenient, as will become apparent. Because of the above, the derivative of a function F of an interval at a point x is defined by $\lim\limits_{I \to x} \frac{F(I)}{|I|}$.

The four Dini derivates and the upper and lower derivates of a function F are defined as follows:

$$\overline{F}^+(x) = \limsup_{t \to x^+} \frac{F(t) - F(x)}{t - x},$$

$$\overline{F}^-(x) = \limsup_{t \to x^-} \frac{F(t) - F(x)}{t - x},$$

$$\underline{F}^+(x) = \liminf_{t \to x^+} \frac{F(t) - F(x)}{t - x},$$

$$\underline{F}^-(x) = \liminf_{t \to x^-} \frac{F(t) - F(x)}{t - x},$$

$$\overline{F}(x) = \limsup_{t \to x} \frac{F(t) - F(x)}{t - x} = \limsup_{I \to x} \frac{F(I)}{|I|},$$

$$\underline{F}(x) = \limsup_{t \to x} \frac{F(t) - F(x)}{t - x} = \liminf_{I \to x} \frac{F(I)}{|I|}.$$

Clearly, $\overline{F}(x) = \max(\overline{F}^+(x), \overline{F}^-(x))$ and $\underline{F}(x) = \min(\underline{F}^+(x), \underline{F}^-(x))$. Also $F^+(x)$ is said to exist if $\overline{F}^+(x) = \underline{F}^+(x)$ and then $F^+(x)$ equals their common value; similarly $F^-(x)$ equals the common value of $\overline{F}^-(x)$ and $\underline{F}^-(x)$ when they are equal.

Some limitations on the possibilities for the Dini derivates of a real function are given by the next theorem. Its proof uses the interesting fact that a real function cannot have uncountably many strict relative maxima (or minima). To see this, fix $f(x)$ and let

$$A_n = \{x\colon\ 0 < |t - x| < 1/n \ \text{ implies } \ f(t) < f(x)\}.$$

Then $\bigcup_{n=1}^{\infty} A_n$ is the set of strict relative maxima of f. But each set A_n consists of isolated points and hence each A_n is at most countable as is the set of strict relative maxima of f.

(8.1) *Theorem. If* $F(x)$ *is an arbitrary real function, then*

$$\{x\colon\ \overline{F}^{+}(x) < \underline{F}^{-}(x)\} \quad \text{and} \quad \{x\colon\ \overline{F}^{-}(x) < \underline{F}^{+}(x)\}$$

are at most countable sets.

Proof. For each rational number r, let

$$E_r = \{x\colon\ \overline{F}^{+}(x) < r < \underline{F}^{-}(x)\}.$$

Then $\{x\colon\ \overline{F}^{+}(x) < \underline{F}^{-}(x)\} = \bigcup_{r} E_r$. Let $F_r(x) = F(x) - rx$. Then each $x \in E_r$ is a strict relative maximum of $F_r(x)$ because at each $x \in E_r$,

$$\overline{F}_r^{+}(x) < 0 < \underline{F}_r^{-}(x).$$

Since the set of strict relative maxima of a function is at most countable, each E_r is at most countable and so is $\bigcup_{r} E_r$. That $\{x\colon\ \overline{F}^{-}(x) < \underline{F}^{+}(x)\}$ is at most countable is proved similarly.

Note. In order to show that a given function $F(x)$ is differentiable a.e., it suffices to show that

$$\{x\colon \overline{F}^+(x) > \underline{F}^-(x)\} \quad \text{and} \quad \{x\colon \overline{F}^-(x) > \underline{F}^+(x)\}$$

are of measure 0. For then, almost everywhere

$$\overline{F}^+ \le \underline{F}^- \le \overline{F}^- \le \overline{F}^+ \quad \text{and} \quad \overline{F}^- \le \underline{F}^+ \le \overline{F}^+ \le \overline{F}^-.$$

A function $F(x)$ defined on $[a,b]$ is said to be <u>non-decreasing</u> on $[a,b]$ if $x_1 < x_2$ implies $F(x_1) \le F(x_2)$ for all $x_1, x_2 \in [a,b]$. A function $F(I)$ is said to be <u>non-decreasing</u> if $I_1 \subset I_2 \subset [a,b]$ implies $F(I_1) \le F(I_2)$. A non-decreasing function F of a point gives rise to a non-negative, non-decreasing, additive $F(I)$.

After Weierstrass constructed a continuous nowhere differentiable function, it was natural historically to try to construct a monotone nowhere differentiable function. That this cannot be done is a consequence of the following theorem. The theorem begins a description of the relationship between differentiation and Lebesgue integration.

(8.2) <u>Theorem</u>. <u>If</u> $F(x)$ <u>is non-decreasing on</u> $[a,b]$, <u>then</u> $F(x)$ <u>is differentiable</u> a.e. <u>on</u> $[a,b]$, $F'(x)$ <u>is Lebesgue integrable on</u> $[a,b]$ <u>and</u>

$$\int_a^b F'(x)\,dx \le F(b) - F(a).$$

<u>Proof</u>. Let $F(I)$ be the associated function of an interval. Suppose that $E = \{x\colon \overline{F}(x) > \underline{F}(x) \ge 0\}$ has positive outer measure. Then there are two rational numbers r_1 and r_2 with $0 \le r_1 < r_2$ such that

$$E' = \{x\colon \overline{F}(x) > r_2 > r_1 > \underline{F}(x)\}$$

has positive outer measure. Given $\varepsilon > 0$, let $G \supset E'$, G open and $|G| - m^*(E') < \varepsilon$. Let

$$\mathcal{V}_1 = \{I\colon I \subset G \text{ and } F(I) < r_1|I|\}.$$

Then $\mathcal{V}_1$ is a Vitali cover of E' and there exists a sequence of pairwise disjoint intervals I_m such that $|E' \setminus \cup I_m| = 0$. Let

$$\mathcal{V}_2 = \{J\colon J \text{ is contained in some } I_m \text{ and } F(J) > r_2|J|\}.$$

Then $\mathcal{V}_2$ is a Vitali cover of $E' \cap \bigcup_m \mathring{I}_m$ and so there exist $J_{m,n}$ a sequence of pairwise disjoint intervals with $J_{m,n} \subset I_m$ and $|E' \setminus \cup J_{m,n}| = 0$. But then $r_2(|G| - \varepsilon) \le \sum_{m,n} r_2|J_{m,n}| \le \sum_{m,n} F(J_{m,n}) \le \sum_m F(I_m) \le \sum_m |I_m| r_1$. And thus $r_2(|G| - \varepsilon) \le r_1|G|$. Since $\varepsilon > 0$ is arbitrary, it cannot be that $|G| > 0$ for each such G. Now, if $F(x)$ is a non-decreasing function of a point on $[a,b]$, let $F(x) = F(b)$ when $x > b$. Since $F'(x)$ exists a.e. (possibly ∞), it follows that

$$\frac{F(x + 1/n) - F(x)}{1/n}$$

is non-negative and approaches $F'(x)$ a.e. By Fatou's Lemma,

$$\begin{aligned}
\int_a^b F'(x)\,dx &\le \underline{\lim} \int_a^b n(F(x + 1/n) - F(x))dx \\
&\le \underline{\lim}\, n\left(\int_b^{b+(1/n)} F(x)dx - \int_a^{a+(1/n)} F(x)dx \right) \\
&\le F(b) - \overline{\lim}\, F(a + 1/n) \\
&\le F(b) - F(a).
\end{aligned}$$

Since $\int_a^b F'(x)\,dx < \infty$ it follows that $F(x)$ has a finite derivative a.e.

Note. We define a function $F(I)$ to be super additive if, whenever $I_1 \cup I_2 = I$, with I_1, I_2 pairwise disjoint, $F(I) \geq F(I_1) + F(I_2)$. Then the first part of the theorem shows that the derivative of a non-negative, super additive function of an interval exists a.e. Note, however, that a non-negative, non-decreasing function of an interval can have an infinite derivate everywhere as in the case $F(I) = |I|^{1/2}$.

A function $F(x)$ is said to be of bounded variation on $[a,b]$ if there is a number M such that for each collection of pairwise non-overlapping intervals $\{I_k\}$ contained in $[a,b]$, $\Sigma|F(I_k)| < M$. We write $F \in BV$ on $[a,b]$ and call $V(F;a,x) = \sup \Sigma|F(I_k)|$ the variation of F on $[a,x]$; here the I_k are pairwise non-overlapping and contained in $[a,x]$. Similarly, $\overline{V}(F;a,x) = \sup \Sigma F(I_k)$ and $\underline{V}(F;a,x) = \sup \Sigma(-F(I_k))$ with I_k non-overlapping, are the upper and lower variations of F on $[a,x]$. These are abbreviated $V(x)$, $\overline{V}(x)$, $\underline{V}(x)$.

Note that if F and G are BV on $[a,b]$, so are constant multiples of F and so are $F + G$ and, less obviously, $F \cdot G$. To see that $F \cdot G \in BV$, note first that $F \in BV$ implies F is bounded and thus $F^2 \in BV$ since

$$\Sigma|F^2(x_k) - F^2(x_k')| = \Sigma|F(x_k) + F(x_k')| \cdot |F(x_k) - F(x_k')| \leq 2M \cdot V(F;a,b)$$

where $|F(x)| \leq M$ on $[a,b]$. That $F \cdot G \in BV$ follows from the equation $4F \cdot G = (F + G)^2 - (F - G)^2$.

(8.3) Theorem. If $F \in BV$ on $[a,b]$, then $F(x) - F(a) = \overline{V}(x) - \underline{V}(x)$ and $V(x) = \overline{V}(x) + \underline{V}(x)$; $\overline{V}$ and $\underline{V}$

<u>are monotone non-decreasing. Furthermore,</u> $F(x)$ <u>is differentiable</u> a.e. <u>and</u> $F'(x) \in \mathcal{L}$.

<u>Proof</u>. Let $x \in [a,b]$ and P partition $[a,x]$ and let $\overline{V}_P = \Sigma' F(I_k)$ and $\underline{V}_P = \Sigma'' -F(I_k)$ where Σ' is over those I_k with $F(I_k) \ge 0$ and Σ'' over those I_k with $F(I_k) \le 0$. Then $F(x) - F(a) = \overline{V}_P - \underline{V}_P$. Thus $F(x) - F(a) \le \overline{V}(x) - \underline{V}_P$ and $F(x) - F(a) \ge \overline{V}_P - \underline{V}(x)$. Since this is true for all partitions P, $F(x) - F(a) = \overline{V}(x) - \underline{V}(x)$. Clearly $V(x) \le \overline{V}(x) + \underline{V}(x)$ and $V(x) \ge \overline{V}_P + \underline{V}_P = 2\overline{V}_P - (F(x) - F(a))$. Since this is true for all P, $V(x) \ge 2\overline{V}(x) - (F(x) - F(a)) = \overline{V}(x) + \underline{V}(x)$. Therefore $V(x) = \overline{V}(x) + \underline{V}(x)$. That $\overline{V}(x)$ and $\underline{V}(x)$ are monotone is clear and it follows from Theorem 8.2 that $F(x)$ is differentiable a.e. and $F'(x) \in \mathcal{L}$.

A function $F(x)$ is said to be <u>absolutely continuous</u> on $[a,b]$ if, given any $\varepsilon > 0$ there is a $\delta > 0$ so that if $\{I_k\}$ is any collection of non-overlapping intervals contained in $[a,b]$, $\Sigma|I_k| < \delta$ implies $\Sigma|F(I_k)| < \varepsilon$. Then F is said to belong to AC on $[a,b]$. A function $F(x)$ is said to satisfy <u>Lusin's condition</u> (N) on $[a,b]$ if, whenever $Z \subset [a,b]$, $|Z| = 0$ implies $|F(Z)| = 0$.

Note that the sum, product, and constant multiples of AC functions can be shown to be AC in the same way as was done for functions of bounded variation. Theorem 8.4 begins to show the relationship between these concepts.

(8.4) Theorem. If $F \in AC$ on $[a,b]$, then F is continuous, BV, and satisfies condition (N). (The converse is true and will be shown shortly.)

Proof. Continuity follows from the definition using a single interval I; that is, for each $\varepsilon > 0$ there is a $\delta > 0$ so that $|I| < \delta$ implies $|F(I)| < \varepsilon$. To see that $F \in BV$, let $\varepsilon = 1$ and select a natural number N so that $(b - a)/N < \delta = \delta(\varepsilon)$. Then if $[a,b]$ is partitioned into N equal length intervals $J_1, J_2, \ldots, J_N$, for any non-overlapping intervals $\{I_k\}$,

$$\sum_k |F(I_k)| \le \sum_k \sum_{n=1} |F(I_k \cap J_n)| \le \sum_{n=1}^{N} \sum_k |F(I_k \cap J_n)| \le N$$

since $\sum_k |F(I_k \cap J_n)| \le 1$ for each $n = 1, 2, \ldots, N$. To see that F satisfies condition (N), let Z be a set of measure 0 and G be an open set with $Z \subset G$ and $|G| < \delta = \delta(\varepsilon)$. Then if $G = \cup I_k$,

$$|F(Z)| \le |F(G)| \le \Sigma m(F(I_k)) \le \varepsilon$$

and ε being arbitrary, it follows that $|F(Z)| = 0$ and $F \in (N)$.

In fact, absolute continuity is precisely the concept needed to characterize those functions F which satisfy for some integrable f, $F(I) = \int_I f(t)\,dt$. Such functions F are called "primitives" for the Lebesgue integral. Theorem 8.5 shows that these primitives are absolutely continuous and that, if $F(I) = \int_I f(t)\,dt$, then $F'(x) = f(x)$ a.e.

(8.5) Theorem. If $f(x)$ is Lebesgue integrable on $[a,b]$ and $F(x) = \int_a^x f(t)\,dt$ then $F(x) \in AC$ on $[a,b]$ and $F'(x) = f(x)$ a.e.

Proof. Fix $\varepsilon > 0$. Let

$$E_n = \{x \in [a,b]: |f(x)| > n\}.$$

Then $\{E_n\}$ is a non-increasing sequence of sets and $m(\cap E_n) = 0$. By the monotone convergence theorem

$$\int_{E_n} |f(t)|dt = \int_a^b |f(t)|\cdot C_{E_n}(t)\,dt$$

approaches 0. Thus there is an N so that $\int_{E_N} |f(t)|dt < \varepsilon/2$. Let $\delta = \varepsilon/2N$. Then if $\Sigma|I_k| < \delta$ and the I_k are non-overlapping,

$$\begin{aligned}\sum_k |F(I_k)| &= \sum_k \left|\int_{I_k} f(t)\,dt\right| \\ &\le \sum_k \int_{I_k\cap E_N} |f(t)|dt + \sum_k \int_{I_k\setminus E_N} |f(t)|dt \\ &\le \frac{\varepsilon}{2} + N\cdot\Sigma|I_k| < \varepsilon.\end{aligned}$$

Thus F is absolutely continuous.

To see that $F'(x) = f(x)$ a.e., suppose $\{x: F'(x) \ne f(x)\}$ has positive measure. Let $E_{pq} = \{x: F'(x) > q > p > f(x)\}$ for rational numbers p, q with $p < q$. Given $\varepsilon > 0$, as shown above there is $\delta > 0$ so that $|E| < \delta$ implies $\int_E |f(t)|dt < \varepsilon$. Let G be an open set containing E_{pq} so that $|G\setminus E_{pq}| < \delta$. If $x \in E_{pq}$ there is $h' > 0$ so that for $0 < h < h'$, $(F(x+h) - F(x))/h > q$. Thus, these intervals $[x, x+h]$ which are contained in G cover E_{pq} in the sense of Vitali. Thus, there exist $I_1, I_2, \ldots, I_n, \ldots$ pairwise disjoint intervals so that $|E_{pq} \setminus \cup I_n| = 0$. Then

$$\int_{I_n} f(t)\,dt = F(I_n) \geq q|I_n|$$

so that

$$\begin{aligned}\int_{E_{pq}} f(t)\,dt &\geq \int_{\cup I_n} f(t)\,dt - \varepsilon \\ &\geq q\cdot\Sigma|I_n| - \varepsilon \\ &\geq q|E_{pq}| - 2\varepsilon.\end{aligned}$$

However $f(t) < p$ on E_{pq} implies $\int_{E_{pq}} f(t)\,dt \leq p|E_{pq}|$. Since $\varepsilon > 0$ is arbitrary, $q|E_{pq}| \leq p|E_{pq}|$ and thus $|E_{pq}| = 0$. Thus $\{x: F'(x) > f(x)\}$ has measure 0. That $\{x: F'(x) < f(x)\}$ has measure 0 follows from the above argument applied to $G(x) = -F(x)$ and $g(x) = -f(x)$.

Before proving that absolutely continuous functions are the Lebesgue integral of their a.e. derivatives (the converse of the previous theorem), we develop three theorems which facilitate the proof and which will be used later.

(8.6) Theorem. *If F is a continuous function on a measurable set D, a necessary and sufficient condition that for each measurable $E \subset D$, $F(E)$ is measurable is that F satisfy (N) in D.*

Note. It is clear from the proof and from Lusin's Theorem that continuity can be replaced by measurability in the statement of Theorem 8.6.

Proof. If $F \in (N)$ and $E \in \mathcal{M}$, then $E = \cup E_n \cup Z$ where the E_n are compact and $|Z| = 0$. Since F is continuous on D, $F(E_n)$ is compact and $F \in (N)$ implies $m(F(Z)) = 0$. Thus $F(E) = \bigcup_n F(E_n) \cup F(Z)$ is measurable. Conversely, if $F \notin (N)$ on D, F takes a set $Z \subset D$ with $|Z| = 0$ onto a set P of positive outer measure. Since P contains a non-measurable subset P', F takes $F^{-1}(P') \cap Z$ into P' and hence does not take measurable sets into measurable sets.

(8.7) Theorem. If D is a measurable set and F is any function which is differentiable on D such that $|F'(x)| \le M$ at each point of D, then $F(D) \in \mathcal{M}$ and $m(F(D)) \le M|D|$.

Proof. Given $\varepsilon > 0$, let
$D_n = \{x \in D: x \in I \text{ and } |I| < 1/n \Rightarrow |F(I)/|I|| < M + \varepsilon\}$.
Then $D_n \subset D_{n+1}$ and $D = \cup D_n$. Fix n. Then D_n is contained in a sequence of non-overlapping intervals $\{I_k\}$ with $|I_k| < 1/n$ so that $\Sigma|I_k| < m^*(D_n) + \varepsilon$. These can be obtained by partitioning the component intervals of an open set containing D_n. Then $m^*(F(D_n)) \le \sum_k m^*(F(D_n \cap I_k)) \le \sum_k m^*(F(I_k)) \le (M + \varepsilon)\Sigma|I_k| \le (M + \varepsilon)(m^*(D_n) + \varepsilon)$. Letting $\varepsilon \to 0$ and then $n \to \infty$, we have $m^*(F(D)) \le M|D|$. It follows that if $|Z| = 0$, $Z \subset D$, $m(F(Z)) = 0$ and thus F satisfies (N) in D. Since F is continuous on D, $F(D)$ is measurable and $m(F(D)) \le M|D|$.

(8.8) Corollary. Differentiable functions satisfy Lusin's condition (N).

(8.9) <u>Corollary</u>. <u>If</u> $F'(x) = 0$ <u>on a set</u> E, $m(F(E)) = 0$.

(8.10) <u>Theorem</u>. <u>If</u> F <u>is derivable on</u> $D \in \mathcal{M}$, $m(F(D)) \le \int_D |F'(x)|dx$.

<u>Proof</u>. Given $\varepsilon > 0$ and $|D| < \infty$, let $D = \cup D_n$ where $D_n = \{x \in D: (n - 1)\varepsilon \le |F'(x)| < n\varepsilon\}$. Then

$$\begin{aligned} m(F(D)) &\le \Sigma m(F(D_n)) \le \Sigma n\varepsilon|D_n| \\ &\le \Sigma \int_{D_n} (\varepsilon + |F'(x)|)dx \\ &\le \int_D |F'(x)|dx + \varepsilon|D|. \end{aligned}$$

Since $\varepsilon > 0$ is arbitrary, $m(F(D)) \le \int_D |F'(x)|dx$. Since this holds on sets of finite measure, it also holds on sets of infinite measure.

We now proceed to the converse of Theorem 8.5.

(8.11) <u>Theorem</u>. <u>If</u> $F(x)$ <u>and</u> $G(x)$ <u>are</u> AC <u>on</u> $[a,b]$ <u>and</u> $F'(x) = G'(x)$ a.e. <u>on</u> $[a,b]$, <u>then</u> $F(x) - G(x) \equiv C$ <u>on</u> $[a,b]$. <u>Moreover</u>, <u>if</u> $f(t) = F'(t)$ a.e., <u>then</u> $F(x) - F(a) = \int_a^x f(t)\, dt$.

<u>Proof</u>. If $F, G \in AC$, then $H = F - G$ does too and $H'(x) = 0$ a.e. on $[a,b]$ since $H'(x) = F'(x) - G'(x)$ a.e. on $[a,b]$. Let D be the set of points of differentiability of H and N the set of points of non-differentiability. Then by Corollary 8.9 $|H(D)| = 0$ and since H satisfies (N), $|H(N)| = 0$. Thus $|H([a,b])| \le |H(D)| + |H(N)| = 0$. Since H is continuous, H is a constant. Suppose $F(x)$ is AC

and $f(t) = F'(t)$ a.e. Let $G(x) = \int_a^x f(t)\,dt$. Then G is AC and $G'(x) = f(x)$ a.e. Hence $G - F \equiv C$ on $[a,b]$. But then $G(x) = \int_a^x f(t)\,dt = G(x) - G(a) = F(x) - F(a)$.

We now provide the proof of the converse of Theorem 8.4.

(8.12) _Theorem_. _If_ $F(x)$ _is a continuous function of bounded variation which satisfies_ (N) _on_ $[a,b]$, _then_ F _is_ AC _on_ $[a,b]$.

Proof. Let $D = \{x\colon F'(x) \text{ exists}\}$ and

$$N = \{x\colon F'(x) \text{ does not exist}\}.$$

For $[c,d] \subset [a,b]$, $|F(d) - F(c)| \leq |F(D \cap [c,d])| + |F(N)|$. But $|N| = 0$ and F satisfies (N) implies $|F(N)| = 0$. Let $G(I) = \int_I |F'(x)|dx$. Then $G(I)$ is AC. By Theorem 8.10 and the above, for each $I = [c,d] \subset [a,b]$, $|F(I)| \leq |G(I)|$. It follows that F is also AC on $[a,b]$.

A function S is said to be _singular_ if S is of bounded variation and $S'(x) = 0$ a.e. Functions of bounded variation can be written as the sum of a singular function and an absolutely continuous one. This is known as the _Lebesgue decomposition of a function_ $F \in BV$.

(8.13) _Theorem_. _Each function of bounded variation on_ $[a,b]$ _is the sum of an absolutely continuous function_

<u>and a singular function. This decomposition is unique up to additive constants.</u>

<u>Proof</u>. For $F \in BV$, let $F_1(x) = \int_a^x F'(t)\,dt$ and $S_1(x) = F(x) - F_1(x)$. Then $F_1 \in AC$ and since $F_1'(t) = F'(t)$ a.e., $S_1'(x) = 0$ a.e. and S_1 is singular. If $F = F_2 + S_2$ with $F_2 \in AC$ and S_2 singular, then $F_2 - F_1 = S_1 - S_2$; $F_2 - F_1 \in AC$ and has derivative 0 a.e. Thus $F_2 - F_1 \equiv C$ and hence $S_1 - S_2 \equiv C$.

A function is absolutely continuous if and only if it is the Lebesgue integral of its almost everywhere derivative. Moreover, absolute continuity, besides characterizing the primitives for the Lebesgue integral, is precisely the condition needed to make several formulas hold. Namely, if F is a function of bounded variation, the variation of F on I is given by $\int_I |F'(x)|dx$ and the arc length is given by

$$\int_I \sqrt{1 + F'(x)^2}\,dx$$

if and only if F is absolutely continuous.

(8.14) <u>Theorem</u>. <u>If</u> F <u>is</u> BV <u>on</u> $[a,b]$, <u>then</u> $V'(x) = |F'(x)|$ a.e. <u>on</u> $[a,b]$. <u>Moreover</u>, <u>a function</u> F <u>is</u> AC <u>on</u> $[a,b]$ <u>if and only if</u>

$$V(F;a,b) = \int_a^b |F'(x)|dx < \infty.$$

<u>Proof</u>. Suppose F is BV on $[a,b]$ and recall that $F(x) - F(a) = \overline{V}(x) - \underline{V}(x)$. Let $\overline{V} = U_1 + S_1$, $\underline{V} = U_2 + S_2$ be the Lebesgue decompositions of $\overline{V}$ and $\underline{V}$. Then

$T = S_1 + S_2$ is the singular part of V. Note that S_1 and S_2 are non-negative functions of an interval. Since for each interval $J \subset I$, $F(J) = \int_J F'(t)\,dt + S_1(J) - S_2(J)$, it follows by summing over a partition of I that

$$V(F;I) \leq \int_I |F'(t)|dt + T(I)$$

and by differentiating both sides of this inequality that

$$V'(x) \leq |F'(x)| \text{ a.e.}$$

Since $V(I) = \overline{V}(I) + \underline{V}(I) \geq |F(I)|$, $V'(x) \geq |F'(x)|$ a.e. Hence, $V'(x) = |F'(x)|$ a.e. and since $F \in BV$, applying Theorem 8.2 to $V(x)$ yields,

$$V(F;a,b) \geq \int_a^b |F'(x)|dx$$

with equality holding if and only if V is AC, that is, if and only if the Lebesgue decomposition of V has no singular part. Then, if F is AC, V is AC and

$$V(b) - V(a) = V(F;a,b) = \int_a^b V'(x)dx = \int_a^b |F'(x)|dx < \infty.$$

Conversely, if $V(F;a,b) = \int_a^b |F'(x)|dx < \infty$ then the Lebesgue decomposition of V has no singular part; that is, $V(x)$ is AC on $[a,b]$ and, since on each interval $I \subset [a,b]$, $|F(I)| \leq V(I)$, it follows that F is AC on $[a,b]$.

Now consider $F(x)$ defined on $[a,b]$ and let

$$H(I) = \left(F(I)^2 + |I|^2\right)^{1/2}$$

(H is not, in general, an additive function of an interval.) Let $L(J) = \sup \Sigma H(I_k)$ where Σ is over a partition $P = \{x_o < x_1 < \ldots < x_n\}$, $J = [x_o, x_n]$, $I_k = [x_{k-1}, x_k]$ and the supremum is over all partitions P of J. By noting that the sum Σ is increased by the addition to P of a single point, it is clear that $L(J) = L(J_1) + L(J_2)$ when $J = J_1 \cup J_2$ and $J_1 \cap J_2$ consists of a single point. Since

$$\begin{aligned} V(J) &= \sup \Sigma |F(I_k)| \le L(J) \\ &\le \sup \Sigma\big(|F(I_k)| + |I_k|\big) = V(J) + |J|, \end{aligned}$$

it is clear that $L(I)$ is finite iff $F \in BV$ on I and $L \in AC$ on I iff $F \in AC$ on I. If $F \in BV$ and F is continuous, $L(I)$ is the definition of the length of the graph of F on I.

(8.15) Theorem. *Given* $F(x)$ *of bounded variation on* $[a,b]$,

$$\int_a^b (F'(x)^2 + 1)^{1/2} dx \le L([a,b])$$

with equality holding iff $F \in AC$ *on* $[a,b]$.

Proof. If $F \in BV$, $L(I)$ is a finite, additive, non-negative function of an interval and thus is differentiable a.e. Since $L(I) \ge H(I)$ and

$$H'(x) = (F'(x)^2 + 1)^{1/2} = \lim_{I \to x} \sqrt{\left(\frac{F(I)}{|I|}\right)^2 + 1}$$

holds at every point where $F'(x)$ exists, $L'(x) \ge H'(x)$ a.e. To see that $L'(x) = H'(x)$ a.e., let $E = \{x\colon L'(x)$ and $H'(x)$ exist and $L'(x) > H'(x)\}$ and let

$$E_n = \{x \in E\colon \text{if } x \in I \text{ and } |I| < 1/n, \text{ then } L(I) > H(I) + |I|/n\}.$$

Then $E = \cup E_n$. Fix $\varepsilon > 0$ and let P be a partition of $[a,b]$ with $|P| < 1/n$ such that $\sum_P H(I_k) > L([a,b]) - \varepsilon$. Let Σ' be the summation over all $I_k = [x_{k-1}, x_k]$ with $x_k \in P$ and $I_k \cap E_n \ne \emptyset$. Then

$$|E_n| \le \Sigma'|I_k| \le \Sigma' n(L(I_k) - H(I_k))$$
$$\le n\cdot\sum_P (L(I_k) - H(I_k)) \le n\varepsilon.$$

Since $\varepsilon > 0$ was arbitrary, $|E_n| = 0$ and hence $|E| = 0$. Now

$$L([a,b]) \ge \int_a^b L'(x)\,dx = \int_a^b \sqrt{F'(x)^2 + 1}\,dx$$

with equality holding iff L is absolutely continuous; that is, iff $F \in AC$ on $[a,b]$.

There are differentiable functions defined on an interval $[a,b]$ which are not of bounded variation. A standard example of such a function is $F(x) = x^a\cos(x^{-b})$ for $x \in (0,1]$, $F(0) = 0$. If $1 < a \le b$, then F is differentiable because $|F(x)| \le x^a$ and hence $F'(0) = 0$. However, since $|\cos(n\pi)| = 1$ and $\cos(n + 1/2)\pi = 0$, if $x_n = (n\pi)^{-1/b}$, $|F(x_n)| = (n\pi)^{-a/b}$ and $V(F;(0,1)) = \Sigma|F(x_n)| = \pi^{-a/b}\cdot\Sigma n^{-a/b} = \infty$ because $a/b \le 1$. Since such an F is not of bounded variation, F is not the integral of its derivative. Historically, this fact, along with the fact that two differentiable functions which have the same derivative must differ by a constant, suggested the existence of a larger integral which integrated both Lebesgue integrable functions and derivatives. The resulting extensions of the Lebesgue integral will be the subject matter of the next chapter. At this point, note that if $F(x)$ is differentiable on $[a,b]$ and $F'(x) \in \mathcal{L}_{[a,b]}$, then $F(x) - F(a) = \int_a^b F'(x)\,dx$. We can actually prove the following better result:

(8.16) <u>Theorem</u>. <u>Suppose</u> $F(x)$ <u>is continuous on</u> $[a,b]$ <u>and differentiable except on an at most countable set or except on a set</u> N <u>such that</u> $|F(N)| = 0$. <u>If</u> $D = [a,b]\backslash N$ <u>and</u> $F'(x) \in \mathcal{L}$ <u>on</u> D, <u>it follows that</u> $F \in AC$ <u>and hence</u> $F'(x)$ <u>exists</u> a.e. <u>and</u>

$$F(x) - F(a) = \int_a^b F'(t)\,dt.$$

(That AC functions satisfy the hypotheses of this theorem is clear and hence the converse of the theorem is also true.)

<u>Proof</u>. Since F is continuous, in order to show that $F \in AC$, it suffices to show that $F \in BV$ and F satisfies (N). Since $F'(x) \in \mathcal{L}$ on D, by Theorem 8.7 it follows that for each $I \subset [a,b]$

$$m(F(I)) \le m(F(I \cap D)) + m(F(I \cap N)) \le \int_D |F'(x)|dx < \infty.$$

But then, for $\{I_k\}$ pairwise non-overlapping

$$\Sigma|F(I_k)| \le \Sigma m(F(I_k)) \le \int_D |F'(x)|dx < \infty.$$

Thus $F \in BV$. If Z is any set of measure 0, by Corollary 8.9 $m(F(Z \cap D)) = 0$ and by hypothesis $m(F(Z \cap N)) = 0$. Hence $m(F(Z)) = 0$ and it follows that F also satisfies (N).

Having provided several characterizations of absolute continuity, we now consider some important subclasses of the class of absolutely continuous functions. A function F is said to satisfy a <u>Lipschitz Condition</u> and F is said to be Lipschitz on $[a,b]$ provided that is a number M such that for each x and $t \in [a,b]$, $|F(t) - F(x)| \le M|t - x|$. The number M is called the Lipschitz constant for F.

(8.17) Theorem. A function F is Lipschitz on $[a,b]$ if and only if $F'(x)$ exists a.e., $F'(x)$ is bounded on the set of points where it exists and $F(x) - F(a) = \int_a^x F'(t)\,dt$.

Proof. If $F(x)$ is a Lipschitz function on $[a,b]$, it is absolutely continuous on $[a,b]$. This is because given $\varepsilon > 0$ whenever $\{I_n\}$ is a sequence of pairwise non-overlapping intervals with $\Sigma|I_n| < \varepsilon/M$, where M is the Lipschitz constant for F, it follows that $\Sigma|F(I_n)| \le \Sigma M|I_n| < \varepsilon$. Thus Lipschitz functions are differentiable a.e. and $F(x) - F(a) = \int_a^x F'(t)\,dt$. Furthermore $\left|\frac{F(t) - F(x)}{t - x}\right| \le M$ and thus $F'(x)$ is bounded by M on the set of points on which it exists. Conversely, if $F'(x)$ is bounded and exists a.e. and $F(x) - F(a) = \int_a^x F'(t)\,dt$ then $|F(I)| = |\int_I F'(t)\,dt| \le M|I|$. That is, $|F(x) - F(t)| \le M|x - t|$ and F is Lipschitz.

A function F is said to be convex on an interval I if, for each $x_1 < x_2 < x_3$ in I, $F(x_2) \le F(x_1)\frac{x_3 - x_2}{x_3 - x_1} + F(x_3)\frac{x_2 - x_1}{x_3 - x_1}$. That F is convex means that $(x_2,F(x_2))$ is below or on the line segment connecting $(x_1,F(x_1))$ and $(x_3,F(x_3))$. If this line segment is called a secant line, F convex means that all secant lines are above or on the graph of F. If F is also differentiable on I, F is frequently referred to as being concave up on I. Convex functions on an interval I can be discontinuous at the endpoints of the interval I and at these points only. The following theorem characterizes convex functions.

(8.18) Theorem. *If F is convex on $[a,b]$, then F is continuous on (a,b) and differentiable except at an at most countable set of points. The right and left derivates exist at every point of (a,b) and are monotone non-decreasing functions of x. Conversely, if $f(x)$ is monotone non-decreasing on (a,b) and if $x_o \in (a,b)$ and $F(x) - F(x_o) = \int_{x_o}^{x} f(t)\,dt$ for $x \in (a,b)$ then $F(x)$ is convex on (a,b).*

Proof. For $x_1 < x_2 < x_3$ in (a,b) and F convex on (a,b), it follows from the definition of convex that

$$\frac{F(x_2) - F(x_1)}{x_2 - x_1} \le \frac{F(x_3) - F(x_1)}{x_3 - x_1} \le \frac{F(x_3) - F(x_2)}{x_3 - x_2}.$$

From these inequalities, we observe that as $x = x_1$ is approached from the right, the difference quotients are non-increasing and bounded below by $\dfrac{F(x_1) - F(x_o)}{x_1 - x_o}$ where x_o is any point to the left of x_1. Thus, $\overline{F}^+(x) = \underline{F}^+(x)$ at every point $x \in (a,b)$; similarly $\overline{F}^-(x) = \underline{F}^-(x)$ at every point x in (a,b). Clearly, $\overline{F}^+(x) = \underline{F}^+(x) \ge \underline{F}^-(x) = \overline{F}^-(x)$ and each of the left and right sided derivates is monotone non-decreasing. Thus the function F is continuous on the right and on the left at each point x in (a,b) and thus is continous on (a,b). By Theorem 8.1 the set of points of non-differentiability of F is at most countable. Now suppose $f(x)$ is a monotone non-decreasing function on (a,b). Let $x_o \in (a,b)$ and

$$F(x) - F(x_o) = \int_{x_o}^{x} f(t)\,dt$$

for each $x \in (a,b)$. The existence of the integral of f is guaranteed by the monotonicity of f. Let $x_1 < x_2 < x_3$ belong to (a,b). Then, by the monotonicity of f, it follows that

$$(1): \quad f(x_2)(x_2 - x_1) \geq F(x_2) - F(x_1)$$

and

$$(2): \quad f(x_2)(x_3 - x_2) \leq F(x_3) - F(x_2).$$

This is because for x in (x_1,x_2), $f(x) \leq f(x_2)$ and $F(x_2) - F(x_1) = \int_{x_1}^{x_2} f(t)\,dt \leq f(x_2)(x_2 - x_1)$. The second inequality is due to the fact that for $x \in (x_2,x_3)$, $f(x) \geq f(x_2)$. From (2) it follows that $F(x_3) - F(x_1) \geq F(x_2) - F(x_1) + f(x_2)(x_3 - x_2)$. Combining this with (1) yields $F(x_3) - F(x_1) \geq F(x_2) - F(x_1) + (F(x_2) - F(x_1))\cdot\dfrac{x_3 - x_2}{x_2 - x_1}$. That is,

$$\frac{F(x_3) - F(x_1)}{x_3 - x_1} \geq \frac{F(x_2) - F(x_1)}{x_2 - x_1}.$$

This last inequality implies that $(x_2, F(x_2))$ lies on or below the line segment connecting $(x_1, F(x_1))$ to $(x_3, F(x_3))$ and hence that $F(x)$ is convex.

8.1 Exercises

1. If $f(x)$ is not of bounded variation on $[0,1]$, show that there is a monotone sequence $x_i \in [0,1]$ so that $\Sigma|f(x_{i+1}) - f(x_i)| = \infty$.
2. Construct an example of a function which is continuous and of bounded variation on $[0,1]$ but is not monotone on any interval.
3. Suppose $\{F_n\}$ is a non-decreasing sequence of non-decreasing absolutely continuous functions defined on $[0,1]$. Prove that if $\sum_{n=1}^{\infty} F_n(x)$ converges to $F(x)$, then $F(x)$ is a non-decreasing absolutely continuous function.
4. Suppose $F'(x) > 0$ on $E \in \mathcal{M}$ with $m(E) > 0$. Prove that $m(F(E)) > 0$.
5. If $f \in \mathcal{L}_{[a,b]}$ a point $x \in (a,b]$ is called a <u>Lebesgue point</u> of f if

$$\lim_{h \to 0} \frac{1}{h} \int_x^{x+h} |f(t) - f(x)|dt = 0.$$

   Show that if $f \in \mathcal{L}_{[a,b]}$ then almost every point of $[a,b]$ is a Lebesgue point of f.
6. Suppose F and G are additive functions of an interval on $[a,b]$. Show that if G is of bounded variation on $[a,b]$ and $F(I) \geq G(I)$ on each $I \subset [a,b]$, then F is of bounded variation.
7. Let F, G, H be additive functions of an interval and suppose that for each I, $F(I) \leq G(I) \leq H(I)$. Prove that if F and H are continuous, then G is continuous and if F and H are absolutely continuous, then G is also absolutely continuous.

8. If $f(x)$ is increasing on $[0,1]$, show that $F(x) = \int_0^x f(t)dt$ is concave up on $[0,1]$.
9. Suppose $F(x)$ is continuous on $[0,1]$. Show F is of bounded variation iff $\int_{-\infty}^{\infty} S(y)dy < \infty$ where $S(y) = \|\{x: f(x) = y\}\|_c$; here $S(y) = \infty$ if the cardinality is infinite. The function $S(y)$ is called the <u>Banach indicatrix</u> of F.
10. Let $\{x_n\}$ be a sequence of points and $\{a_n\}$, $\{b_n\}$ satisfy $\Sigma|a_n| < \infty$, $\Sigma|b_n| < \infty$. A function of the form $j(x) = \sum_{x_n<x} a_n + \sum_{x_n \le x} b_n$ is called a <u>jump function</u>. Show that every jump function is a singular function.
11. Show that every singular function $S(x)$ on $[0,1]$ can be written as $j(x) + S(x)$ where $j(x)$ is a jump function and $S(x)$ is a continuous singular function.

8.2 Differentiation of Functions of a Set

We begin this section with further consideration of completely additive functions of a set. Suppose that Φ is such a function and that Φ is defined on the σ-algebra of subsets of a set X which are also the measurable sets for a given measure m. Then Φ is said to be <u>absolutely continuous with respect to</u> m provided that for each set Z with $m(Z) = 0$, we have $\Phi(Z) = 0$. Due to the Jordan decomposition theorem, in considering absolute continuity it will frequently only be necessary to study a measure m_1 which is absolutely continuous with respect to a given measure m. This is because $\Phi = \bar{V}_\Phi + \underline{V}_\Phi$, both $\bar{V}_\Phi$ and $-\underline{V}_\Phi$ are measures and it is easily seen that Φ is absolutely continuous with respect to m if and only if both $\bar{V}_\Phi$ and $\underline{V}_\Phi$ are absolutely continuous with respect to m.

The definition of absolute continuity which was just given for completely additive functions of a set is analogous to Lusin's condition (N) for functions of a real variable. If our concern is restricted to functions Φ with $V_\Phi(X) < \infty$, an alternate definition can be given which is analogous to that of absolute continuity for functions of a real variable which are of bounded variation.

Alternate definition: A countably additive function of a set Φ with $V_\Phi(X) < \infty$ is absolutely continuous with respect to m provided that for each $\varepsilon > 0$ there is a $\delta > 0$ so that, if $\{E_i\}$ is a sequence of pairwise disjoint measurable sets, then $\Sigma m(E_i) < \delta$ implies $\Sigma|\Phi(E_i)| < \varepsilon$; due to the countable additivity of Φ and m, this is equivalent to $m(E) < \delta$ implies $|\Phi(E)| < \varepsilon$.

Note that the alternate definition implies the original one; for if $m(Z) = 0$, then for any $\delta > 0$ it follows that $m(Z) < \delta$ and thus for every $\varepsilon > 0$ it follows that $|\Phi(Z)| < \varepsilon$ and $\Phi(Z) = 0$. To see that the definitions are equivalent when $|\Phi(X)| < \infty$, we need only consider the case where Φ is non-negative; that is, the case where Φ is a bounded measure. Suppose then that the original definition holds but the alternate does not. Then there is a number $\varepsilon > 0$ and a sequence of sets $\{E_n\}$ such that $m(E_n) < 2^{-n}$ but $\Phi(E_n) \geq \varepsilon$. Since $\Phi(X) < \infty$, we have $\Phi(\overline{\lim}\, E_n) \geq \overline{\lim}\, \Phi(E_n) \geq \varepsilon$. However, $m(\overline{\lim}\, E_n) = 0$ because, for each natural number N, $\overline{\lim}\, E_n \subset \bigcup_{n=N}^{\infty} E_n$ and thus $m(\overline{\lim}\, E_n) \leq \sum_{n=N}^{\infty} m(E_n) \leq 2^{-N+1}$. If $Z = \overline{\lim}\, E_n$, then $m(Z) = 0$ and $\Phi(Z) > 0$ contrary to the assumption that Φ satisfies the original definition of absolute continuity with respect to a measure m. It follows that the two definitions agree on sets X whenever Φ satisfies $V_\Phi(X) < \infty$.

Note that, if a measure is defined by $m_2(E) = \int_E f(x)\, dm$ where f is a non-negative measurable function, then m_2 is absolutely continuous with respect to m; this is because if $m(Z) = 0$, then $m_2(Z) = \int_Z f(x)\, dm = 0$. The Radon-Nikodym theorem, which will be proved shortly, implies that any m_2 is absolutely continuous with respect to m on a space X of σ-finite m-measure if and only if there is a non-negative measurable function f such that $m_2(E)$ is given by $\int_E f(x)\, dm$. The proof of this theorem is more easily understood if it is preceeded by a few observations. Suppose $m(X) < \infty$ and m_2 is defined by $m_2(E) = \int_E f(x)\, dm$. Consider the problem of finding a function equal to f almost everywhere using only the measures m and m_2. First of all, if for each pair of non-negative integers k and n the sets

$$A_{k,n} = \{x\colon f(x) \geq k/2^n\}$$

were known, then f would be determined. Indeed, $f(x)$ would equal $\lim f_n(x)$ where $f_n(x) = k/2^n$

when $x \in A_{k,n} \setminus A_{k+1,n}$ and $f_n(x) = \infty$ when $x \in \bigcap_k A_{k,n}$. Moreover, if $\Phi_{k,n}(E) = \int_E (f(x) - k/2^n)dm$ then $A_{k,n}$ and $A_{k,n}^c$ would be a Hahn decomposition of $\Phi_{k,n}$. Since $\Phi_{k,n}(E)$ can be written as $m_2(E) - m(E) \cdot k/2^n$, each $\Phi_{k,n}$ is determined by m_2 and m. Thus m_2 and m determine additive set functions $\Phi_{k,n}$ which have Hahn decompositions $A_{k,n}$ and $A_{k,n}^c$ which are unique up to sets of measure 0 and these in turn determine a function f as the limit of a monotone sequence of functions $\{f_n\}$ each of which is defined almost everywhere in terms of the $A_{k,n}$. The proof of the theorem given below uses these key ideas.

(8.19) <u>Theorem</u>. (Radon-Nikodym Theorem) <u>If m_2 and m are two measures defined on the same σ-algebra of subsets of a set X, if m_2 is absolutely continuous with respect to m and if $m(X) < \infty$, or X is of σ-finite m-measure, then there exists a function f such that for each measurable set E, $m_2(E) = \int_E f(x)\, dm$.</u>

<u>Proof</u>. First suppose $m(X) < \infty$. For each pair of non-negative integers k,n, let $\Phi_{k,n}$ be defined on the measurable sets $E \subset X$ by $\Phi_{k,n}(E) = m_2(E) - m(E) \cdot k/2^n$. Let $A_{0,n} = X$ and $B_{0,n} = \emptyset$ and let $A_{k,n}$ and $B_{k,n}$ be Hahn decompositions of each $\Phi_{k,n}$. Since, for each k and n, $\Phi_{k,n} = \Phi_{2k,n+1}$, the decompositions can be chosen so that $A_{2k,n+1} = A_{k,n}$.

Fix k, n, k' and n' with $k/2^n < k'/2^{n'}$. Let $C = A_{k',n'} \backslash A_{k,n} = A_{k',n'} \cap B_{k,n}$. Because $A_{k',n'} \supset C$, $0 \le \Phi_{k',n'}(C) = m_2(C) - m(C) \cdot k'/2^{n'}$. Because $B_{k,n} \supset C$, $0 \ge \Phi_{k,n}(C) = m_2(C) - m(C) \cdot k/2^n$. Thus $m_2(C) < \infty$ and $m(C) \cdot k'/2^{n'} \le m(C) \cdot k/2^n$. This implies that $m(C) = 0$ because $k/2^n < k'/2^{n'}$. Let Z be the union of all sets of the form $A_{k',n'} \backslash A_{k,n}$ which satisfy $k/2^n < k'/2^{n'}$. Then $m(Z) = 0$. By the absolute continuity of m_2 with respect to m, $m_2(Z) = 0$. Also, if g is any function, $\int_Z g \, dm = 0$. In effect, Z can be removed from the space X; that is, let $X' = X \backslash Z$, $A'_{k,n} = A_{k,n} \backslash Z$ and $B'_{k,n} = B_{k,n} \backslash Z$. Then the sets $A'_{k,n}$ satisfy for each k and n

i) $A'_{0,n} = X'$

ii) $A'_{2k,n+1} = A'_{k,n}$

iii) $A'_{k+1,n} \subset A'_{k,n}$.

Here iii) is due to the removal of the set Z from each $A_{k,n}$ from which it follows that, if $k/2^n < k'/2^{n'}$, then $A_{k',n'} \backslash A_{k,n} \subset Z$ and $A_{k',n'} \backslash Z \subset A_{k,n} \backslash Z$. Now let $f_n(x) = k/2^n$ if $x \in A'_{k,n} \backslash A'_{k+1,n}$ and let $f_n(x) = \infty$ if $x \in \bigcap_k A_{k,n}$. Since

$$f_{n+1}(x) = 2k/2^{n+1} \quad \text{if} \quad x \in A'_{2k,n+1} \backslash A'_{2k+1,n+1}$$

and

$$f_{n+1}(x) = (2k+1)/2^{n+1} \quad \text{if} \quad x \in A'_{2k+1,n+1} \backslash A'_{2k+2,n+1}$$

and

$$f_n(x) = k/2^n \quad \text{if} \quad x \in A'_{2k,n+1} \setminus A'_{2k+2,n+1},$$

it follows that for each $x \in X'$, $f_n(x) \le f_{n+1}(x)$. Thus $\{f_n\}$ is a monotone non-decreasing sequence of non-negative functions, f_n converges to a function f, and, by the monotone convergence theorem, $\int_E f_n(x)\, dm$ approaches $\int_E f(x)\, dm$ for each measurable set $E \subset X'$. Let E be any measurable set. Note that, because of ii) and iii) above, for each m and n, $\bigcap_k A'_{k,n} = \bigcap_k A'_{k,m}$. Let $A = E \cap \bigcap_k A'_{k,m}$. Suppose $m(A) > 0$. Since $f(x) = \infty$ at each $x \in A$, $\int_E f(x)\, dm = \infty$ when $m(A) > 0$. However, since for each k,n,

$$0 \le \Phi_{k,n}(A) = m_2(A) - m(A)\cdot k/2^n,$$

letting k approach ∞, it follows that $m_2(A) = \infty = m_2(E)$ when $m(A) > 0$. The theorem is thus true for E if $m(A) > 0$. If $m(A) = 0$, then $\int_A f(x)\, dm = 0$ and $m_2(A) = 0$ by the absolute continuity of m_2. It remains to consider the set $B = E \setminus A$. Let

$$E_{k,n} = B \cap (A'_{k,n} \setminus A'_{k+1,n}) = B \cap A'_{k,n} \cap B'_{k+1,n}.$$

Since $A'_{k,n} \subset E_{k,n}$, $\Phi_{k,n}(E_{k,n}) \ge 0$ and $m_2(E_{k,n}) \ge m(E_{k,n})\cdot k/2^n$. Since $B'_{k+1,n} \subset E_{k,n}$, $\Phi_{k+1,n}(E_{k,n}) \le 0$ and $m_2(E_{k,n}) \le m(E_{k,n})\cdot(k+1)/2^n$. Consequently,

$$\begin{aligned}
\int_E f_n &= \sum_k m(E_{k,n})\cdot k/2^n \\
&\le \sum_k m_2(E_{k,n}) = m_2(B) \\
&\le \sum_k m(E_{k,n})(k+1)/2^n \\
&= \int_E f_n(x)\, dm + m(E)/2^n.
\end{aligned}$$

Thus $\lim_{n\to\infty} \int_E f_n(x)\, dm = m_2(B)$. Since $\lim_{n\to\infty} \int_B f_n(x)\, dm$ $= \int_B f(x)\, dm$ it follows that $\int_B f(x)\, dm = m_2(B)$. Since $E = A \cup B$, $\int_E f(x)\, dm = m_2(E)$. The theorem is thus true for any measurable set E when $m(X) < \infty$. If X is of σ-finite m-measure, then $X = \cup X_i$ where the X_i are measurable, pairwise disjoint and, for each i, $m(X_i) < \infty$. If m_2 is absolutely continuous with respect to m then there are functions f_i on X_i so that $m_2(E\cap X_i) = \int_{E\cap X_i} f_i(x)\, dm$. Clearly, if $f(x) = f_i(x)$ when $x \in X_i$, then, for any measurable set E, $m_2(E) = \Sigma m_2(E\cap X_i) = \sum \int_{E\cap X_i} f(x)\, dm =$ $\int_E f(x)\, dm$. Thus the theorem holds true for spaces of σ-finite m-measure.

A function f which relates Φ and m by the formula $\Phi(E) = \int_E f(x)\, dm$ is called the Radon-Nikodym derivative of Φ with respect to m. This is indicated by writing $f = d\Phi/dm$. The function f behaves in many ways like a derivative. For example, if m_1 and m_2 are measures which are absolutely continuous with respect to m, then for non-negative numbers a and b it is apparent that the measure $am_1 + bm_2$ is absolutely continuous with respect to m. Furthermore, $d(am_1 + bm_2)/dm = adm_1/dm + bdm_2/dm$. This is because $m_1(E) = \int_E f_1(x)\, dm$ and $m_2(E) =$ $\int_E f_2(x)\, dm$ for appropriately chosen $f_1 = dm_1/dm$ and $f_2 = dm_2/dm$, and hence

$$(am_1 + bm_2)(E) = am_1(E) + bm_2(E)$$

is equal to $\int_E (af_1(x) + bf_2(x))dm$ and

$$d(am_1 + bm_2)/dm = af_1 + bf_2 = adm_1/dm + bdm_2/dm.$$

This formula clearly also holds for completely additive set functions Φ_1 and Φ_2 provided Φ_1 does not take on the value $+\infty$ and Φ_2 the value $-\infty$ or vice-versa.

If m_1 is a measure which is absolutely continuous with respect to m and g is a function which is integrable with respect to m_1, then

$$\int_E g(x)\, dm_1 = \int_E (g(x)\cdot dm_1/dm)dm.$$

To see this, let $f = dm_1/dm$ and E be an m_1-measurable set. Note that $\int C_E(x)\, dm_1 = m_1(E) = \int_E f(x)\, dm = \int C_E(x)\cdot f(x)\, dm$. Thus $\int g(x)\, dm_1 = \int g(x)\cdot f(x)\, dm$ holds when g is the characteristic function of a set. By the additivity of the integral, it also holds for simple functions; that is, if $s(x) = \sum_{i=1}^{n} a_i C_{E_i}(x)$, then $\int s(x)\, dm_1 = \sum_{i=1}^{n} a_i \cdot \int C_{E_i}(x)\, dm_1 = \sum_{i=1}^{n} a_i \cdot \int C_{E_i}(x)\cdot f(x)\, dm = \int s(x)\cdot f(x)\, dm$. By the monotone convergence theorem, the formula holds for non-negative measurable functions. For if $g(x) = \lim s_n(x)$ where $s_n(x)$ is a non-decreasing sequence of non-negative simple functions, then $s_n(x)\cdot f(x)$ is a non-decreasing sequence of non-negative functions and

$$\int g(x)\, dm_1 = \lim_{n\to\infty} \int s_n(x)\, dm_1$$

$$= \lim_{n\to\infty} \int s_n(x)\cdot f(x)\, dm$$

$$= \int g(x)\cdot f(x)\ dm.$$

Finally, by considering the positive and negative parts of g, the formula also holds for functions g which are integrable with respect to m_1.

It follows from the above that, if m_2 is absolutely continuous with respect to m_1 and m_1 is absolutely continuous with respect to m, then $dm_2/dm = (dm_2/dm_1)(dm_1/dm)$. For if $g = dm_2/dm_1$ and $f = dm_1/dm$, then

$$m_2(E) = \int_E g(x)\ dm_1 = \int_E g(x)\cdot f(x)\ dm$$

and hence $dm_2/dm = g(x)\cdot f(x)$. The obvious generalizations with measures replaced by completely additive functions of a set can be shown by considering the positive and negative variations of the functions of a set.

The last property of the Radon-Nikodym derivative can be applied to functions of a real variable. If $F(x)$ is non-decreasing on $[a,b]$, F is AC on $[a,b]$ if and only if the Lebesgue-Stieltjes measure m_F is absolutely continuous with respect to Lebesgue measure. To see this, suppose F is AC. Then for each $\varepsilon > 0$ there is a $\delta > 0$ so that if $\{I_n\}$ is a sequence of pairwise disjoint intervals with $\Sigma|I_n| < \delta$ then $\Sigma F(I_n) < \varepsilon$. If Z is a set of measure 0, Z is contained in an open set G for which $\Sigma|I_n| < \delta$ where $\{I_n\}$ is the collection of component intervals of G. Hence $\Sigma F(I_n) < \varepsilon$ and since $\varepsilon > 0$ was arbitrary, $m_F(Z) = 0$ and m_F is absolutely continuous with respect to Lebesgue measure. Conversely, if m_F is absolutely continuous with respect to Lebesgue measure, F is continuous. For if

F were not continuous at x_o then $m_F(\{x_o\}) = F(x_o+) - F(x_o-) > 0$ contrary to the fact that $m(\{x_o\}) = 0$. The absolute continuity of m_F also implies that F satisfies Lusin's condition (N). Thus F is AC if and only if m_F is absolutely continuous with respect to Lebesgue measure. Thus we have:

(8.20) <u>Theorem</u>. <u>If a non-decreasing function</u> F <u>is</u> AC <u>on</u> $[a,b]$, <u>then every function</u> g <u>which is integrable with respect to</u> m_F <u>satisfies</u> $\int_a^t g(x)\, dm_F = \int_a^t g(x)\cdot f(x)\, dm$ <u>and here</u> $f(x)$, <u>the Radon-Nikodym derivative of</u> m_F <u>with respect to</u> m, <u>equals</u> $F'(x)$ a.e.

<u>Proof</u>. From the above discussion, m_F is absolutely continuous with respect to Lebesgue measure and hence the Radon-Nikodym derivative $f = dm_F/dm$ exists on $[a,b]$ and satisfies $\int_E g(x)\, dm_F = \int_E g(x)\cdot f(x)\, dx$ for each measurable set E and function g which is integrable with respect to m_F. But then $F(t) - F(a) = \int_a^t 1\, dm_F = \int_a^t f(x)\, dm$ and thus $F'(x) = f(x)$ a.e.

If F is a continuous increasing function on $[a,b]$, then any function g which is integrable on $[F(a), F(b)]$ satisfies $\int_{F(a)}^{F(b)} g(u)\, du = \mathcal{LS}\int_a^b g(F(x))\, dF$. To see this note that for any Lebesgue measurable set E, $m(E) = m_F(F^{-1}(E))$. This follows by considering a cover of E with intervals and the corresponding cover of $F^{-1}(E)$. Thus, if $[y_{i-1}, y_i)$ is an interval in the range of g, then

$$m_F\left((g \circ F)^{-1}([y_{i-1}, y_i))\right) = m_F\left(F^{-1}(g^{-1}([y_{i-1}, y_i)))\right)$$
$$= m\left(g^{-1}([y_{i-1}, y_i))\right).$$

The equality of the integrals follows from the equality of approximations using partitions of the range of g. Now, if F is an increasing, absolutely continuous function, the previous theorem shows that

$$\mathcal{LS}\int_a^b g(F(x))\, dF = \int_a^b g(F(x))F'(x)\, dx.$$

That is, for g integrable on $[F(a), F(b)]$ and F increasing and AC on $[a,b]$ the familiar substitution formula holds; that is,

$$(*) \qquad \int_{F(a)}^{F(b)} g(u)\, du = \int_a^b g(F(x))F'(x)\, dx.$$

If F is increasing but not absolutely continuous, then for $g(x) = 1$,

$$\int_{F(a)}^{F(b)} 1\, dm = F(b) - F(a) > \int_a^b F'(x)\, dx.$$

Thus the formula (*) is valid in general for integrable g and increasing F if and only if F is absolutely continuous.

We now consider those additive functions of a set which are complementary to the absolutely continuous ones. A completely additive function of a set θ defined on the σ-algebra of measurable sets of a measure m is said to be <u>singular with respect to</u> m provided there is a set Z with $m(Z) = 0$ such that any measurable set E contained in Z^c satisfies $\theta(E) = 0$. If a function Φ is both absolutely continuous and singular with respect to m, then Φ must be identically 0. The agreement of this definition with the definition of singular for functions of bounded variation of a real variable is the import of the following theorem. Here it suffices to consider a non-decreasing function F which is singular in the sense that $F'(x) = 0$ a.e.

(8.21) Theorem. *Suppose $F(x)$ is a non-decreasing function defined on $[a,b]$. Then m_F is a singular measure with respect to Lebesgue measure if and only if $F'(x) = 0$ a.e.*

Proof. First suppose $F'(x) = 0$ a.e. If

$$Z = \{x\colon F'(x) \neq 0 \text{ or } F'(x) \text{ does not exist}\},$$

then $E \subset Z^c$ implies $m_F(E) = m(F(E)) \le \int_E F'(x)\,dx = 0$ by Theorem 8.10. Thus m_F is singular with respect to Lebesgue measure. To see the converse, suppose F is non-decreasing and Z is a set satisfying $m(Z) = 0$ and $m_F(E) = 0$ whenever $E \subset Z^c$. Since $m_F(Z^c) = 0$, given $\varepsilon > 0$ there is an open set G with $Z^c \subset G$ and $\Sigma F(I_n) < \varepsilon$ where $\{I_n\}$ is the sequence of component intervals of G. Since $\int_{I_n} F'(x)\,dx \le F(I_n)$ it follows that $\int_G F'(x)\,dx \le \Sigma F(I_n) < \varepsilon$ and because $F'(x) \ge 0$ a.e. it follows that $\int_E F'(x)\,dx = 0$ and $F'(x) = 0$ a.e. on E. Thus

$$Z_o = Z \cup \{x\colon F'(x) \neq 0 \text{ or } F'(x) \text{ does not exist}\}$$

is a set of measure 0 and for each $x \in Z_o^c$, $F'(x) = 0$ and thus $F'(x) = 0$ a.e.

The next theorem shows that the decomposition of a function of bounded variation of a real variable into the sum of an AC function and a singular function has an exact analogue for σ-finite measures defined on the same σ-algebra of subsets of a set; this is also known as the Lebesgue decomposition.

(8.22) Theorem. (Lebesgue Decomposition Theorem) *If m and m_1 are two σ-finite measures defined on the same σ-algebra of subsets of a set X, then m_1 can be written as the sum of two measures; the first is absolutely continuous with respect to m and the second is singular with respect to m. Moreover, this decomposition into the sum of an absolutely continuous and a singular measure is unique.*

Proof. Suppose that m and m_1 are given and let $m_2 = m_1 + m$. Then both m and m_1 are absolutely continuous with respect to m_2. By the Radon-Nikodym Theorem there are non-negative functions f and g such that for each measurable set E, $m(E) = \int_E f\, dm_2$ and $m_1(E) = \int_E g\, dm_2$. Let $A = \{x: f(x) > 0\}$ and $B = \{x: f(x) = 0\}$ and let $m_A(E) = m_1(E \cap A)$ and $m_B(E) = m_1(E \cap B)$. Then $m_1 = m_A + m_B$. Also $m(B) = \int_B f\, dm_2 = \int_B 0\, dm_2 = 0$. That is, B is a set of m-measure 0 and, since $E \subset B^c$ implies $m_B(E) = 0$, it follows that m_B is singular with respect to m. To see that m_A is absolutely continuous with respect to m, let Z be a set with $m(Z) = 0$. Since $0 = m(Z) = \int_Z f\, dm_2$ and $f > 0$ on A, it follows that $m_2(Z \cap A) = 0$. But then $m_A(Z) = m_1(A \cap Z) = \int_{A \cap Z} g\, dm_2 = 0$. Since an arbitrary set Z with $m(Z) = 0$ has $m_A(Z) = 0$, it follows that m_A is absolutely continuous with respect to m. To see that this decomposition is unique, suppose that $m_1 = m'_A + m'_B$ where m'_A is absolutely continuous and m'_B is singular with respect to m. Since the space X is of σ-finite m_1 and m measure, X can be written as a countable union of pairwise disjoint sets of finite m and m_1 measure. On each such set, $m_1 = m_A + m_B = m'_A + m'_B$ and the additive function of a set Φ defined on the measurable subsets of that set by $\Phi = m'_A - m_A = m_B - m'_B$ is both absolutely continuous and singular with respect to m. It follows that Φ is identically zero and that $m_A(E) = m'_A(E)$ and $m_B(E) = m'_B(E)$ holds for all m-measurable sets; that is, the Lebesgue decomposition is unique.

An alternate, direct, set-theoretic proof of a slight generalization of this theorem sheds additional light on the Lebesgue decomposition. It requires the following preliminary result.

(8.23) Theorem. *If X is a measure space of σ-finite m-measure, X cannot be written as the pairwise disjoint union of uncountably many measurable sets each of positive measure.*

Proof. If X is of σ-finite m-measure, X can be written as the union of a sequence $\{X_i\}$ of pairwise disjoint sets each of finite measure. Suppose $X = \cup E_\alpha$ where each E_α is measurable and $m(E_\alpha) > 0$. Suppose further that $\cup E_\alpha$ is an uncountable union of sets.

Since each set E_α meets one of the sets X_i in a set of positive measure, there is a natural number k so that $\{E_\alpha: m(E_\alpha \cap X_k) > 0\}$ is an uncountable collection of sets. Then there is a natural number n so that $\{E_\alpha: m(E_\alpha \cap X_k) > 1/n\}$ is an uncountable collection. Since the E_α were supposed measurable and pairwise disjoint, this contradicts the fact that $m(X_k) < \infty$.

The next theorem, which implies the Lebesgue decomposition theorem, shows that the decomposition can be achieved in a slightly more general setting.

(8.24) Theorem. *Suppose m_1 and m are two measures defined on the same σ-algebra of subsets of X. If m_1 is a σ-finite measure, then $m_1 = m_A + m_B$ where m_A and m_B are measures with m_A absolutely continuous and m_B singular with respect to m.*

Proof. Presumably, there are sets Z with $m(Z) = 0$ and $m_1(Z) > 0$; for otherwise, m_1 is absolutely continuous with respect to m. Well order the collection of sets Z_α of m-measure 0 which have positive m_1 measure. Let $E_o = Z_o$. If $A_\alpha = Z_\alpha \setminus \bigcup_{\beta<\alpha} Z_\beta$ is measurable and has positive m_1 measure, let $E_\alpha = A_\alpha$; otherwise, let $E_\alpha = \emptyset$. It follows from the previous theorem that for every ordinal α there are at most countably many non-empty sets in $\{E_\beta: \beta < \alpha\}$. Thus each $A_\alpha = Z_\alpha \setminus \bigcup_{\beta<\alpha} Z_\beta$ is measurable and the E_α which are not empty form an at most countable collection of sets each of m-measure 0. Thus $B = \cup E_\alpha$ is measurable and $m(B) = 0$. Let $m_A(E) = m_1(E \setminus B)$ and

$m_B(E) = m_1(E \cap B)$. Clearly, $m_B(E) = 0$ if $E \subset B^c$ and since $m(B) = 0$, m_B is singular with respect to m. To see that m_A is absolutely continuous with respect to m, let Z be a set with $m(Z) = 0$. If $m_A(Z)$ were greater than 0, since $m_A(Z \cap B) = m_1(\emptyset) = 0$, there would be an ordinal number α_o so that $Z_{\alpha_o} = Z \backslash B$. But it is not possible that $m_A(Z_{\alpha_o}) = m_1(Z_{\alpha_o})$ be positive; for if it were positive, then $Z_{\alpha_o} \subset \bigcup_{\beta \le \alpha_o} E_\beta \subset B$ and $m_A(Z_{\alpha_o}) = 0$. Consequently, for each Z with $m(Z) = 0$ so is $m_A(Z) = 0$ and thus m_A is absolutely continuous with respect to m.

Note that the σ-finite condition on m is not present in the above theorem; this condition cannot be deleted from the statement of the Radon-Nikodym theorem. An example showing this is as follows: Let m_1 be Lebesgue measure on the line. Define m on the Lebesgue measurable sets by $m(E) = n$ if there are n elements in E and $m(E) = \infty$ if E is an infinite set. Since $m(E) = 0$ only if $E = \emptyset$ and then $m_1(E) = 0$, m_1 is absolutely continuous with respect to m. However, there is no function f such that for each set E, $m_1(E) = \int_E f\, dm$. For if there were, letting $E_x = \{x\}$, since $m_1(E_x) = 0$ and $m(E_x) = 1$, f would have to be 0 at x. That is, f would be identically 0 and if $m_1(E) > 0$, then $m_1(E) \neq \int_E f\, dm = 0$.

The comparison of measures in a general setting will be completed with the following result.

(8.25) <u>Theorem</u>. <u>Suppose</u> m_1 <u>and</u> m <u>are</u> σ-<u>finite measures on the same</u> σ-<u>algebra of subsets of</u> X. <u>If for each measurable set</u> E, $m_2(E) = \int_E f\,dm + m_1(E)$ <u>with</u> $f(x) \geq 0$, <u>then any function</u> g <u>which is integrable with respect to</u> m_2 <u>also satisfies</u>

$$\int g(x)\,dm_2 = \int g(x)\cdot f(x)\,dm + \int g(x)\,dm_1.$$

<u>Proof</u>. This formula clearly holds for characteristic functions of a set since

$$\int C_E(x)\,dm_2 = m_2(E) = \int_E f\,dm + m_1(E)$$
$$= \int C_E(x)\cdot f(x)\,dm + \int C_E(x)\,dm_1.$$

By the additivity properties of the integral the formula also holds for simple functions. By the monotone convergence theorem it holds for non-negative measurable functions. Finally, by considering the positive and negative parts of an integrable function, the formula holds in general for functions integrable with respect to m_2.

We now consider two topics which involve the Radon-Nikodym derivative. The first of these is differentiation of functions of a set in Euclidean k-dimensional space. In the general setting of a σ-finite measure space on which a completely additive function of a set is defined on the collection of measurable sets, it is frequently not possible to define a pointwise derivative. The possibility of doing so requires a collection of sets with which the derivative is to be computed at each point and a limit process with which the computation is performed. If $\Phi(E) = \int_E f\,dm$, the differentiation of $\Phi(E)$ to $f(x)$ at almost every point x cannot in general be obtained without there being a theorem like the Vitali covering theorem which applies to the collection of all sets with which the derivative is computed. With some restrictions on Φ and an appropriate definition of the derivative of Φ at a point, differentiation of completely additive functions of a set with respect to Lebesgue measure in Euclidean k-dimensional space does occur a.e.; so does the differentiation of the integral to its integrand.

Suppose a function of a set Φ is defined on the Borel subsets of $\mathbb{R}^k$. Then the *general upper derivate* of Φ, denoted by $\overline{D}\Phi(x)$, is defined to be the supremum of the set of all numbers of the form $\lim \Phi(E_n)/m(E_n)$ where $x \in E_n$, the sets E_n are compact, $\lim(\operatorname{diam}(E_n)) = 0$ and, if I_n is the smallest k-cell with equal length sides which is centered at x and contains E_n, there is a number $r > 0$ such that $m(E_n)/|I_n| \geq r$. This number r is called a *parameter of regularity* for the sequence $\{E_n\}$. The infimum of the set of all such numbers $\lim \Phi(E_n)/m(E_n)$ is the *general lower derivate* of Φ at x, denoted by $\underline{D}\Phi(x)$. If $\underline{D}\Phi(x) = \overline{D}\Phi(x)$, then Φ is said to have a *general derivative* at x equal to the common value of $\underline{D}\Phi(x)$ and $\overline{D}\Phi(x)$; this general derivative is denoted by $D\Phi(x)$. Restricting the sets E_n to be closed intervals of the form

$$\{(x_1, x_2, \ldots, x_n) : a_1 \leq x_1 \leq b_1, \ldots, a_n \leq x_n \leq b_n\},$$

defines the *ordinary upper derivate* $\overline{\Phi}(x)$ and the *ordinary lower derivate* $\underline{\Phi}(x)$; if these are equal Φ is said to have an *ordinary derivative* at x and this is denoted by $\Phi'(x)$. Sometimes the restriction that there be a parameter of regularity for the sequences of intervals is dropped; the resulting derivates are called the *strong upper derivate*, *strong lower derivate* and the resulting derivative, when it exists, is called the *strong derivative*. Since it is clear that $\overline{D}\Phi(x) \geq \overline{\Phi}(x) \geq \underline{\Phi}(x) \geq \underline{D}\Phi(x)$, it follows that $\Phi'(x)$ exists and equals $D\Phi(x)$ at any point x where the general derivative of Φ exists.

Suppose Φ_1 and Φ_2 are functions of a set and x is a point at which $D\Phi_1$ and $D\Phi_2$ exist and do not take on both $+\infty$ and $-\infty$. If $\Phi = \Phi_1 + \Phi_2$ can be defined, then $D\Phi(x)$ exists and equals $D\Phi_1(x) + D\Phi_2(x)$. To see this, let $\{E_n\}$ be a sequence of closed sets each of which satisfies the same parameter of regularity. If $x \in E_n$ for each natural number n and $\lim \text{diam}(E_n) = 0$, then $\lim_{n\to\infty} \Phi_i(E_n)/m(E_n) = D\Phi_i(x)$, $i = 1, 2$. Clearly, for each such sequence

$$\lim_{n\to\infty} [\Phi_1(E_n) + \Phi_2(E_n)]/m(E_n) = D\Phi_1(x) + D\Phi_2(x).$$

Thus $D\Phi(x)$ exists and equals $D\Phi_1(x) + D\Phi_2(x)$. The same comment applies to the ordinary derivative and to the strong derivative.

The next theorem gives sufficient conditions for a completely additive function defined on the Borel subsets of Euclidean n-dimensional space to be derivable almost everywhere in the general sense.

(8.26) *Theorem*. (Lebesgue's Theorem) *Let* Φ *be a completely additive function of a set defined on the Borel subsets of n-dimensional space. If* Φ *is finite valued on bounded sets, then* $D\Phi$ *exists and is finite almost everywhere.*

Proof. Because of the comment above, we may consider only the non-negative part of Φ or, what amounts to the same thing, we may assume that Φ is non-negative. We begin with the proof of the following two assertions which hold for non-negative completely additive functions of a set Φ which are finite valued on bounded sets and defined on a σ-algebra which contains the Borel sets:

i) If $\overline{D}\Phi(x) \geq a$ at each point x of a set E, then each Borel set A containing E satisfies $\Phi(A) \geq a \cdot m^*(E)$.

ii) The set of points x where $\overline{D}\Phi(x) = \infty$ is of measure 0.

To see that i) holds, suppose $\overline{D}\Phi(x) \geq a$ at each point $x \in E$. Let A be a Borel set containing E. If $\Phi(A) = \infty$, there is nothing to prove. So suppose $\Phi(A) < \infty$. Fix $\varepsilon > 0$. For each $x \in E$, there is a sequence of closed sets $\{E_{n,x}\}$ which contain x each satisfying the same parameter of regularity such that $\lim \operatorname{diam}(E_{n,x}) = 0$ and $\lim \Phi(E_{n,x})/m(E_{n,x}) \geq a$. Let G be an open set containing A with $\Phi(G \backslash A) < \varepsilon$ which is possible due to Theorem 5.6 since Φ is non-negative. Then the collection of sets

$$\mathcal{V} = \{E_{n,x}\colon x \in E \text{ and } E_{n,x} \subset G \text{ and } \frac{\Phi(E_{n,x})}{m(E_{n,x})} > a - \varepsilon\}$$

covers E in the sense of Vitali. Thus there is a sequence of pairwise disjoint sets $\{A_n\}$ contained in $\mathcal{V}$ such that $m(E \backslash \cup A_n) = 0$. Since $\cup A_n \subset G$, $\Phi(A) + \varepsilon \geq \Phi(G) \geq \Phi(\cup A_n) = \Sigma\Phi(A_n) \geq (a-\varepsilon) \cdot \Sigma m(A_n)$. Thus $\Phi(A) \geq (a-\varepsilon) \cdot m^*(E) - \varepsilon$ and since $\varepsilon > 0$ was arbitrary it follows that $\Phi(A) \geq a \cdot m^*(E)$. Since A was an arbitrary Borel set containing E, it follows that i) holds. To see that ii) holds, suppose that $\overline{D}\Phi(X) = \infty$ at each point of a bounded set E. Let A be a Borel set containing E with $m(A) = m^*(E)$. Then, for each natural number n, since $\overline{D}\Phi(x) \geq n$ at each $x \in E$, it follows that $\Phi(A) \geq n \cdot m^*(E)$. Since Φ is finite on bounded sets, $m^*(E) = 0$. But then $\{x: \overline{D}\Phi(x) = \infty\}$ is of measure 0 because each of its bounded subsets is of measure 0. Now suppose that Φ is a non-negative function which satisfies the conditions of the theorem. Suppose, if possible, that $\{x: \overline{D}\Phi(x) > \underline{D}\Phi(x)\}$ has positive outer measure. Then there would exist rational numbers r_1 and r_2 such that

$$\{x: \underline{D}\Phi(x) < r_1 < r_2 < \overline{D}\Phi(x)\}$$

has positive outer measure. To show that this is impossible, it will be shown that every bounded subset E of this set is of measure 0. So let E be such a set. For each $x \in E$, since $\underline{D}\Phi(x) < r_1$, there is a sequence $\{E_{n,x}\}$ of closed sets each satisfying the same parameter of regularity with $x \in E_{n,x}$, $\lim_{n\to\infty} \text{diam}(E_{n,x}) = 0$ and $\lim_{n\to\infty} \Phi(E_{n,x})/m(E_{n,x}) < r_1$. Given $\varepsilon > 0$, let G be an open set containing E with $m(G) < m^*(E) + \varepsilon$. Then the collection of sets

$$\mathcal{V} = \{E_{n,x}: E_{n,x} \subset G \text{ and } \Phi(E_{n,x})/m(E_{n,x}) < r_1\}$$

covers E in the sense of Vitali. Thus there is a sequence $\{A_n\}$ of pairwise disjoint sets contained in $\mathcal{V}$ with $m(E\setminus\cup A_n) = 0$. Then $\Phi(\cup A_n) = \Sigma\Phi(A_n) < r_1 \cdot \Sigma m(A_n) \leq r_1 \cdot m(G) < r_1(m^*(E) + \varepsilon)$. However, since $\overline{D}\Phi(x) > r_2$ at each $x \in E$, it follows from i) that $\Phi(A_n) \geq r_2 \cdot m^*(E \cap A_n)$. Since $m(E\setminus\cup A_n) = 0$, it follows that $r_2 \cdot m^*(E) \leq r_2 \cdot \Sigma m^*(E \cap A_n) \leq r_2 \cdot \Sigma m(A_n) \leq \Sigma \Phi(A_n) \leq r_1(m^*(E) + \varepsilon)$. Since $r_1 < r_2$ and $\varepsilon > 0$ is arbitrary, $m^*(E) = 0$; that is, $m(E) = 0$.

Now suppose that f is an integrable function defined in $\mathbb{R}^k$, that $\Phi(E) = \int_E f\, dm$ and that Φ is finite valued on bounded sets. We wish to show that $D\Phi(x) = f(x)$ almost everywhere; that is, that the Radon-Nikodym derivative of such a Φ equals the general derivative a.e. This will be accomplished as follows: First, it will be shown (Theorem 8.27) that, if A is a measurable set, $\Phi(E) = \int_E C_A(x)\, dm = m(E \cap A)$ satisfies $D\Phi(x) = 1$ at a.e. point $x \in A$ and $D\Phi(x) = 0$ at a.e. point $x \in A^c$. Thus, the equality of the two derivatives holds for characteristic functions. It then follows that $D\Phi(x) = f(x)$ whenever $f(x)$ is a non-negative simple function. Next, suppose that $\{\Phi_n\}_{n=0}^{\infty}$ is a sequence of non-negative completely additive set functions which are defined on the Borel sets and are finite valued on bounded sets. Then it will be shown (Theorem 8.28) that, if for each Borel set E, $\{\Phi_n(E)\}_{n=1}^{\infty}$ is non-decreasing and $\lim \Phi_n(E) = \Phi_o(E)$, then $\lim D\Phi_n(x) = D\Phi_o(x)$ a.e. Using this along with the monotone convergence theorem and the fact that each non-negative measurable f is the limit of a non-decreasing sequence of simple functions, it will follow that $D\Phi(x) = f(x)$ whenever f is non-negative and $\Phi(E) = \int_E f(x)\, dm$, provided Φ takes on a finite value on each bounded Borel set. Finally, each completely additive function of a set Φ with $\Phi(E) = \int_E f\, dm$, where Φ is finite valued on bounded sets, is the sum of $\Phi_1(E) = \int_E f^+ \, dm$ and $\Phi_2(E) = -\int_E f^- \, dm$. Then $D\Phi(x) = D\Phi_1(x) + D\Phi_2(x) = f(x)$ a.e. We proceed with the theorems referred to above.

(8.27) <u>Theorem</u>. <u>If</u> A <u>is a measurable set in</u> $\mathbb{R}^k$ <u>and</u> $\Phi(E) = \int_E C_A(x) \, dm = m(A \cap E)$, <u>then at almost every</u> $x \in A$, $D\Phi(x) = 1$ <u>and at almost every</u> $x \in A^c$, $D\Phi(x) = 0$.

<u>Proof</u>. (Note that this is a type of density theorem and its proof is analogous to that of the Lebesgue density theorem on the line. Indeed, if $x \in A$ and $\Phi'_s(x) = 1$, x is called a point of density of A.) Let A be a measurable set and $\Phi(E) = \int_E C_A(x) \, dm$. If $m(E) > 0$, then $\Phi(E)/m(E) \le 1$ and thus $\overline{D}\Phi(x) \le 1$ at each point x. Suppose there is a set $B \subset A$ of positive outer measure such that $\underline{D}\Phi(x) < 1$ at each $x \in B$. If $B_n = \{x \in B: \underline{D}\Phi(x) < 1 - 1/n\}$, one of the sets B_n must have positive outer measure, say B_N. Let G be an open set containing B_N with $m(G) < m^*(B_N)(1 + 1/N)$. Then at each $x \in B_N$ there is a sequence of closed sets $\{E_{n,x}\}$ each satisfying the same parameter of regularity, containing x and contained in G such that $\lim \operatorname{diam}(E_{n,x}) = 0$ and $\Phi(E_{n,x})/m(E_{n,x}) < 1 - 1/N$. The collection of all such sets $E_{n,x}$ covers B_N in the sense of Vitali and hence it contains a sequence of pairwise disjoint sets $\{E_j\}$ which satisfies $m(B_N \setminus \cup E_j) = 0$. But then

$$\begin{aligned} m^*(B_N) &\le \Sigma\, m^*(B_N \cap E_j) \le \Sigma\,(1 - 1/N) m(E_j) \\ &\le (1 - 1/N) m(G) \le (1 - 1/N)(1 + 1/N) m^*(B_N) \\ &= (1 - 1/N^2) m^*(B_N) \end{aligned}$$

and this is impossible if $m^*(B_N) > 0$. It follows that $m(B) = 0$ and $D\Phi(x) = 1$ at a.e. point of A. Let $\Phi_C(E) = \int_E C_{A^C}(x)\, dm$. Then $D\Phi_C(x) = 1$ at a.e. point $x \in A^C$. Let x be a point of A^C such that $D\Phi_C(x) = 1$. Let $\{E_n\}$ be a sequence of sets each satisfying the same parameter of regularity with $x \in E_n$ and $\lim \operatorname{diam}(E_n) = 0$. Since $\Phi(E_n) + \Phi_C(E_n) = m(E_n)$ it follows that $\lim \Phi(E_n)/m(E_n) = \lim(1 - \Phi_C(E_n)/m(E_n)) = 0$. That is, at almost every $x \in A^C$, $D\Phi(x) = 0$. This completes the proof.

(8.28) <u>Theorem</u>. <u>Let $\{\Phi_n\}_{n=0}^{\infty}$ be a sequence of non-negative completely additive functions of a set which are defined on the Borel sets and are finite valued on bounded sets. Suppose that, for each bounded Borel set E, $\{\Phi_n(E)\}$ is non-decreasing and $\Phi_o(E) = \lim \Phi_n(E)$ is finite valued on bounded sets, then</u> $\lim_{n\to\infty} D\Phi_n(x) = D\Phi_o(x)$ a.e.

<u>Proof</u>. Suppose $\{\Phi_n\}_{n=0}^{\infty}$ satisfies the hypotheses of the theorem. Then, for each Borel set E and natural number n, $\Phi_o(E) - \Phi_n(E) \geq 0$ and thus $\overline{D}(\Phi_o - \Phi_n)(x)$ is a non-increasing sequence of non-negative numbers at each point x. It remains to show that $\overline{D}(\Phi_o - \Phi_n)$ approaches 0 a.e. For then $D(\Phi_o - \Phi_n)$ approaches 0 almost everywhere; that is, at almost every point x each $D\Phi_n$ exists and $\lim D(\Phi_o - \Phi_n)(x) = D\Phi_o(x) - \lim D\Phi_n(x) = 0$. To see that this occurs, let

$$B_N = \{x\colon \|x\| < N \text{ and for each natural number } n,\ D(\Phi_o - \Phi_n)(x) \geq 1/N\}.$$

Then, for each n and $x \in B_N$, $\overline{D}(\Phi_o - \Phi_n)(x) \geq 1/N$.

If A is a bounded Borel set with $B_N \subset A$, this inequality involving the upper derivate of $\Phi_o - \Phi_n$ implies that $(\Phi_o - \Phi_n)(A) \geq 1/N \cdot m^*(B_N)$. Since $\lim(\Phi_o - \Phi_n)(A) = 0$, it follows that $m^*(B_N) = 0$. Because this is true for each N, $m(\cup B_N) = 0$ and $\lim_{n\to\infty} D\Phi_n(x) = D\Phi(x)$ a.e.

From these last two theorems and the comments preceeding them, we now have:

(8.29) <u>Theorem</u>. <u>If</u> $f(x)$ <u>is integrable and has a finite integral on every bounded measurable set in</u> $\mathbb{R}^k$, <u>then</u> $\Phi(E) = \int_E f(x)\,dm$ <u>satisfies</u> $D\Phi(x) = f(x)$ a.e.

A converse to this theorem completes the Lebesgue decomposition theorem for those Φ which are finite valued on bounded sets.

(8.30) <u>Theorem</u>. <u>Let</u> Φ <u>be a completely additive function of a set which is defined on the Borel sets and is finite valued on bounded sets</u>. <u>If</u> $\Phi(E) = \int_E f\,dm + \theta(E)$, <u>where</u> $f = d\Phi/dm$ <u>and</u> θ <u>is singular, is the Lebesgue decomposition of</u> Φ, <u>then</u> $D\theta(x) = 0$ a.e. <u>and</u> $D\Phi(x) = f(x)$ a.e.

<u>Proof</u>. Clearly, $\int_E f\,dm$ and $\theta(E)$ obtained from such a Φ must both be finite valued on bounded sets. There is a set Z of measure 0 such that $\theta(E) = 0$ if $E \subset Z^c$. Let θ_1 be the non-negative part of θ. For each natural number N let

$$B_N = \{x \in Z^c\colon |x| < N \text{ and } \overline{D}\theta_1(x) > 1/N\}.$$

Then each Borel set $A \subset Z^c$ with $B_N \subset A$ satisfies $0 = \theta_1(A) \geq 1/N \cdot m^*(B_N)$. Hence, for each N, $m^*(B_N) = 0$ and thus $m(\cup B_N) = 0$ and $D\theta_1 = 0$ a.e. By considering θ_2, the non-positive part of θ, we obtain $D\theta_2(x) = 0$ a.e. and hence $D\theta(x) = 0$ a.e. Then $D(\Phi - \theta)(x) = D\Phi(x)$ a.e. But $(\Phi - \theta)(E) = \int_E f(x)\, dm$ and hence $D(\Phi - \theta)(x) = f(x)$ a.e. and thus $D\Phi(x) = f(x)$ a.e.

In the event that V_Φ is a σ-finite measure and Φ takes on infinite values on bounded sets, it is not necessary that the derivative of Φ exist a.e. For example, let P be a nowhere dense perfect subset of $[0,1]$ of positive measure with contiguous intervals $I_n = [a_n, b_n]$. Let $f(x) = 1/(b_n - a_n)$ when $x \in I_n$ and, if $x \in P$, let $f(x) = 0$. Then $\Phi(E) = \int_E f(x)\, dx$ is absolutely continuous with respect to Lebesgue measure. However, if I is an interval which contains points of P in its interior, then I contains countably many contiguous intervals I_n in its interior and hence $\Phi(I) = \infty$ because for each $I_n \subset I$, $\Phi(I_n) = 1$. Thus at each $x \in P$, $\overline{D}\Phi(x) = \infty$ and, since $\underline{D}\Phi(x) = 0$ at each point of density of P, Φ does not have a general derivative a.e. The same situation can occur for a singular function Φ. For example, let $E_n \subset [0,1]$ be a collection of pairwise disjoint closed sets each of measure 0. If $\lim \operatorname{diam}(E_n) = 0$, $\cup E_n$ is dense, Φ is a measure on each E_n with $\Phi(E_n) = 1$ and $\Phi(E) = \Sigma\Phi(E \cap E_n)$, then Φ is singular. However, given any point x and interval I containing x, the interval I contains countably many E_n and hence $\Phi(I) = \infty$. Thus, even though Φ is singular, $\overline{D}\Phi(x) = \infty$ at every point $x \in$

[0,1].

In spite of such examples as those just given, the Radon-Nikodym derivative of any Φ defined on the Borel subsets of $\mathbb{R}^k$ for which V_Φ is σ-finite can be determined from the general derivates of Φ.

(8.31) Theorem. *Suppose Φ is a countably additive function of a set defined on the Borel subsets of $\mathbb{R}^k$ such that V_Φ is a σ-finite measure. If Φ takes on the value $+\infty$ (respectively, $-\infty$) the Radon-Nikodym derivative* $d\Phi/dm = \underline{D}\Phi$ a.e. (*respectively,* $\overline{D}\Phi$ a.e.).

Proof. Suppose Φ takes on the value $+\infty$. Then the negative part of Φ is bounded and its general derivative exists, is finite a.e. and equals its Radon-Nikodym derivative a.e. Consider the non-negative part of Φ or, without loss of generality, suppose that Φ is non-negative. Then $\Phi(E) = \int_E f\,dm + \theta(E)$ where θ is non-negative and singular and f is a non-negative measurable function. For such a θ, there is a set Z of measure 0 such that for $E \subset Z^c$, $\theta(E) = 0$. But each point x of Z^c is a point of density of Z^c and by the Lusin-Menchoff theorem there is a compact set $E_x \subset Z^c$ so that x is a point of density of E_x. Then the sequence of intervals $\{I_n\}$ each centered at x with diameters $1/n$ satisfies $\inf\{m(E_x \cap I_n)/|I_n|\} > 0$ because

$$\lim m(E_x \cap I_n)/|I_n| = 1.$$

Using the intervals I_n it follows that $\underline{D}\theta(x) = 0$; indeed, $\theta(E_x \cap I_n)/m(E_x \cap I_n) = 0$ for each natural number n and $\{E_x \cap I_n\}$ satisfies a parameter of regularity. Since $\underline{D}\theta = 0$ a.e. it remains to show that $\Phi_o(E) =$

$\int_E f(x)\, dm$ has a lower derivate equal to f almost everywhere. To see this, for each natural number n let $f_n(x) = \min(f(x), n)$ and let $\Phi_n(E) = \int_E f_n(x)\, dm$. Then $\{f_n\}$ is a non-decreasing sequence of functions which approaches f. For each Borel set E, the monotone convergence theorem implies that $\lim \Phi_n(E) = \Phi_o(E)$. Moreover, since each Φ_n is finite valued on bounded sets, $D\Phi_n$ exists a.e. Clearly, at any point x where $D\Phi_n$ exists, $\underline{D}\Phi_o(x) \geq D\Phi_n(x)$ because whenever $x \in E_i$, where each E_i satisfies the same parameter of regularity and $\lim \operatorname{diam}(E_i) = 0$, then $\Phi_o(E_i)/m(E_i) \geq \Phi_n(E_i)/m(E_i)$. Since Φ_o is a σ-finite measure, it follows that f is finite valued a.e. For otherwise, if f were infinite on a set E of positive measure, Φ_o would be infinite on each set $E' \subset E$ of positive measure and Φ_o would be zero on each set $Z \subset E$ of measure 0 and E would not be a countable union of sets of finite measure. Let $A_n = \{x\colon f_n(x) = f(x) < n \text{ and } D\Phi_n(x) = f_n(x)\}$. Let x be a point of density of A_n. Again, by the Lusin-Menchoff theorem there is a compact set E_x with density 1 at x such that $E_x \subset A_n$. If I_j is the k-cell centered at x of diameter $1/j$ and equal length sides, then the sets $E_j = E_x \cap I_j$ satisfy the same parameter of regularity since $m(E_j)/|I_j|$ approaches 1. Moreover,

$$\begin{aligned}\underline{D}\Phi_o(x) &\leq \underline{\lim} \int_{E_j} f(x)\, dm \,/\, m(E_j)\\ &= \lim \int_{E_j} f_n(x)\, dm \,/\, m(E_j)\\ &= f_n(x) = f(x).\end{aligned}$$

Since almost every point x is a point of density of

some set A_n, it follows that $\underline{D}\Phi_o(x) = f(x)$ a.e. and hence $\underline{D}\Phi(x) = f(x)$ a.e. Since the case where $\Phi(X) = -\infty$ is completely analogous, the theorem is proved.

The second topic which involves the Radon-Nikodym derivative is a classical result which gives a characterization of linear functionals defined on an $\mathcal{L}^p$ space. The Radon-Nikodym theorem is needed in order to consider this subject in the setting of a general σ-finite measure space. The fact that $f(x) = d\Phi/dm$ is determined even if there is insufficient structure for pointwise differentiation will allow us to suppose that the underlying space X has no other structure. Actually, we will consider the $\mathcal{L}^p$ spaces defined on a measure space X of finite measure; the generalization to spaces of σ-finite measure poses no real difficulty.

Given p with $1 \le p < \infty$, let q satisfy $1/p + 1/q = 1$ or $q = \infty$ in the case that $p = 1$. Fix $g \in \mathcal{L}^q_X$. Then for $f \in \mathcal{L}^p_X$, $\int f \cdot g \, dm < \infty$. Moreover, if $L(f) = \int f \cdot g \, dm$, L is a real-valued function defined on $\mathcal{L}^p_X$ and L satisfies

1) $L(f_1 + f_2) = L(f_1) + L(f_2)$
2) $L(c \cdot f) = c \cdot L(f)$
3) if $f_n \to f$ [mean p], $L(f_n)$ approaches $L(f)$.

That 1) and 2) hold is clear. If $f_n \to f$ [mean p], $f \in \mathcal{L}^p$. By the completeness of $\mathcal{L}^p$ and by Hölder's inequality,

$$\begin{aligned} L(f_n) - L(f) &= L(f_n - f) = \int (f_n - f) g \, dm \\ &\le \left(\int |f_n - f|^p dm\right)^{1/p} \cdot \left(\int |g|^q dm\right)^{1/q} \end{aligned}$$

and the latter approaches 0 as n approaches ∞. This implies that 3) holds.

A real-valued function defined on a vector space is called a <u>functional</u>; 1) and 2) show that the function L is linear and a function satisfying 1) and 2) is called a <u>linear functional</u>; 3) is the statement that L is continuous and hence is a continuous linear functional.

The study of linear spaces and their continuous linear functionals form the subject matter of Functional Analysis. Here we wish to show that all continuous linear functionals defined on $\mathcal{L}_X^p$, $1 \le p < \infty$, are of the form $L(f) = \int f \cdot g \, dm$ for appropriate choice of $g \in \mathcal{L}_X^q$. In general, the collection of linear functionals defined on a linear Banach space B will be shown to be a Banach space B^* under the norm defined by $\|L\| = \sup|L(x)|$; here the supremum is over all $x \in B$ with $\|x\| \le 1$. The space B^* is called the <u>dual</u> of B. In the particular case where L is a linear functional on $\mathcal{L}_X^p$, it will be shown that $\|L\| = \sup|L(f)| = \|g\|_q$ provided $L(f)) = \int f \cdot g \, dm$. Moreover, the space of linear functionals on $\mathcal{L}_X^p$ is isomorphic to $\mathcal{L}_X^q$ because the map that takes L, where $L(f) = \int f \cdot g \, dm$, to g preserves norms.

To begin, consider the general setting where L is a continuous linear functional defined on a linear space B and let B^* denote the collection of all such linear functionals. Note that if L_1 and L_2 belong to B^* and c is a real number, then $L_1 + L_2$ and cL_1 are linear functionals on B and, if $x_n \to x_o$ in B, $(L_1 + L_2)(x_n) = L_1(x_n) + L_2(x_n)$ approaches $L_1(x_o) + L_2(x_o)$ and $cL_1(x_n)$ approaches $cL_1(x_o)$. Thus $L_1 + L_2$ and cL_1 are continuous and belong to B^*. Hence B^* is a linear space.

First note that, if $\overline{0}$ denotes the zero element in B, $L(\overline{0}) = L(\overline{0} + \overline{0}) = L(\overline{0}) + L(\overline{0})$ and thus $L(\overline{0})$ must equal 0 for each linear functional L defined on B.

We now show that if $L \in B^*$ then there is a number M such that whenever $\|x\| \le 1$, $|L(x)| \le M$. To see this suppose there were a linear functional $L \in B^*$ and a sequence $\{x_n\} \subset B$ with each $\|x_n\| \le 1$ and $|L(x_n)| \ge n$. If $y_n = (1/n)x_n$, then $\|y_n\| = (1/n)\|x_n\| \le 1/n$. Then $\|y_n\|$ would approach 0, y_n would approach $\overline{0}$ and since L is continuous $L(y_n)$ would approach $L(\overline{0}) = 0$. But $|L(y_n)| = (1/n)|L(x_n)| \ge 1$. This contradiction implies that, given $L \in B^*$ there is a number M such that if $\|x\| \le 1$, $|L(x)| \le M$.

We can then define $\|L\| = \sup|L(x)|$ where the supremum is over all $x \in B$ with $\|x\| \le 1$. Note that $\|L\|$ also equals the supremum of all $|L(x)|$ with $\|x\| = 1$ because, if $0 \ne \|x\| < 1$, then $\|x/\|x\|\| = 1$ and $|L(x/\|x\|)| = |L(x)\cdot(1/\|x\|)| > |L(x)|$. In addition, if $L \in B^*$ and there is a number M such that for each $x \in B$, $|L(x)| \le M\|x\|$, then $\|L\| \le M$. This is because for each $x \in B$ with $\|x\| = 1$, $|L(x)| \le M$. Now, if L is the linear functional that is identically 0, $\|L\| = \sup_{\|x\|\le 1} |L(x)| = 0$. Also, if $\|L\| = 0$, then $L(\overline{0}) = 0$ and, if $x \ne \overline{0}$, $0 = L\left(\frac{x}{\|x\|}\right) = \frac{1}{\|x\|} L(x)$ and $L(x) = 0$. Thus

$$1)\quad \|L\| = 0 \text{ iff } L \equiv 0.$$

Also $\|cL\| = \sup_{\|x\|\le 1} |c\cdot L(x)| = |c| \sup_{\|x\|\le 1} |L(x)| = |c|\|L\|$. Thus

$$2)\quad \|cL\| = |c|\|L\|.$$

Finally, $\|L_1 + L_2\| = \sup_{\|x\|\le 1} |L_1(x) + L_2(x)| \le \sup_{\|x\|\le 1} |L_1(x)| + \sup_{\|x\|\le 1} |L_2(x)| = \|L_1\| + \|L_2\|$ and

$$3)\quad \|L_1 + L_2\| \le \|L_1\| + \|L_2\|.$$

Thus $\|\ \|$ is a norm on B^*. To show that B^* is a Banach space, it remains to show that B^* is complete. So let L_n be a Cauchy sequence with respect to the norm $\|\ \|$ of B^*. Given $\varepsilon > 0$ there is a natural number N so that $\|L_n - L_m\| < \varepsilon$ whenever $n,m > N$. For each $x \in B$ with $x \ne \overline{0}$, $\varepsilon > \|L_n - L_m\| \ge |(L_n - L_m)(x/\|x\|)| = |(L_n - L_m)(x)|/\|x\|$. Thus

$$|L_n(x) - L_m(x)| \le \|L_n - L_m\|\|x\| \le \varepsilon\|x\|$$

and $\{L_n(x)\}$ is a Cauchy sequence of real numbers. This is also true for $x = \overline{0}$ and indeed, for $\|x\| \le 1$, $L_n(x)$ is uniformly Cauchy and hence L_n converges to a function L and does so uniformly on

$$\{x \in B: \|x\| \le 1\}.$$

It now remains to show that L is a continuous linear functional. But $L(x+y) = \lim L_n(x+y) = \lim L_n(x) + \lim L_n(y) = L(x) + L(y)$. Also $L(cx) = \lim L_n(cx) = \lim c\cdot L_n(x) = c\cdot\lim L_n(x) = c\cdot L(x)$. Since this holds true for each $x,y \in B$ and real number c, L is a linear functional. To see that L is continuous let $x_i \to x_o$ in B. Then $\|x_i\|$ is bounded, say by M and $\dfrac{x_i}{M} \to \dfrac{x_o}{M}$, $\left\|\dfrac{x_i}{M}\right\| \le 1$ and $\left\|\dfrac{x_o}{M}\right\| \le 1$. Since $L_n \to L$ uniformly on all x with $\|x\| \le 1$,

$$\lim_i L\left(\frac{x_i}{M}\right) = \lim_i \lim_n L_n\left(\frac{x_i}{M}\right) = \lim_n \lim_i L_n\left(\frac{x_i}{M}\right)$$

$$= \lim_n L_n\left(\frac{x_o}{M}\right) \quad \text{by the continuity of } L_n$$

$$= L\left(\frac{x_o}{M}\right).$$

Thus $\lim_i L(x_i) = \lim M \cdot L\left(\frac{x_i}{M}\right) = M \cdot \lim L\left(\frac{x_i}{M}\right) = M \cdot L\left(\frac{x_o}{M}\right) = L(x_o)$ and L is continuous. Thus each Cauchy sequence in B^* converges to an element of B^* and B^* is complete.

We are now ready to consider the collection of continuous linear functionals defined on a space $\mathcal{L}_X^p$ where $1 \le p < \infty$.

(8.32) Theorem. (Riesz Representation Theorem) *If X is a measure space of σ-finite measure and $1 \le p < \infty$ and q satisfies $1/p + 1/q = 1$ or $q = \infty$ in the case where $p = 1$, then for every continuous linear functional L defined on $\mathcal{L}_X^p$ there is a unique function g in $\mathcal{L}_X^q$ so that L has a representation of the form $L(f) = \int_X f(x) \cdot g(x)\, dm$. Moreover, $\|L\| = \|g\|_q$.*

Proof. We only consider the case where $m(X) < \infty$. When $m(X) = \infty$ and m is σ-finite the function g can be defined piecewise on each of a collection of pairwise disjoint measurable sets X_i with $X = \cup X_i$. It has already been noted that if $g \in \mathcal{L}_X^q$, then $L(f) = \int f(x) \cdot g(x)\, dm$ is a linear functional. Moreover, distinct $g \in \mathcal{L}_X^q$ correspond to distinct linear functionals. For if $g_1 \ne g_2$ a.e., there is a set E with $m(E) > 0$ such that either $g_1(x) > g_2(x)$ at each $x \in E$ or $g_2(x) > g_1(x)$ at each $x \in E$. Then $f(x) = C_E(x)$ satisfies

$$\int f(x)\cdot g_1(x)\, dm \neq \int f(x)\cdot g_2(x)\, dm.$$

Now let L be any continuous linear functional defined on $\mathcal{L}_X^p$. Let $\Phi(E) = L(C_E(x))$. Since L is linear, if $E = \bigcup_{i=1}^{n} E_i$, where the sets E_i are pairwise disjoint, $\Phi(\bigcup_{i=1}^{n} E_i) = L(C_E(x)) = \sum_{i=1}^{n} L(C_{E_i}(x)) = \sum_{i=1}^{n} \Phi(E_i)$. Let $E = \bigcup_1^\infty E_i$ where again the sets E_i are pairwise disjoint. Let $A_n = \bigcup_1^n E_i$ and note that $\lim m(A_n) = m(E)$. Since $m(X) < \infty$, $\lim m(E\backslash A_n) = 0$ and hence $\lim C_{A_n}(x) = C_E(x)$ [meas.]. Also $C_{A_n}(x)$ is a bounded sequence of functions on X and since $m(X) < \infty$, it follows that $\lim C_{A_n}(x) = C_E(x)$ [mean p]. But since L is continuous on $\mathcal{L}_X^p$,

$$\Phi(E) = L(C_E(x)) = \lim L(C_{A_n}(x))$$
$$= \sum_{n=1}^{\infty} L(C_{E_n}(x)) = \sum_{n=1}^{\infty} \Phi(E_n).$$

Hence Φ is a completely additive function of a set. Moreover, if $m(Z) = 0$, $C_z(x) = 0$ a.e. and hence $\Phi(Z) = L(C_z(x)) = L(\overline{0}) = 0$ and Φ is absolutely continuous with respect to m. By the Radon-Nikodym theorem, there is a function g such that $\Phi(E) = \int C_E(x)\cdot g(x)\, dm = L(C_E(x))$ for each measurable set E contained in X. It then follows that for each non-negative simple function $s(x) = \sum_{i=1}^{n} a_i\cdot C_{E_i}(x)$, L satisfies

$$\begin{aligned} L(s(x)) &= L(\Sigma a_i \cdot C_{E_i}(x)) \\ &= \int \sum_1^n a_i C_{E_i}(x) \cdot g(x)\, dm \\ &= \int s(x) \cdot g(x)\, dm. \end{aligned}$$

Then, if $f(x) \geq 0$, and $f \in \mathscr{L}_X^p$, the usual monotone sequence of simple functions s_n which converge to f, also converge to f [mean p]. By the continuity of L and by considering the monotone convergence theorem for both the positive and the negative parts of g, it follows that

$$\begin{aligned} L(f) &= \lim L(s_n) \\ &= \lim \int s_n(x) \cdot g(x)\, dm \\ &= \int f(x) \cdot g(x)\, dm. \end{aligned}$$

Finally, if f is any function in $\mathscr{L}^p$, $f = f^+ - f^-$ and $\mathscr{L}(f) = \mathscr{L}(f^+) - \mathscr{L}(f^-) = \int f^+(x) \cdot g(x)\, dm - \int f^-(x) \cdot g(x)\, dm = \int f(x) \cdot g(x)\, dm$. To see that the function g belongs to $\mathscr{L}_X^q$, first suppose that $1 < p < \infty$. Let

$$g_n(x) = \begin{cases} |g(x)|^{q-1} \operatorname{sgn}(g(x)) & \text{if } |g(x)| \leq n \\ n \operatorname{sgn}(g(x)) & \text{if } |g(x)| > n \end{cases}.$$

Since each g_n is bounded on X and $m(X) < \infty$, each g_n belongs to $\mathscr{L}_X^p$. Then

$$\begin{aligned} \left|\int g_n(x) \cdot g(x)\, dm\right| &= |L(g_n)| \\ &\leq \|L\| \|g_n\|_p \\ &= \|L\| \left(\int |g_n(x)|^p\, dm\right)^{1/p}. \end{aligned}$$

Also, at each x, $|g_n(x)|^p = |g_n(x)||g_n(x)|^{1/(q-1)} \leq |g_n(x)||g(x)| = g_n(x) \cdot g(x)$. Then $\int |g(x)|^p\, dm \leq \|L\|(\int |g_n(x)|^p\, dm)^{1/p}$ and it follows that

$(\int |g_n(x)|^p\, dm)^{1-1/p} = (\int |g_n(x)|^p\, dm)^{1/q} \le \|L\|$. Since $|g_n|^p$ is a monotone sequence of non-negative functions approaching $|g|^q$, the function g belongs to $\mathcal{L}_X^q$ and $\|g\|_q = (\int |g|^q)^{1/q} \le \|L\|$. To see that $\|L\| = \|g\|_q$, consider $f \in \mathcal{L}^p$. By Hölder's inequality, $|L(f)| = |\int f(x)\cdot g(x)\, dm| \le \int |f(x)\cdot g(x)|dm \le \|f\|_p \|g\|_q$. Thus $\|L\| \le \|g\|_q$ and it follows that $\|L\| = \|g\|_q$. Now consider the case where $p = 1$ and $q = \infty$, we first show that $g \in \mathcal{L}_X^\infty$. If $|g(x)| \ge M$ on a set E of positive measure, let $f(x) = \text{sgn}\, g(x) \cdot \frac{1}{M\cdot m(E)} C_E(x)$. Then $\int |f(x)|dm = 1/M$. However, for this function f, $\|L\|\cdot\int |f|dm \ge \int |f\cdot g|dm \ge \int_E 1/m(E)\, dm = 1$ and hence $\|L\| \ge M$. It follows that $|g(x)| \le \|L\|$ a.e., that is g is essentially bounded and $g \in \mathcal{L}_X^\infty$. It also follows that $\|g\| \le \|L\|$. To see that $\|g\|_\infty = \|L\|$, note that, for any function $f \in \mathcal{L}_X^1$,

$$|L(f)| = |\int f\cdot g\, dm| \le \int |f\cdot g|dm \le \|g\|_\infty \int |f|dm = \|g\|_\infty \|f\|_1$$

and hence $\|L\| \le \|g\|_\infty$ and $\|L\| = \|g\|_\infty$.

8.2 Exercises

1. Let $C(x)$ be the Cantor function. Extend C to $(-\infty,\infty)$ by defining $C(x + 1) = C(x) + 1$. Then if $\{r_n\}$ is a countable dense subset of $[0,1]$ show that $\sum_{n=1}^{\infty} C(x + r_n)\cdot 2^{-n}$ is a strictly increasing singular function on $\mathbb{R}$.

2. For $f, g \in \mathcal{L}^2_{[a,b]}$ let $f \cdot g = \int_a^b f(x) \cdot g(x)\, dx$. Prove that $f \cdot g$ is an inner product on $\mathcal{L}^2_{[a,b]}$.
3. For functions f which are of bounded variation on $[0,1]$ let $\|f\|_V = |f(0)| + V(f;\ 0,1)$. Show that this is a norm and that the resulting linear space is complete.
4. If $f_1(0) = f_2(0) = 0$ and f_1 is absolutely continuous on $[0,1]$ and f_2 is singular on $[0,1]$, prove that $\|f_1 + f_2\|_V = \|f_1\|_V + \|f_2\|_V$ with $\|\cdot\|_V$ as given in Exercise 3.
5. Show that the functions which are monotone on some subinterval of $[0,1]$ are of first category in the space of functions of bounded variation with $\|\cdot\|_V$ as defined in Exercise 3.
6. Show that there is a measure Φ on $[0,1]$ which is σ-finite and absolutely continuous with respect to Lebesgue measure so that $\overline{D}\Phi(x) = \infty$ at each $x \in [0,1]$.

Chapter Nine

9.1 The Denjoy-Perron Integral

The fact that there are measurable functions which have an improper Lebesgue integral but are not Lebesgue integrable has already been pointed out. Indeed, once it is known that the primitive or indefinite integral $F(x)$ of a Lebesgue integrable function $f(x)$ must be absolutely continuous, it becomes apparent that these are derivatives which are not Lebesgue integrable. The derivative of $F(x) = x^2\cos(\pi/x^2)$ with $F(0) = 0$ is one such function. It is easily seen that, on each interval $[\varepsilon,1]$, $F(x)$ is absolutely continuous but that $F(x)$ is not of bounded variation on $[0,1]$. For if $I_n = [(n+1)^{-1/2}, n^{-1/2}]$, then $\int_{I_n} |F'(x)|dx = n^{-1} + (n + 1)^{-1}$ and thus $\int_0^1 |F'(x)|dx = \infty$. Thus, although the improper integral $\lim_{\varepsilon\to 0^+} \int_\varepsilon^1 F'(x)\, dx$ exists, $F'(x)$ is not Lebesgue integrable. This was known to Lebesgue when he developed his integral (1904). There thus arose a problem of finding, if possible, a continuous process of integration which is defined on intervals, agrees with the Lebesgue integral, includes improper integrals and integrates each derivative to its corresponding differentiable function. For such an integral it turns out that integrability of each integrable function on subintervals and additivity of the integral on non-overlapping intervals can be retained, but the integrability of integrable functions on measurable subsets and countable additivity must necessarily be lost. A constructive approach involving a transfinite collection of improper integrals of two types was developed by Denjoy (1912) and this solved the problem. Lusin (1912) recognized that a descriptive definition of the indefinite integrals from Denjoy's solution could be given. That is, for the class of all functions f integrable on an interval I_o in the constructive sense of Denjoy, the class of corresponding $F(x)$ or $F(I)$ were described by Lusin. Moreover, each $f(x)$ which is integrable was shown to be the almost everywhere derivative of the corresponding function $F(x)$. Another integral, that of Perron (1914) was also developed and it turned out to be equivalent to that of Denjoy. The resulting integral was thus called the Denjoy-Perron integral or the restricted sense Denjoy integral. (The wide-sense Denjoy integral or Denjoy-Khinechine integral (1916) was discovered by these two mathematicians almost simultaneously. It will be presented in the second section of this chapter.) A fourth approach to the integral, developed by Henstock and Kurzweil (1947) was named the Riemann-complete integral because it uses Riemann style partitions of the interval on which a function is to be integrated. This integral also turned out to be equivalent to the Denjoy-Perron integral. The equivalence of these four definitions, the agreement of the resulting integral with the Lebesgue integral on the class of summable functions along with its ability to integrate derivatives and its closure under improper integration form the subject of this section.

Before giving the definition of the four integrals, some criteria will be stated for the type of integral under consideration. We will only be concerned with generalizations of the Lebesgue integral on an interval $I_o = [a_o, b_o]$ although generalizations of other integrals are possible. Just as $\mathcal{L}\int f$ denotes the Lebesgue integral of f on its domain I_o, in general, $\mathcal{I}\int f$ will be used to denote an $\mathcal{I}$-integral of a function f; here an $\mathcal{I}$-integral is to be defined on a collection $\mathcal{I}$ of measurable functions f each having domain I_o. A function F is called an $\mathcal{I}$-<u>indefinite</u> <u>integral</u> of f if, for each interval $I \subset I_o$, $f \cdot C_I \in \mathcal{I}$ and $\mathcal{I}\int f \cdot C_I = F(I)$. The properties given below for a class of functions $\mathcal{I}$ and the real-valued functions $\mathcal{I}\int f$ for $f \in \mathcal{I}$ are the ones that we will require in order that $\mathcal{I}$ be called an integral. The statement $0 \in \mathcal{I}$ along with properties i) and ii) are sometimes taken to define the most general concept of an integral on an interval I_o; Denjoy used the term "totalization" for such a process. The properties defining an integral $\mathcal{I}$ are as follows:

i) If $f \in \mathcal{I}$ and I is an interval contained in I_o, then $f \cdot C_I \in \mathcal{I}$.

ii) For each $f \in \mathcal{I}$, $F(I) = \mathcal{I}\int f \cdot C_I$ is a real-valued, continuous additive function of an interval. (The continuity of the function F does not hold for certain more general integrals which will not be considered here.)

iii) The class $\mathcal{I}$ is a vector space of measurable functions and $\mathcal{I}\int f$ is linear on this space; that is, if $f, g \in \mathcal{I}$ and $a, b \in \mathbb{R}$, then $af + bg \in \mathcal{I}$ and $\mathcal{I}\int (af + bg) = a \cdot \mathcal{I}\int f + b \cdot \mathcal{I}\int g$.

Any real-valued function $\mathcal{I}\int f$ defined on a collection $\mathcal{I}$ of measurable functions and satisfying i), ii) and iii) will be called an integral. Note that the same definition could be applied equally well in $\mathbb{R}^k$ where the intervals I are taken to be k-cells.

Since we will be concerned with generalizations of the Lebesgue integral on the line, a fourth condition should also hold for $\mathcal{I}$ if it is to be an extension of the Lebesgue integral; namely,

iv) If $f \in \mathcal{I}$ and $f \geq 0$ a.e., then $f \in \mathcal{L}$ and $\mathcal{L}\int f = \mathcal{I}\int f$.

From this it follows that each $f \in \mathcal{L}$ belongs to $\mathcal{I}$

and for $f \in \mathcal{L}$, $\mathcal{L}\int f = \mathcal{I}\int f$. (If $\mathcal{I}$ were an extension of a different integral, iv) would be replaced by the analogous assertion for that integral.)

Finally, if possible, one would like the following condition to hold for $\mathcal{I}$ when $\mathcal{I}$ is an extension of $\mathcal{L}$ which contains the derivatives of differentiable functions:

v) If $f \in \mathcal{I}$ and $F(I) = \mathcal{I}\int f\cdot C_I$, then $F'(x) = f(x)$ a.e.

(A condition similar to v) is frequently obtainable for more general integrals with the ordinary derivative of $F(I)$ replaced by a more general derivative.) The condition v) will be seen to hold for the Denjoy-Perron integral.

A function f is said to be integrable on $E \subset I_o$ if $f\cdot C_E \in \mathcal{I}$ and one writes, as usual, $\mathcal{I}\int_E f = \mathcal{I}\int f\cdot C_E$. Note that if $\mathcal{I}$ satisfies iv) and $Z \subset I_o$ is a set of measure 0, then $f\cdot C_Z \in \mathcal{I}$ and $\mathcal{I}\int_Z f = 0$. Thus if $f \in \mathcal{I}$ and $g = f$ a.e. then $g \in \mathcal{I}$ and $\mathcal{I}\int g = \mathcal{I}\int f$. It also follows from iv) that if $f,g \in \mathcal{I}$ and $f \le g$ a.e., then $\mathcal{I}\int f \le \mathcal{I}\int g$. Moreover, when applicable, the convergence theorems hold for an integral $\mathcal{I}$ which extends the Lebesgue integral. For example, if $f_n \in \mathcal{I}$ and f_n approaches f a.e. and there are $g_1,g_2 \in \mathcal{I}$ with $g_1 \le f_n \le g_2$, then $f \in \mathcal{I}$ and $\mathcal{I}\int f = \lim \mathcal{I}\int f_n$. This is because $g_2 - g_1 \ge f_n - g_1 \ge 0$ and since $0 \le$

$f_n - g_1 \in \mathcal{I}$, by iv), $f_n - g_1 \in \mathcal{L}$ and $\mathcal{L}\int(f_n - g_1) = \mathcal{I}\int f_n - \mathcal{I}\int g_1$. By the dominated convergence theorem for the Lebesgue integral, $\mathcal{L}\int(f - g_1) = \lim \mathcal{L}\int(f_n - g_1)$. On subtracting $\mathcal{I}\int g_1$ from both sides of this equation, one has $\lim \mathcal{I}\int f_n = \mathcal{I}\int f$ and thus the dominated convergence theorem holds for $\mathcal{I}$.

We need some preliminary notions before proceeding to the original definition of Denjoy for the Denjoy-Perron integral. If $\mathcal{I}$ is an integral on I_o, the standard method of forming an improper integral is ascribed to Cauchy and the resulting integral will be denoted by $\mathcal{I}^C$. Suppose $x_1, \ldots, x_n$ are finitely many points in I_o and f is $\mathcal{I}$-integrable on each closed interval $I \subset I_o$ which does not contain any of these points. If there is a continuous function F such that $F(I) = \mathcal{I}\int_I f$ for such I, then F is said to be an improper integral of f. The collection of all such functions f forms $\mathcal{I}^C$, the Cauchy extension of the integral $\mathcal{I}$. One easily obtains that: if $\mathcal{I}$ merely satisfies i) and ii), so does $\mathcal{I}^C$; if $\mathcal{I}$ satisfies i), ii) and iii), so does $\mathcal{I}^C$; if $\mathcal{I}$ is an extension of $\mathcal{L}$, so is $\mathcal{I}^C$; and if $\mathcal{I}$ also satisfies v), so does $\mathcal{I}^C$.

Another type of improper integral which is ascribed to Harnack is as follows: suppose that $P \subset I_o$ is a perfect set and f is $\mathcal{I}$-integrable on P and on each interval I_n complementary to P and that $\Sigma\, \theta(\mathcal{I}\int_{I_n} f; I_n) < \infty$. Then the function

$$F(x) = \mathscr{I}\int f\cdot C_{P\cap[a_o,x]} + \Sigma\mathscr{I}\int f\cdot C_{I_n\cap[a_o,x]}$$

is defined on I_o and $F(x)$ will be called the H*-type improper integral of f on I_o. We wish to consider an extension of an integral $\mathscr{I}$ which involves such improper integrals. For this purpose, we will need the following definition: If $\mathscr{I}$ is an integral on I_o (or if $\mathscr{I}$ merely satisfies i) and ii) and $0 \in \mathscr{I}$) and f is a function defined on I_o, a point $x \in I_o$ is called a point of singularity for $\mathscr{I}$ if there does not exist an interval I with $x \in \mathring{I} \cap I_o$ for which $f\cdot C_I$ belongs to $\mathscr{I}$. Given f and $\mathscr{I}$, $S(f) = S(f,\mathscr{I})$ will denote the set of points of singularity of f with respect to $\mathscr{I}$.

(9.1) Theorem. For each f defined on I_o and $\mathscr{I}$ satisfying i) and ii), the set $S(f,\mathscr{I})$ is a closed subset of I_o and the function f belongs to $\mathscr{I}$ iff $S(f,\mathscr{I}) = \emptyset$.

Proof. Let x_o be a limit point of $S(f)$. Then there is a sequence $\{x_n\}$ which approaches x_o with each $x_n \in S(f)$. If $x_o \in \mathring{I} \cap I_o$, then there is $x_n \in \mathring{I} \cap I_o$ and since $x_n \in S(f)$, $f\cdot C_I \notin \mathscr{I}$. Since this holds for any I with $x_o \in \mathring{I}$, $x_o \in S(f)$ and hence $S(f)$ is closed. By i), if $f \in \mathscr{I}$, then $S(f) = \emptyset$. Conversely, if $S(f) = \emptyset$, then for each $x \in I_o$ there is an interval I_x with $x \in \mathring{I}_x \cap I_o$ and $f\cdot C_{I_x} \in \mathscr{I}$. This set of intervals $\{\mathring{I}_x: x \in I_o\}$ is an open cover of I_o and contains a finite subcover $\{I_n\}_{n=1}^{N}$. Without loss of generality we assume that no I_n is contained in $\bigcup_{k \neq n} I_k$ and that the intervals I_n are in their natural order. Hence there are x_o, ..., x_N so that $a = x_o$, $b = x_N$ and $x_n \in I_{n-1} \cap I_n$ for $n = 1$, ..., N and thus $[x_{n-1}, x_n] \subset I_n$ and $\{x_o, \ldots, x_N\}$ is a partition of I_o. Then by i) each $f \cdot C_{[x_{n-1}, x_n]} \in \mathscr{I}$ and by ii) it follows that $f \in \mathscr{I}$. Thus $f \in \mathscr{I}$ iff $S(f) = \emptyset$.

Now all f for which $S(f)$ is a perfect set and the H*-type integral of f on I_o is defined with $P = S(f)$ make up the class $\mathscr{I}^{H*}$. If $\mathscr{I}$ satisfies merely i) and ii) then so does $\mathscr{I}^{H*}$. However, the collection $\mathscr{I}^{H*}$ does not in general satisfy iii) when $\mathscr{I}$ does. Indeed, one can show that $\mathscr{L}^{H*}$ does not satisfy iii). Nonetheless, if the extension methods (C) and (H*) are applied transfinitely starting with the Lebesgue integral, the union of all the resulting extensions can be shown to be an integral; that is, to satisfy i), ii) and iii). The extension which results when one starts with $\mathscr{L}$ and applies transfinitely the Cauchy (C) and Harnack (H*) extensions is the constructive definition of the integral which was first given by Denjoy. That the resulting class of functions is an integral will be shown indirectly by proving that it coincides with the class of functions obtained by the second definition of the Denjoy-Perron integral.

The constructive definition is begun by letting $\mathscr{L}_o^* = \mathscr{L}$ and $\mathscr{L}_1^* = \mathscr{L}^{CH*} = (\mathscr{L}^C)^{H*}$. If for each $\beta < \alpha <$

ω_1, $\mathcal{L}^*_\beta$ has been defined, let $\mathcal{L}^*_\alpha = (\bigcup_{\beta<\alpha} \mathcal{L}^*_\beta)^{CH*}$. Finally, $\mathcal{L}^*_{\omega_1} = \bigcup_{\alpha<\omega_1} \mathcal{L}^*_\alpha$ defines the class of Denjoy-Perron integrable functions. One may easily observe that applying (C) or (H*) to $\mathcal{L}^*_{\omega_1}$ does not result in any new functions. This is because for each sequence $\{f_n\} \subset \mathcal{L}^*_{\omega_1}$ there is $\alpha_o < \omega_1$ such that $\{f_n\} \subset \mathcal{L}^*_{\alpha_o}$ and hence an extension using (C) or (H*) which involves these functions must belong to $\mathcal{L}^*_{\alpha_o+1}$.

Without proving anything further concerning $L^*_{\omega_1}$, we turn to the descriptive definition of the Denjoy-Perron integral. Again several preliminary definitions are needed. A function F defined on I_o is said to be <u>absolutely continuous in the restricted sense on a set</u> $E \subset I_o$, written $F \in AC^*$ on E, if F is bounded on an interval containing E and for every $\varepsilon > 0$ there is a $\delta > 0$ such that $\Sigma\, \mathcal{O}(F;I_n) < \varepsilon$ whenever $\{I_n\}$ is a sequence of non-overlapping intervals with their endpoints in E and $\Sigma|I_n| < \delta$. A function F defined on I_o is said to be <u>generalized absolutely continuous in the restricted sense</u> on $E \subset I_o$ if F is continuous on E and E is a countable union of sets on each of which F is AC*. The functions which have this property are said to be ACG* on E or, if $E = I_o$, to be ACG*. We will show shortly that each $F \in ACG^*$ is differentiable a.e. and that if two functions F and G in ACG* have the same derivative almost everywhere then there is a constant C so that $F = G + C$; that is, for $I \subset I_o$, $F(I) = G(I)$. Then the class $\mathcal{D}*$ will consist of the collection of all functions which are almost everywhere the derivative of some function which is ACG* and $\mathcal{D}*\int_I f = F(I)$ if and only if $F \in$ ACG* and $F'(x) = f(x)$ a.e.

Actually, we want to prove more than this. A function F will be said to be BV* or of <u>bounded variation in the restricted sense on</u> $E \subset I_o$ if there is a number $M < \infty$ so that $\Sigma\, \theta(F;I_n) \le M$ whenever $\{I_n\}$ is a sequence of non-overlapping intervals whose endpoints are in E. A function F will be said to be BVG* or of <u>generalized bounded variation in the restricted sense on</u> E if $E = \cup E_n$ and F is BV* on each E_n. The functions which have this property will be said to be BVG* on E or, if $E = I_o$, to be BVG*. Given $E \subset I_o$, the number given by $\sup \Sigma\, \theta(F;I_n)$, where the supremum is taken over all sequences $\{I_n\}$ of pairwise non-overlapping intervals with endpoints in E, is called the <u>strong variation</u> of F on E and is denoted by $V_*(F;E)$. In verifying the properties AC* and BV* and in computing $V_*(F;E)$ it is sufficient to consider the supremum taken over finite sequences of intervals. Note that, if $E = I_o$, then AC* and BV* agree respectively with the concepts of absolute continuity and bounded variation and $V_*(F;I_o) = V_F(I_o)$. Also note that, if F is AC* on E (resp. BV* on E) and $E' \subset E$, then clearly F is AC* on E' (resp. BV* on E').

(9.2) <u>Theorem</u>. <u>Each function</u> F <u>which is</u> AC* <u>on</u> $E \subset I_o$ <u>is necessarily</u> BV* <u>on</u> E. <u>Moreover, if</u> F <u>is</u>

BV* on E, F is BV* on $\overline{E}$ and if F is AC* on E and continuous on $\overline{E}$ then F is also AC* on $\overline{E}$.

Proof. Let F be AC* on E, a = inf E and b = sup E. There is $\delta > 0$ such that, if $|I| < \delta$, $V_*(F;E\cap I) \le 1$. Since F is AC* on E, F is bounded on [a,b]. Let $M = \sup\{|F(x)|: x \in [a,b]\}$. Let $\{x_o, x_1, \ldots, x_n\}$ be a partition of [a,b] of norm less than δ. Then

$$V_*(F;E) \le \Sigma\, V_*(F; E\cap[x_{i-1},x_i]) + 2M\cdot n < \infty.$$

Hence F is necessarily BV* on E. Now suppose F is BV* on E. Let I = [inf E, sup E] and $\{x_o, x_1, \ldots, x_n\}$ be a partition of I with each $x_i \in \overline{E}$. Then

$$\Sigma\, \theta(F;[x_{i-1},x_i]) = \Sigma'\theta(F;[x_{i-1},x_i]) + \Sigma''\theta(F;[x_{i-1},x_i])$$

where Σ' is over all i for which $[x_{i-1},x_i]$ contains a point of E and Σ'' is over all other intervals of the partition. For each interval $[x_{i-1},x_i]$ occurring in Σ', let w_i be a point in $E \cap [x_{i-1},x_i]$. Note that $[x_o,x_1]$ and $[x_{n-1},x_n]$ occur in Σ' and any interval in Σ'' has both of its adjacent intervals occurring in Σ'. It follows from this observation that

$$\Sigma''\theta(F;[x_{i-1},x_i]) \le \Sigma'\theta(F;[w_{i-1},w_i]) \le V_*(F;E).$$

It is also easy to observe that

$$\begin{aligned}\Sigma'\theta(F;[x_{i-1},x_i]) &\le \theta(F;[x_o,x_1]) + \theta(F;[x_{n-1},x_n]) \\ &\quad + 2\Sigma'\theta(F;[w_{i-1},w_i]) \\ &\le 2\theta(F;I) + 2V_*(F;E).\end{aligned}$$

Thus, if $M = 2\theta(F;I) + 3V_*(F;E)$, one has $\Sigma\,\theta(F;[x_{i-1},x_i]) \le M < \infty$ and hence $V_*(F;\bar{E}) \le M < \infty$ and F is BV* on $\bar{E}$. Finally, suppose that F is AC* on E and continuous on $\bar{E}$. Given $\varepsilon > 0$ by definition there is $\delta > 0$ such that, if $\{I_n\}$ is a sequence of pairwise non-overlapping intervals with endpoints in E, then $\Sigma\,\theta(F;I_n) < \varepsilon/4$. Let $\{I_n\}_{n=1}^N$ be non-overlapping with endpoints in $\bar{E}$ with $\Sigma|I_n| < \delta$. Let $\eta = \min\{|I_n|;\ n = 1, \ldots, N\}$. By the uniform continuity of F on $\bar{E}$, for each n there are points w_n and w'_n in E so that, if $I_n = [x_n,x'_n]$, I'_n is the interval with endpoints w_n and x_n and I''_n is the interval with endpoints x'_n and w'_n, then $|x_n - w_n| < \eta/2$, $|x'_n - w'_n| < \eta/2$, $\theta(F;I'_n) < \varepsilon/4N$ and $\theta(F;I''_n) < \varepsilon/4N$. Then

$$\begin{aligned}\Sigma\,\theta(F;I_n) \le\ & \Sigma'\theta(F;[w_n,w'_n]) + \Sigma''\theta(F;[w_n,w'_n]) \\ & + \Sigma\,\theta(F;I'_n) + \Sigma\,\theta(F;I''_n)\end{aligned}$$

where Σ' is the sum over odd indices n and Σ'' is over even ones. Assuming that the intervals I_n occur in their natural order, the intervals in Σ' are non-overlapping as are those in Σ''. Hence, $\Sigma\,\theta(F;I_n) \le \varepsilon/2 + \varepsilon/2 + N\cdot\varepsilon/4N + N\cdot\varepsilon/4N = \varepsilon$. It follows that F is AC* on $\bar{E}$ whenever F is AC* on E and continuous on $\bar{E}$.

Further properties of functions which are BV* on E or AC* and continuous on E or ACG* on I_o are given in the next theorem.

(9.3) Theorem. *Every function F which is BV* on E is equal on $\bar{E}$ with a function G which is of bounded variation on I_o; every F which is AC* on E and continuous on $\bar{E}$ agrees with a function G which is absolutely continuous on I_o. Furthermore, if F is ACG* on I_o, then F satisfies Lusin's condition (N) on I_o.*

Proof. In the first case F is BV* on $\bar{E}$ and in the second case F is AC* on $\bar{E}$. In either case, one may construct the function G which agrees with F on $\bar{E}$ and is linear on intervals contiguous to $\bar{E}$ with $G(x) = F(b)$ when $x \geq b = \sup E$ and $G(x) = F(a)$ when $x \leq a = \inf E$. Then if I is a contiguous interval to $\bar{E}$, $\inf_{x \in I} F(x) \leq \inf_{x \in I} G(x) \leq \sup_{x \in I} G(x) \leq \sup_{x \in I} F(x)$. From this it follows that for any $I' \subset I_o$ with endpoints in E, $\theta(G;I') \leq \theta(F;I')$. Thus if F is BV* on E then G is of bounded variation on I_o and if F is AC* on $\bar{E}$ then G is absolutely continuous on I_o. Finally, if F is ACG* on I_o then $I_o = \cup E_n$ where F is AC* on each E_n. Since by definition F is continuous, F is AC* on each $\bar{E}_n$. Thus there is a sequence of functions $\{G_n\}$ each absolutely continuous on I_o so that $F(x) = G_n(x)$ whenever $x \in E_n$. But then if $Z \subset I_o$ is a set of measure 0, then $F(Z) = \bigcup_{n=1}^{\infty} F(Z \cap E_n) = \bigcup_{n=1}^{\infty} G_n(Z \cap E_n)$. Since each E_n is absolutely continuous, for each n, $|G(Z \cap E_n)| = 0$ and hence $|F(Z)| = 0$. Thus, if F is ACG* on I_o, F satisfies Lusin's condition (N).

Theorem 9.1 also gives rise to characterizations of the classes of functions which are ACG* and BVG* on I_o which will be needed later; these characterizations are sometimes taken as the definitions of ACG* and BVG*.

(9.4) Theorem. A function F is ACG* on I_o iff F is continuous on I_o and each non-empty closed subset of I_o contains a portion on which F is AC*. A function F is BVG* on I_o iff each non-empty closed subset of I_o contains a portion on which F is BV*.

Proof. The proof of this theorem can be given in a very general setting. Suppose a property $\mathcal{P}$ holds on each subset of a set E and on the closure of the set E whenever it holds on E. We will show that $I_o = \bigcup_{n=1}^{\infty} E_n$ with $\mathcal{P}$ holding on each E_n iff each non-empty closed subset of I_o contains a portion on which $\mathcal{P}$ holds. To see this, first suppose $I_o = \cup E_n$ where $\mathcal{P}$ holds on each E_n. Then $\mathcal{P}$ holds on each $\overline{E}_n$. If $K \neq \emptyset$ is a closed subset of I_o, then $K = \bigcup_{n=1}^{\infty} (K \cap \overline{E}_n)$ and by the Baire category theorem one of the sets $K \cap \overline{E}_n$ is dense in a portion of K. But since K is closed, this set $K \cap \overline{E}_n$ contains a portion of K and hence $\mathcal{P}$ holds on this portion. Conversely, if each non-empty closed subset K of I_o contains a portion on which $\mathcal{P}$ holds, first let J_o be a portion of I_o on which $\mathcal{P}$ holds. If $J_o \neq I_o$, then let J_1 be a portion of $K_o = I_o \backslash \mathring{J}_o$ on which $\mathcal{P}$ holds. In general, if for every ordinal number $\beta<\alpha$, K_β has been defined and $\bigcap_{\beta<\alpha} K_\beta \neq \emptyset$, let J_α be a closed interval so that $\mathring{J}_\alpha \cap \bigcap_{\beta<\alpha} K_\beta \neq \emptyset$ and $\mathcal{P}$ holds on $J_\alpha \cap \bigcap_{\beta<\alpha} K_\beta$. Let $K_\alpha = \bigcap_{\beta<\alpha} K_\beta \backslash \mathring{J}_\alpha$. Since the K_α then form a non-increasing sequence of closed subsets of I_o, there is $\alpha_o < \omega_1$ so that for $\alpha > \alpha_o$, $K_\alpha = K_{\alpha_o}$. But if K_{α_o} were not empty then it would have a portion on which $\mathcal{P}$ holds and then K_{α_o+1} would be properly contained in K_{α_o}. It follows that $K_{\alpha_o} = \emptyset$, $I_o = \bigcup_{\alpha<\alpha_o} (J_\alpha \cap \bigcap_{\beta<\alpha} K_\beta)$ and $\mathcal{P}$ holds on each of the sets in the union. Now when $\mathcal{P}$ is interpreted for a function F defined on I_o as "F is BV* on E", the first part of the theorem follows; when $\mathcal{P}$ is interpreted for F continuous on I_o as "F is AC* on E", the second part follows.

In order to show that the functions in $\mathcal{D}*$ form an integral, we need the results of the next two theorems.

(9.5) Theorem. *Each function F which is BVG* on I_o (and, in particular, each F which is ACG* on I_o) is differentiable a.e.*

Proof. Suppose F is BVG* on $I_o = [a_o,b_o]$. Then there is a sequence of sets $\{E_n\}$ with $I_o = \cup E_n$ such that F is BV* on each E_n and hence is BV* on each $\overline{E}_n$. Consider a fixed natural number n and let $a = \inf E_n$ and $b = \sup E_n$. Define $G(x)$ and $H(x)$ on $[a,b]$ to be equal to $F(x)$ on $\overline{E}_n$ and on each I contiguous to $\overline{E}_n$ let

$$G(x) = \inf\{F(x)\colon x \in I\} \quad \text{and} \quad H(x) = \sup\{F(x)\colon x \in I\}.$$

Then, since F is BV^* on $\overline{E}_n$, both G and H are of bounded variation on $[a,b]$ and $G(x) \le F(x) \le H(x)$ at each $x \in [a,b]$. Let t be a limit point of E which is also a point of differentiability of both G and H. Since t is a limit point of E, $G'(t)$ must equal $H'(t)$. Since $G(t) = F(t) = H(t)$ and for all x, $G(x) \le F(x) \le H(x)$, for $x > t$,

$$\frac{G(t) - G(x)}{t - x} \le \frac{F(t) - F(x)}{t - x} \le \frac{H(t) - H(x)}{t - x}$$

and for $x < t$,

$$\frac{H(t) - H(x)}{t - x} \le \frac{F(t) - F(x)}{t - x} \le \frac{G(t) - G(x)}{t - x}.$$

Letting x approach t one has by the standard argument that $G'(t) = F'(t) = H'(t)$. Since all but an at most countable set of points of E_n are limit points of E_n and since G and H are of bounded variation and thus differentiable a.e., F is differentiable at almost every point of E_n. Since this is true on each set E_n it follows that F is differentiable at almost every point of $I_o = \cup E_n$.

(9.6) <u>Theorem</u>. <u>If</u> $F(x)$ <u>is</u> ACG^* <u>on</u> I_o <u>and</u> $F'(x) = 0$ a.e. <u>on</u> I_o, <u>then</u> F <u>is identically constant on</u> I_o.

<u>Proof</u>. Let $F(x)$ be ACG^* on I_o and

$$A = \{x \in I_o\colon F'(x) = 0\}.$$

Suppose $F'(x) = 0$ a.e. on I_o. Then $m(I_o \backslash A) = 0$ and, since F satisfies Lusin's condition (N), $m(F(I_o \backslash A)) = 0$. Since $F'(x) = 0$ at each $x \in A$, one also has $m(F(A)) = 0$. Hence, $m(F(I_o)) = 0$. Since F is a continuous function whose image on I_o is of measure 0, F must be identically constant. For otherwise, if F took on two different values at two points if I_o, being continuous, F would have to take on all intermediate values between them and then the image of I_o under F would not be of measure 0.

We are now in a position to show that $\mathcal{D}*$ is an integral.

(9.7) Theorem. The class of functions $\mathcal{D}*$ consisting of those f which are equal almost everywhere to the derivative of a function which is ACG* satisfy the properties i) - v) where for $f \in \mathcal{D}*$ and $F'(x) = f(x)$ a.e. with $F \in$ ACG*, $\mathcal{D}*\int f$ is defined to be $F(I_o)$.

Proof. If $F \in$ ACG* on $I_o = [a_o, b_o]$ then $I_o = \cup E_n$ and, on each of the sets E_n, F agrees with a function G_n which is absolutely continuous on I_o. Also $F'(x) = G_n'(x)$ at almost every point of E_n. Since each $G_n'(x)$ is measurable, it follows that $F'(x)$ is a measurable function because at almost every x, $F'(x) = \sum_{n=1}^{\infty} G_n'(x) \cdot C_{E_n'}$ where $E_n' = \overline{E}_n \backslash \bigcup_{i=1}^{n-1} \overline{E}_i$. If $f \in \mathcal{D}*$ then $f(x) = F'(x)$ a.e. for some $F \in$ ACG*. If $f(x) = G'(x)$ a.e. for some $G \in$ ACG*, then $(F-G)'(x)$

$= 0$ a.e. and by the previous theorem $F - G$ is identically constant. Then $F(I_o) = F(b_o) - F(a_o) = G(b_o) - G(a_o) = G(I_o)$ and $\mathcal{D}{*}\int f = F(I_o)$ is well-defined. Again, if $f \in \mathcal{D}{*}$ and $f(x) = F'(x)$ a.e. when $F \in ACG{*}$, then for $I = [a,b] \subset I_o$ one has $\mathcal{D}{*}\int f \cdot C_I = F(I)$. This follows by simply letting the function G be given by $G(x) = F(x)$ if $x \in I$, $G(x) = F(b)$ if $x > b$ and $G(x) = F(a)$ if $x < a$. Then G is $ACG{*}$ on I_o and $G'(x) = f(x) \cdot C_I(x)$ at almost every point of I_o and thus $\int f \cdot C_I = G(I) = F(I)$. Thus $\mathcal{D}{*}$ satisfies i). Clearly, one also has for each such $f \in \mathcal{D}{*}$ and $I = I_1 \cup I_2$ where I_1 and I_2 are non-overlapping, that $F(I) = F(I_1) + F(I_2)$; that is, $F(I) = \mathcal{D}{*}\int f \cdot C_I$ is an additive function of an interval and since all such $F \in ACG{*}$ are continuous, ii) holds. Now if f and g are in $\mathcal{D}{*}$ and F and G are in $ACG{*}$ with $F'(x) = f(x)$ a.e. and $G'(x) = g(x)$ a.e., then for $a,b \in \mathbb{R}$, $(aF + bG)'(x) = af(x) + bg(x)$ a.e. Since F is $ACG{*}$ on I_o, $I_o = \cup E_n$ and on each E_n, F is $AC{*}$. Similarly, $I_o = \cup E'_m$ and on each E'_m, G is $AC{*}$. But then $I_o = \cup E_{n,m}$ where $E_{n,m} = E_n \cap E'_m$ and both F and G are $AC{*}$ on each $E_{n,m}$. Then $aF + bG$ is readily seen to be $AC{*}$ on each $E_{n,m}$ and thus $aF + bG \in ACG{*}$. Hence the class $\mathcal{D}{*}$ is a vector space of measurable functions and for $f,g \in \mathcal{D}{*}$ and $a,b \in \mathbb{R}$ one has $\mathcal{D}{*}\int(af + bg) = a \cdot \mathcal{D}{*}\int f + b \cdot \mathcal{D}{*}\int g$; that is, iii) holds. Since the absolutely continuous functions with domain I_o are in $ACG{*}$, it follows that if $f \in \mathcal{L}$ on I_o then $f \in \mathcal{D}{*}$ and $\mathcal{D}{*}\int f = \mathcal{L}\int_{I_o} f$; that is, iv) holds. Finally, v) holds, by definition. That is, if $f \in \mathcal{D}^*$ and $F \in ACG^*$ with $F(I) = \mathcal{D}^*\int f \cdot C_I$, then $F'(x) = f(x)$ a.e. by the definition of the $\mathcal{D}^*$-integral.

We now proceed to establish the equivalence of the $\mathcal{L}^*_{\omega_1}$ process and the D*-integral, thereby also establishing indirectly the fact that $\mathcal{L}^*_{\omega_1}$ is an integral. For this purpose, we first show that $(\mathcal{D}^*)^C = (\mathcal{D}^*)^{H^*} = \mathcal{D}^*$; this implies the inclusion relationship $\mathcal{L}^*_{\omega_1} \subset \mathcal{D}^*$ because $\mathcal{L} \subset \mathcal{D}^*$ and $\mathcal{L}^*_{\omega_1}$ is the smallest class of functions closed under (C) and (H*).

(9.8) <u>Theorem</u>. <u>The class of functions</u> $\mathcal{D}^*$ <u>is closed under both the Cauchy</u> (C) <u>and the Harnack</u> H*-<u>type of improper integral</u>.

<u>Proof</u>. Let $\{x_1, \ldots, x_n\} \subset I_o$ and suppose F is ACG* on each closed interval I which contains none of the points x_i and suppose F is continuous on I_o. It is then almost immediate that F is ACG* on I_o. Indeed, this follows from the fact that $I_o = [a_o, x_1] \cup [x_1, x_2] \cup \ldots \cup [x_{n-1}, x_n]$ and on each of these intervals F is ACG*; F is trivially ACG* at their endpoints and each open interval (x_{i-1}, x_i) is a countable union of intervals on each of which F is ACG*. Now let P be a closed subset of I_o and f be $\mathcal{D}^*$-integrable on P and in each I_n contiguous to P. Because $\mathcal{D}^*$ is closed under Cauchy extensions, without loss of generality, one may assume that there

are no isolated points in P and thus that P is a perfect subset of I_o. Let $c_n = \theta(\mathcal{D}*\int f\cdot C_{I_n};I_n)$ and assume $\Sigma c_n < \infty$. Let $F_n(x) = \mathcal{D}*\int f\cdot C_{I_n\cap[a_o,x]}$ and $F(x) = \mathcal{D}*\int_{a_o}^{x} f(t)\cdot C_P(t)\,dt + \sum_{n=1}^{\infty} F_n(x)$. Because $\Sigma c_n < \infty$, the sum $\Sigma F_n(x)$, involved in the definition of F converges absolutely. Clearly $F(x)$ is ACG* on each I_n. Also, since the sum, $\Sigma F_n(x)$, is a sum of continuous functions and converges uniformly, $F(x)$ is also a continuous function. It remains to show that F is ACG* on P. In order to show this, we show that $G(x) = \sum_{n=1}^{\infty} F_n(x)$ is AC* on P. Let $\varepsilon > 0$ be given. Since $\Sigma c_n < \infty$, there is N so that $\sum_{n=N}^{\infty} c_n < \varepsilon/2$. Let $\delta_1 = \min\{|I_n|,\ n = 1, 2, \ldots, N\}$. Let $h(x) = 0$ if $x \in P$ and $h(x) = (1/|I_n|)\mathcal{D}*\int f\cdot C_{I_n}$ if $x \in I_n$. Then $h(x)$ is Lebesgue integrable. Let $H(x) = \int_{a_o}^{x} h(t)dt$. Then $H(x) = G(x)$ on P and since $H(x)$ is absolutely continuous, there is δ_2 so that if $\{I'_n\}$ is a sequence of non-overlapping intervals and $\Sigma|I'_n| < \delta_2$. then $\Sigma m(H(I'_n)) < \varepsilon/2$. Let $\delta = \min(\delta_1,\delta_2)$. Now let $\{I'_n\}$ consist of pairwise non-overlapping intervals with end points in P with $\Sigma|I'_n| < \delta$. Then, for each I'_n,

$$\begin{aligned}\theta(G;I'_n) &= m(G(I'_n)) \\ &\le \Sigma' m(G(I_n)) + m(G(P\cap I'_n))\end{aligned}$$

where Σ' is over all $I_n \subset I'_n$ with I_n contiguous to P. Then $\theta(G;I'_n) \le \Sigma'\theta(G;I_n) + m(H(I_n))$ and summing both sides of this inequality yields

$$\Sigma\, \theta(G;I'_n) \le \varepsilon/2 + \varepsilon/2 = \varepsilon.$$

Thus G is AC* on P and, consequently F is ACG* on P. Hence F is ACG* on I_o and we have that $\mathcal{D}*$ is closed under the Harnack H*-type improper integrals.

The fact that $\mathcal{L}^*_{\omega_1} \subset \mathcal{D}*$ and that the integrals agree follows by induction on the classes $\mathcal{L}^*_\alpha$. We have $\mathcal{L}^*_o = \mathcal{L} \subset \mathcal{D}*$ and $\mathcal{L}\int f = \mathcal{D}*\int f$. If for each $\beta < \alpha$ and $f \in \mathcal{L}^*_\beta$, $\mathcal{L}^*_\beta\int f = \mathcal{D}*\int f$, since $\mathcal{L}^*_\alpha$ is formed by applying (C) and (H*) to $\bigcup_{\beta<\alpha} \mathcal{L}^*_\beta$, if $f \in (\bigcup_{\beta<\alpha} \mathcal{L}^*_\beta)^C$ then $\mathcal{L}^*_\alpha\int f = F(I_o)$ where F, the continuous function given by the Cauchy process, is ACG*. Also, when (H*) is applied with $f\cdot C_P \in (\mathcal{L}^*_{\beta_o})^C$ and $f\cdot C_{I_n} \in (\mathcal{L}^*_{\beta_n})^C$ for each I_n contiguous to P with $\beta_n < \alpha$, then

$$\mathcal{L}^*_\alpha\int f = (\mathcal{L}^*_{\beta_o})^C \int f\cdot C_P + \Sigma(\mathcal{L}^*_{\beta_k})^C \int f\cdot C_{I_n}$$
$$= \mathcal{D}*\int f\cdot C_P + \Sigma\mathcal{D}*\int f\cdot C_{I_n} = \mathcal{D}*\int f.$$

The next theorem shows that every $f \in \mathcal{D}*$ belongs to $\mathcal{L}^*_{\omega_1}$; that is, that each $f \in \mathcal{D}*$ can be integrated by applying (C) and (H*) repeatedly up to an ordinal which depends on f and is less than ω_1.

(9.9) <u>Theorem</u>. <u>The class</u> $\mathcal{D}^*$ <u>is identical with</u> $\mathcal{L}^*_{\omega_1}$ <u>and for each</u> $f \in \mathcal{L}^*_{\omega_1}$, $\mathcal{D}^*\int f = \mathcal{L}^*_{\omega_1}\int f$.

<u>Proof</u>. After Theorem 9.8 all that remains is to show that $\mathcal{D}^* \subset \mathcal{L}^*_{\omega_1}$. So suppose $f \in \mathcal{D}^*$ and let $S = S(f, \mathcal{L}^*_{\omega_1})$. By Theorem 9.1, S is closed and in order to show that $f \in \mathcal{L}^*_{\omega_1}$ it suffices to show that S is empty. For this purpose note that S can not have any isolated points. This is because $F(x) = \mathcal{D}^*\int_{a_o}^{x} f(t)dt$ is a continuous function and if x_1 were the only point in $S \cap \mathring{I}$, by applying the Cauchy extension $f \cdot C_I$ would belong to $(\mathcal{L}^*_{\omega_1})^C = \mathcal{L}^*_{\omega_1}$. Thus S is a perfect set. The same argument as that which was just given also shows that $f \in \mathcal{L}^*_{\omega_1}$ on each interval contiguous to S. Suppose $S \neq \emptyset$. Since $F \in ACG^*$, there is a portion of S, $S \cap I$, on which F is AC^*. But $f \in \mathcal{L}^*_{\omega_1}$ on each interval I_n contiguous to $S \cap I$ in I. Also $F \in AC^*$ on $S \cap I$ which implies that F agrees with an absolutely continuous function G on $S \cap I$ and $F'(x) = G'(x) = f(x)$ a.e. on $S \cap I$ implies that $f \cdot C_{S \cap I} \in \mathcal{L}$. Since $\Sigma\, \mathcal{O}(\mathcal{L}^*_{\omega_1}\int f \cdot C_{I_n}; I_n) = \Sigma\, \mathcal{O}(F; I_n) < \infty$, we have $f \cdot C_I \in (\mathcal{L}^*_{\omega_1})^{H*} = \mathcal{L}^*_{\omega_1}$. This contradiction shows that $S = \emptyset$ and thus that $\mathcal{D}^* = \mathcal{L}^*_{\omega_1}$.

We now turn to the Perron approach to generalizing the Lebesgue integral and to its equivalence to the $\mathcal{D}*$ integral. For this we will need some additional definitions. A function $U(x)$ is said to be a *majorant* of a function $f(x)$ on I_o if at each $x \in I_o$, $-\infty \neq \underline{U}(x) \geq f(x)$; $V(x)$ is said to be a *minorant* of $f(x)$ if $\infty \neq \overline{V}(x) \leq f(x)$. If U is a majorant and V a minorant of a function $f(x)$, then $\underline{(U - V)}(x) \geq \underline{U}(x) - \overline{V}(x) \geq 0$. We begin by showing that this implies that $U - V$ is a non-negative function of an interval on I_o; equivalently, $(U - V)(x)$ is monotone non-decreasing.

(9.10) *Theorem*. *If* $\underline{F}(x) \geq 0$ *at each* x *in* I_o, *then* $F(I)$ *is non-negative on each* $I \subset I_o$.

Proof. Suppose not. Then there is an F with $\underline{F}(x) \geq 0$ at each $x \in I_o$ and $I_1 \subset I_o$ with $F(I_1) < 0$. Let $\varepsilon > 0$ satisfy $F(I_1) < -\varepsilon|I_1|$ and let $G(I) = F(I) + \varepsilon|I|$ for each $I \subset I_o$. Then $\underline{G}(x) \geq \varepsilon$ at each $x \in I_o$ and $G(I_1) < 0$. Since G is an additive function of an interval, by repeated subdivision of I_1 one can determine a nested sequence of intervals $\{I_n\}$ on each of which G is negative and this can be done so that $\cap I_n$ contains a single point x_o. But then $\underline{G}(x_o) \leq 0$ contradicting the fact that $\underline{G}(x) \geq \varepsilon > 0$ at each $x \in I_o$.

Now a function $f(x)$ is said to be *Perron integrable* on I_o if for each $\varepsilon > 0$ there is a majorant $U(x)$ of $f(x)$ and a minorant $V(x)$ of $f(x)$ so that $U(I_o) - V(I_o) < \varepsilon$. Since for each majorant U and minorant V of f one has $U(I_o) \geq V(I_o)$, one may define the <u>Perron integral</u> of f on I_o as $\mathcal{P}\int f = \inf U(I_o) = \sup V(I_o)$. One may clearly also define upper and lower Perron integrals and f will be Perron integrable when they are equal; in this case we will write $f \in \mathcal{P}$.

It is worth noting from the start that, in the case where $F(x)$ is differentiable on I_o, then $F(x)$ is both a majorant and a minorant for $f(x) = F'(x)$ and hence the derivative of a differentiable function is easily seen to belong to $\mathcal{P}$; also, if $f(x) = F'(x)$ on I_o, then $\mathcal{P}\int f = F(I_o)$.

We proceed to check that the $\mathcal{P}$-integral satisfies i) - v). Some of this information so obtained will be used in the proof that the $\mathcal{P}$ and $\mathcal{D}*$ integrals are equivalent; in any case, checking this has value in that it illuminates some methods applicable to the Perron approach.

To see that i) holds for the Perron integral, let $f \in \mathcal{P}$ on I_o and $[a,b] = I \subset I_o$. Then given $\varepsilon > 0$ there is a majorant $U(x)$ and a minorant $V(x)$ so that $U(I_o) - V(I_o) < \varepsilon/2$. Note that U and V restricted to I are almost what is required for a majorant and minorant for $f \cdot C_I$; but care has to be taken near the endpoints of I to insure that $\underline{U}$ and $\overline{V}$ are respectively above and below the value of f. Thus, let

$$U_1(x) = \begin{cases} U(x) & \text{if } x \in [a,b] \\ U(b) + \frac{\varepsilon}{8}\left(\frac{x - b}{b_o - a_o}\right)^{1/3} & \text{if } x > b \\ U(a) + \frac{\varepsilon}{8}\left(\frac{x - a}{b_o - a_o}\right)^{1/3} & \text{if } x < a \end{cases}$$

$$V_1(x) = \begin{cases} V(x) & \text{if } x \in [a,b] \\ V(b) - \frac{\varepsilon}{8}\left(\frac{x-b}{b_o - a_o}\right)^{1/3} & \text{if } x > b \\ V(a) - \frac{\varepsilon}{8}\left(\frac{x-a}{b_o - a_o}\right)^{1/3} & \text{if } x < a. \end{cases}$$

Then it is easy to check that $-\infty \neq \underline{U}_1(x) > f(x)\cdot C_I(x)$ and $\infty \neq \overline{V}_1(x) < f(x)\cdot C_I(x)$ at each $x \in I_o$ and $U_1(I_o) - V_1(I_o) \leq U(I_o) - V(I_o) + \varepsilon/2 \leq \varepsilon$. Hence $f\cdot C_I \in \mathcal{P}$ and $\mathcal{P}$ satisfies property i).

To see that $P(I) = \mathcal{P}\int f\cdot C_I$ is an additive function of an interval for each $f \in \mathcal{P}$, the same care is needed. Let $I = I_1 \cup I_2$ with $\{c\} = I_1 \cap I_2$ and let $f \in \mathcal{P}$ and $\varepsilon > 0$ be given. Let U_1, U_2 and V_1, V_2 be respectively majorants and minorants for $f\cdot C_{I_1}$ and $f\cdot C_{I_2}$ so that

$$(U_1 - V_1)(I_o) < \varepsilon/4 \quad \text{and} \quad (U_2 - V_2)(I_o) < \varepsilon/4.$$

Then

$$U(x) = U_1(x) + U_2(x) + \frac{\varepsilon}{4}\cdot\left(\frac{x-c}{x_o - a_o}\right)^{1/3}$$

and

$$V(x) = V_1(x) + V_2(x) - \frac{\varepsilon}{4}\cdot\left(\frac{x-c}{x_o - a_o}\right)^{1/3}$$

are majorants and minorants of $f\cdot C_I$ and $(U - V)(I_o) < \varepsilon$. It is then clear that $\mathcal{P}\int f\cdot C_I = \mathcal{P}\int f\cdot C_{I_1} + \mathcal{P}\int f\cdot C_{I_2}$. To see that $P(I) = \mathcal{P}\int f\cdot C_I$ is also continuous, first note that each majorant U and minorant V satisfy on each $I \subset I_o$, $U(I) \geq P(I) \geq V(I)$ and since $(U - V)(x)$ is non-decreasing, if $U(I_o) - V(I_o) < \varepsilon$, then $0 \leq U(I) - P(I) < \varepsilon$ and $0 \leq P(I) - V(I) < \varepsilon$ and each corresponding function of a point $(U - P)(x)$ and $(P - V)(x)$ are non-decreasing. Since $\underline{U} \ge -\infty$, $\underline{\lim}_{I \to x} U(I) \ge 0$ and hence $U(I) \le \varepsilon + P(I)$ which implies that $\underline{\lim}_{I \to x} U(I) \le \varepsilon + \underline{\lim}_{I \to x} P(I)$. Since $\varepsilon > 0$ was arbitrary, $0 \le \underline{\lim}_{I \to x} P(I)$. A similar argument using minorants shows that $\overline{\lim}_{I \to x} P(I) \le 0$ and hence P is continuous at each $x \in I_o$ and property ii) of integrals holds.

Note that if U and V are a majorant and, respectively, a minorant for f, then $-V$ and $-U$ are a majorant and, respectively, a minorant for $-f$. With this in mind, a similar argument to that given for property i) shows that $\mathcal{P}$ is a vector space of functions and that if $f, g \in \mathcal{P}$ and $a, b \in \mathbb{R}$ then $af + bg \in \mathcal{P}$ and $\mathcal{P}\int (af + bg) = a \cdot \mathcal{P}\int f + b \cdot \mathcal{P}\int g$. To complete the proof that iii) holds, it is necessary to show that if $f \in \mathcal{P}$, then f is measurable. This will follow from the proof of v). That is, once it is shown that if $P(x) = \mathcal{P}\int_{a_o}^{x} f(t)dt$, then $P'(x) = f(x)$ a.e., then f is measurable because it is the almost everywhere derivative of a continuous function.

That iv) holds and thus the Lebesgue integral is contained in the Perron integral is a consequence of the next two theorems.

(9.11) *Theorem*. (Vitali-Caratheodory) *If* $f(x)$ *is Lebesgue integrable on* I_o, *then for each* $\varepsilon > 0$ *there exist extended real-valued functions* $g(x)$ *and*

$h(x)$ <u>on</u> I_o <u>with</u> $g(x)$ <u>upper semi-continuous</u>, $h(x)$ <u>lower semi-continuous</u>, $g(x) \leq f(x) \leq h(x)$,

$$\int_{I_o} (h(x) - f(x))dx < \varepsilon \quad \underline{\text{and}} \quad \int_{I_o} (f(x) - g(x))dx < \varepsilon.$$

<u>Proof</u>. By symmetry, it clearly suffices, given $\varepsilon > 0$ and $f \in \mathcal{L}$, to find $h(x)$ which is lower semi-continuous on I_o so that $f(x) \leq h(x)$ and $\int_{I_o} (h(x) - f(x))dx < \varepsilon$. For simplicity, assume $I_o = [0,1]$. If $f_m(x) = \max(f(x), -m)$ then $\lim f_m(x) = f(x)$ and by the monotone convergence theorem $\lim_{m\to\infty} \int_{I_o} f_m(x)dx = \int_{I_o} f(x)dx$. Thus there is $M > 0$ so that if $E_M = \{x: f(x) < -M\}$ then $\int_{E_M} f(x)dx > -\varepsilon$. Let $f_o(x) = f_M(x) + M$ and note that $f_o(x) \geq 0$. For each natural number n, let $A_n = \{x: f_o(x) > (n-1)\varepsilon\}$ and let G_n be an open set with $A_n \subset G_n$ such that $m(G_n \backslash A_n) < \varepsilon/2^n$. Then $C_{G_n}(x)$ is lower semi-continuous as is $\sum_{i=1}^{N} C_{G_n}(x)$ and, since the increasing limit of lower semi-continuous functions is lower semi-continuous, so is $h_o(x) = \sum_{n=1}^{\infty} \varepsilon \cdot C_{G_n}(x)$. Clearly, if $f_o(x) \in ((n-1)\varepsilon, n\varepsilon]$ then $h_o(x) \geq n\varepsilon$ and if $f_o(x) = \infty$, $h_o(x) = \infty$. Thus $h_o(x) \geq f_o(x)$ at each x. Finally, $\int_0^1 (h_o(x) - f_o(x))dx \leq \varepsilon + \sum_{n=1}^{\infty} m(G_n \backslash A_n) \leq 2\varepsilon$. Then $h(x) = h_o(x) + M$ is lower semi-continuous,

$h(x) \geq f(x)$ at each x and

$$\int_0^1 (h(x)-f(x))dx \leq \int_0^1 (h_o(x)-f_o(x))dx - \int_{E_M} f(x)dx \leq 3\varepsilon.$$

Since $\varepsilon > 0$ was arbitrary this proves the theorem.

Note: The complete Vitali-Caratheodory theorem is slightly more general than the above but is proved similarly. It asserts the existence of functions g_n and h_n for a given $f(x)$ which is measurable with respect to a Lebesgue-Stieltjes measure m so that each h_n is lower semi-continuous, each g_n is upper semi-continuous, $g_n \leq f \leq h_n$ and $\int \lim g_n(x)\, dm = \int \lim h_n(x)\, dm = \int f(x)\, dm$.

(9.12) <u>Theorem</u>. <u>If</u> $f(x)$ <u>is Lebesgue integrable on</u> I_o, <u>then for each</u> $\varepsilon > 0$ <u>there is a majorant</u> U <u>and minorant</u> V <u>of</u> f <u>on</u> I_o <u>so that</u> U <u>and</u> V <u>are absolutely continuous and</u> $(U - V)(I_o) < \varepsilon$ <u>and consequently</u> f <u>is Perron integrable and</u> $\mathcal{L}\int f = \mathcal{P}\int f$. <u>Moreover,</u> U <u>can be taken to be the indefinite integral of a lower semi-continuous function, and</u> V <u>the indefinite integral of an upper semi-continuous function</u>.

<u>Proof</u>. Let f be Lebesgue integrable on I_o and for simplicity assume $I_o = [0,1]$. Given $\varepsilon > 0$, let h be lower semi-continuous and g upper semi-continuous with $g(x) \leq f(x) \leq h(x)$, $\int(h - f) < \varepsilon$ and $\int(f - g) < \varepsilon$ as guaranteed by the last theorem. Then $U(x) = \int_{a_o}^{x} h(t)dt$ and $V(x) = \int_{a_o}^{x} g(t)dt$ are the majorant and

minorant functions with the required properties. By symmetry, it suffices to consider $U(x)$. Recall that there is a number $M > 0$ so that $h(x) > -M > -\infty$ and that $h(x) \geq f(x)$. By the lower semi-continuity of h, given $x \in I_o$ and $\varepsilon > 0$,

$$\frac{U(I)}{|I|} = \frac{\int_I h(t)dt}{|I|} \geq \frac{\int_I (h(x) - \varepsilon)dt}{|I|} = h(x) - \varepsilon$$

provided I is a small enough interval containing x. Letting I approach x yields $\underline{U}(x) \geq -M - \varepsilon$ and $\underline{U}(x) \geq f(x) - \varepsilon$ and, since $\varepsilon > 0$ was arbitrary, $\underline{U}(x) \geq -\infty$ and $\underline{U}(x) \geq f(x)$. Since this holds at each $x \in I_o$, U is a majorant of f. Clearly, if one desires, U can be defined to the left of a_o and right of b_o to guarantee that $\underline{U}(a_o) \geq f(a_o)$ and $\underline{U}(b_o) \geq f(b_o)$. Then, for the corresponding V, one has $(U - V)(I_o) = \int_{I_o} (h(x) - g(x)) = \int_{I_o} (h(x) - f(x)) + \int_{I_o} (f(x) - g(x)) \leq 2\varepsilon$. Finally, it is apparent that $\inf U(I_o) = \sup V(I_o) = \int_{I_o} f(x)dx$ and thus f is Perron integrable and $\mathcal{P}\int f = \mathcal{L}\int f$.

It now remains to show that v) holds for the Perron integral. Fix $\varepsilon > 0$ and $f \in \mathcal{P}$ and select U a majorant for f and V a minorant so that $U(I_o) - V(I_o) < \varepsilon^2$. Let $P(x) = \mathcal{P}\int_{a_o}^{x} f(t)dt$ and $H(x) = U(x) - P(x)$. Then $H(x)$ is non-decreasing, $H'(x) \geq 0$ a.e. and $\int_{I_o} H'(x)dx \leq H(I_o) \leq \varepsilon^2$. Since $U(x) = H(x) + P(x)$, $\underline{U}(x) = H'(x) + \underline{P}(x)$ a.e. and hence $\underline{P}(x) > -\infty$ a.e. and $\underline{U}(x) - \underline{P}(x) \geq 0$ a.e. Consider $E = \{x\colon \underline{U}(x) - \underline{P}(x) \leq \varepsilon\}$. Then $|E| = |\{x\colon H'(x) \leq \varepsilon\}|$

and, since $\varepsilon^2 \geq H(I_o) \geq \int_{I_o} H'(x)dx \geq \int_E H'(x)dx \geq \varepsilon \cdot m(E)$, it follows that $m(E) \leq \varepsilon$. Then

$$m^*\{x: f(x) - \underline{P}(x) \geq \varepsilon\} \leq \varepsilon$$

since this set is contained in E. Since $\varepsilon > 0$ is arbitrary, $m\{x: f(x) > \underline{P}(x)\} = 0$ and $m\{x: f(x) = \infty\} = 0$. In an analogous fashion using the minorants one obtains $m\{x: f(x) < \overline{P}(x)\} = 0$ and $m\{x: f(x) = -\infty\} = 0$. Hence, if $f \in \mathcal{P}$ on I_o, then f is finite a.e. and $P'(x) = f(x)$ a.e.

The mathematician Hake (1921) proved that $\mathcal{D}^* \subset \mathcal{P}$ and that the $\mathcal{D}^*$-integral agrees with the $\mathcal{P}$-integral for each $\mathcal{D}^*$-integrable function. Alexandroff (1924) and Looman (1925) independently showed that $\mathcal{P} \subset \mathcal{D}^*$ and thus that the two integrals are equivalent. The resulting equivalence is known as the theorem of Hake, Alexandroff and Looman. This equivalence will be established by the next three theorems. In order to do this we need a definition of a seemingly more restrictive integral. A function f will be said to be <u>$\mathcal{P}_o$-integrable</u> on I_o if, for every $\varepsilon > 0$, there is a continuous majorant U and a continuous minorant V of f so that $(U - V)(I_o) < \varepsilon$. The natural definition of the $\mathcal{P}_o$-integral of f, $\mathcal{P}_o\int f = \sup V(I_o) = \inf U(I_o)$, clearly agrees with the $\mathcal{P}$-integral when f is $\mathcal{P}_o$-integrable. Indeed, the previous discussion of the $\mathcal{P}$-integral also shows that the $\mathcal{P}_o$-integral satisfies i) - v). In any event, the class of functions $\mathcal{P}_o$ is contained in $\mathcal{P}$ and the equivalence of $\mathcal{P}$ and $\mathcal{D}^*$ will be obtained when it is shown that $\mathcal{D}^* \subset \mathcal{P}_o$, with the $\mathcal{D}^*$-integral agreeing with the $\mathcal{P}_o$-integral, and that $\mathcal{P} \subset \mathcal{D}^*$. It will then follow that the Perron and narrow-sense Denjoy integrals are equivalent.

(9.13) <u>Theorem</u>. <u>The class</u> $\mathcal{P}_o$ <u>is closed under Cauchy (C) and Harnack H*-type improper integrals and the</u> $\mathcal{P}_o$<u>-integral agrees with the</u> $\mathcal{P}_o^C$ <u>and</u> $\mathcal{P}_o^{H*}$ <u>integrals; consequently,</u> $\mathcal{L}^*_{\omega_1} = \mathcal{D}^* \subset \mathcal{P}_o \subset \mathcal{P}$ <u>and the Perron integral includes the</u> $\mathcal{D}^*$<u>-integral</u>.

<u>Proof</u>. To see that $\mathcal{P}_o$ is closed under the Cauchy extension, it suffices to consider a single point $c \in I_o$ and to show that, if $f\cdot C_I \in \mathcal{P}_o$ for each $I = [a_o,t] \subset [a_o,c)$ and if $\lim_{t\to c^-} P(t)$ exists, then $f\cdot C_{[a_o,c]} \in \mathcal{P}_o$ and this limit of $P(t)$ as $t \to c^-$ equals $\mathcal{P}_o\int f\cdot C_{[a_o,c]}$. Closure of $\mathcal{P}_o$ under the Cauchy extension then follows by also considering right hand limits and then extending to finite sets of points by induction. Thus, let $\{t_n\}_{n=1}^{\infty}$ be an increasing sequence of points with $\lim t_n = c$. Let $t_o = a_o$ and U_n and V_n be continuous majorants and minorants respectively of $f\cdot C_{[t_{n-1},t_n]}$ so that $(U_n - V_n)(I_o) < \varepsilon/2^n$. Let $\overline{U}_o$ be the additive function of an interval defined so that $U_o(I) = 0$ if $I \subset [c,b_o]$ and $U_o(I) = U_n(I)$ if $I \subset [t_{n-1},t_n]$ and let V_o be defined similarly. Since for each $I \subset [t_{n-1},t_n]$, $U_o(I) - P(I) \le \varepsilon/2^n$, it follows that $\lim_{t\to c^-}\theta(U_o;(t,c)) = 0$ and hence U_o is continuous and well-defined on I_o, as

is V_o. It is necessary to adjust U_o and V_o slightly to create a majorant U and minorant V which satisfy $\infty = \underline{U}(c) \geq f(c) \geq \overline{V}(c) = -\infty$. This can be accomplished by choosing $\delta > 0$ so that $\theta(U;[c-\delta,c]) < \varepsilon$, letting

$$G_1(x) = -\theta(U_o; [x,c]\cap[c-\delta,c]),$$

$$G_2(x) = \varepsilon\left(\frac{x - c}{b_o - a_o}\right)^{1/3}$$

and letting $U(I) = U_o(I) + G_1(I) + G_2(I)$; one obtains V in a similar fashion. Then $\underline{(U_o + G_1)}(c) \geq 0$ and since $\underline{G}_1(c) = \infty$, $\underline{U}(c) = \infty$; the analogous construction yields V with $\overline{V}(c) = -\infty$. Since U and V are clearly continuous and $(U - V)(I_o) \leq (U_o - V_o)(I_o) + 4\varepsilon \leq 5\varepsilon$, it follows that $f\cdot C_{[a_o,c]} \subset \mathcal{P}_o$ and, if $P(x) = P(c-)$ for $x > c$, then $P(I) = \mathcal{P}_o\int f\cdot C_I$ for each $I \subset I_o$. Thus the $\mathcal{P}_o$-integral is closed under the Cauchy extension. To see that it is also closed under the H*-type improper integrals extension suppose P is a perfect set, $f\cdot C_P \in \mathcal{P}_o$ and $f\cdot C_{I_n} \in \mathcal{P}_o$ for each I_n contiguous to P and that $\Sigma\, \theta(\mathcal{P}_o\int f\cdot C_{I_n}; I_n) < \infty$. Let $\varepsilon > 0$ be given and choose N so that

$$\sum_{n=N}^{\infty} \theta(\mathcal{P}_o\int f\cdot C_{I_n}; I_n) < \varepsilon.$$

Since $f\cdot C_P \in \mathcal{P}_o$ and each $f\cdot C_{I_n} \in \mathcal{P}_o$, $f\cdot C_P + \sum_{n=1}^{N-1} f\cdot C_{I_n}$ belongs to $\mathcal{P}_o$ and hence has a continuous majorant U_o and minorant V_o so that $(U_o - V_o)(I_o) < \varepsilon$. It will suffice to find a continuous majorant U and a continuous minorant V for $\sum_{n=N}^{\infty} f\cdot C_{I_n}^{\circ}$ so that $(U - V)(I_o) < 2\varepsilon$. For each $n \geq N$, let U_n and V_n be respectively continuous majorants and minorants for $f\cdot C_{I_n}^{\circ}$ such that $(U_n - V_n)(I_o) < \varepsilon/2^{n+1}$. For each $I_n = [a_n, b_n]$ let positive numbers $\delta_n < (b_n - a_n)/2$ be chosen so that

$$\theta(U_n;[a_n,a_n+\delta] \leq \varepsilon/2^{n+1}, \quad \theta(U_n;[b_n-\delta,b_n]) < \varepsilon/2^{n+1}$$

and so that the analogous restrictions on V_n also hold. Let

$$G_n(x) = \theta(U_n;\ [a_n,x]\cap[a_n,a_n+\delta_n]) + \theta(U_n;\ [b_n-\delta_n,b_n]) - \theta(U_n;\ [x,b_n]\cap[b_n-\delta_n,b_n]).$$

Then each $G_n(x)$ is continuous, non-decreasing and satisfies $0 \leq G_n(x) \leq \varepsilon/2^n$. Let $G(x) = \sum_{n=N}^{\infty} G_n(x)$ and $U(I) = G(I) + \sum_{n=N}^{\infty} U_n(I\cap I_n)$. When the minorant V is constructed similarly $(U - V)(I_o) < \varepsilon$. The uniform convergence of ΣG_n and ΣU_n guarantees the continuity of U; the definition of each G_n guarantees that $\underline{U}(a_n)) \geq 0$ and $\underline{U}(b_n) \geq 0$ and since $\sum_{n=N}^{\infty} f\cdot C_{I_n}^{\circ}$ is 0 at each a_n and b_n, U is a continuous majorant and, similarly, V is a continuous minorant for $\sum_{n=N}^{\infty} f\cdot C_{I_n}^{\circ}$. Finally, since by Theorem 9.12

the Lebesgue integral is contained in the $\mathcal{P}_o$-integral and for $f \in \mathcal{L}$, $\mathcal{P}_o\int f = \mathcal{L}\int f$, it follows that the $\mathcal{P}_o$-integral, being closed under the (C) and (H*) extensions, contains the integral $\mathcal{D}* = \mathcal{L}^*_{\omega_1}$.

In order to show the inclusion of the Perron integral in the $\mathcal{D}*$-integral, we will need a consequence of the next theorem; namely, that if U is a majorant (resp., V a minorant) of a function f on I_o, then U (resp., V) is necessarily BVG* on I_o.

(9.14) Theorem. *Suppose that F is a real-valued function defined on I_o and that at each point x of a set $E \subset I_o$, except for at most countably many points of E, either $\overline{F}(x) < \infty$ or $\underline{F}(x) > -\infty$. Then F is BVG* on E.*

Proof. Clearly F is BV* at each point of the at most countable exceptional set. Let

$$E^+ = \{x \in E: \overline{F}(x) < \infty\}$$

and for each natural number k let

$$E_k = \{x \in E^+: |t-x| \le 1/k \text{ implies } \frac{F(t)-F(x)}{t-x} < k\}.$$

For each integer j let $E_{j,k} = E_k \cap [(j-1)/k, j/k]$. Then to show that F is BVG* on E^+ it suffices to show that F is BV* on each interval $[a,b]$ where a and b belong to $E_{j,k}$. Consider two such points a and b and let $G(x) = F(x) - kx$ on $[a,b]$. We first show that G is monotone non-increasing on $E_{i,k} \cap [a,b]$. But, if $c,d \in E_{j,k} \cap [a,b]$ with $c < d$, then $G(d) - G(c) = F(d) - F(c) - k(d - c) \le 0$ since

$F(d) - F(c) \le k(d - c)$. Now note that if $t_1, t_2 \in [c,d]$, then $G(t_1) - G(c) = F(t_1) - F(c) - k(t_1 - c) \le 0$ and

$$G(d) - G(t_2) = F(d) - F(t_2) - k(d - t_2) \le 0.$$

Thus adding these inequalities yields that $G(t_1) - G(t_2) \le G(c) - G(d)$ and thus $\theta(G;[c,d]) \le G(c) - G(d)$. Consequently, if $\{I_n\}$ is a sequence of pairwise non-overlapping intervals with endpoints in $E_{i,k} \cap [a,b]$ and $I_n = [a_n, b_n]$, then

$$\Sigma\, \theta(G;I_n) \le \Sigma\, G(a_n) - G(b_n) \le G(a) - G(b).$$

It follows that $G(x)$ is BV* on $E_{j,k} \cap [a,b]$ and that $F(x) = G(x) + kx$ is also BV* on $E_{j,k}$. Hence G is BVG* on E^+ and similar considerations for $E^- = \{x: \underline{F}(x) > -\infty\}$ show that F is BVG* on the entire set E.

We will need the following result: If $F(x)$ is BV* on $E \subset I_o$ and $F(x)$ agrees with an absolutely continuous function $G(x)$ on $\bar{E}$, then F is AC* on $\bar{E}$. To see this, let I be an interval whose endpoints belong to $\bar{E}$. If x_1 belongs to I, one has $F(x_1) \le \sup_{x \in I} G(x) + \theta(F;J_i)$ provided $x \in J_i$ contiguous to $\bar{E} \cap I$ or $x \in \bar{E}$. Thus, for any $x_1 \in I$,

$$F(x_1) \le \sup_{x \in I} G(x) + \Sigma_I\, \theta(F;J_i)$$

where Σ_I is over all J_k contiguous to $E \cap I$. Similarly, if $x_2 \in I$, then

$$F(x_2) \ge \inf_{x \in I} G(x) - \Sigma_I\, \theta(F;J_i).$$

Thus, $\theta(F;I) \le \theta(G;I) + 2\Sigma_I\, \theta(F;J_i)$. Since $F \in$ BV* on E, $\Sigma\, \theta(F;I'_n) < \infty$ where $\{I'_n\}$ is the set of intervals contiguous to $\bar{E}$. Thus, given $\varepsilon > 0$, there is N so that $\sum_{n>N} \theta(F;I'_n) < \varepsilon/4$. Suppose $\delta < \min(|I_1|, \ldots, |I_n|)$ and δ is such that $\Sigma|G(I_k)| < \varepsilon/2$ whenever the I_k are pairwise non-overlapping and $\Sigma|I_k| < \delta$. Then for such a sequence of I_k with endpoints in $\bar{E}$, one has

$$\Sigma\,\theta(F;I_k) \le \Sigma\,\theta(G;I_k) + 2\cdot\sum_k\left(\Sigma_{I_k}\theta(F;J_i)\right) < \varepsilon.$$

This implies that F must be AC^* on $\bar{E}$.

We are now ready to show that the $\mathcal{D}^*$ and Perron integrals are equivalent.

(9.15) <u>Theorem</u>. <u>Each</u> $f \in \mathcal{P}$ <u>also belongs to</u> $\mathcal{D}^*$ <u>and thus</u> $\mathcal{P} = \mathcal{D}^*$ <u>and the narrow-sense Denjoy integral and the Perron integral are one and the same</u>.

<u>Proof</u>. Let $f \in \mathcal{P}$ and $P(x) = \mathcal{P}\int f\cdot C_{[a_o,x]}$. To show that $f \in \mathcal{D}^*$, since it is already known that $P'(x) = f(x)$ a.e., it suffices to show that the function $P(x)$ is ACG^*. To do this we need only show that each closed set F contains a portion on which $P(x)$ is AC^*. Fix such a closed set F and let U_o and V_o be respectively a majorant and minorant of f. Since U_o is BVG^*, F contains a portion on which U_o is BV^* and this portion in turn contains a portion $F \cap I$ on which V_o is also BV^*. We may assume that $I = [a,b]$ and $a,b \in F$. Now, $U_o - P$ is monotone non-decreasing and hence $P = U_o - (U_o - P)$ is BV^* on $F \cap I$. It remains to show that P is also AC^* on $F \cap I$. Given $\varepsilon > 0$, let U,V be any majorant and minorant of f so that $(U - V)(I_o) < \varepsilon$. Let $U_1(x) = U(x)$ if $x \in F \cap I$, $P_1(x) = P(x)$ if $x \in F \cap I$, $V_1(x) = V(x)$ if $x \in F \cap I$ and let U_1, P_1, V_1 each

be linear on intervals contiguous to $F \cap I$. Since P is $BV*$ on I, since $U - P$ and $P - V$ are non-decreasing and since U and V are $BV*$ on $F \cap I$, it follows that U_1, P_1, and V_1 are of bounded variation on I. Also, at a.e. point x of $F \cap I$, $P_1'(x)$ and $P'(x)$ exist and the two are equal. Let $f_1(x) = f_2(x) = P_1'(x)$ at each such point and at each interior point of each I_k' contiguous to $F \cap I$. Let $f_1(x) = -\infty$ and $f_2(x) = \infty$ at all other points of I. Then $-\infty \neq \underline{U}_1(x) \geq f_1(x)$ and $f_2(x) \geq \overline{V}_1(x) \neq \infty$ and thus U_1 is a majorant of f_1 and V_1 a minorant of f_2. Since f_1 and f_2 are almost everywhere the derivative of P_1 and P_1' is Lebesgue integrable on I, for each interval $J \subset I$,

$$U_1(J) \geq \int_J f_1' = \int_J P_1' = \int_J f_2' \geq V_1(J).$$

Now let $S(J)$ be the singular function associated with P_1. Then $P_1(J) = \int_J P_1' + S(J)$ and hence, for each $J \subset I$,

$$V_1(J) + S(J) \leq P_1(J) \leq U_1(J) + S(J);$$

that is, for each $J \subset I$,

$$-S(J) \leq U_1(J) - P_1(J) \leq \varepsilon$$

$$S(J) \leq P_1(J) - V_1(J) \leq \varepsilon$$

and hence $|S(J)| \leq \varepsilon$ for each $J \subset I$ and $\varepsilon > 0$. Then $S(x)$ is constant and $P_1(x)$ is absolutely continuous. Then since $P(x)$ is $BV*$ on $F \cap I$ and agrees with an absolutely continuous function on $F \cap I$, as noted earlier, $P(x)$ must be $AC*$ on $F \cap I$. Thus $P(x)$ is $ACG*$. Since this holds for each $f \in \mathcal{P}$, $\mathcal{P} \subset \mathcal{D}*$ and thus $\mathcal{P} = \mathcal{D}*$.

Having the equivalence of the integrals $\mathcal{L}^*_{\omega_1}$, $\mathcal{D}^*$ and $\mathcal{P}$, we turn now to the fourth approach to the integral. First note that a function which is either Lebesgue or Denjoy-Perron integrable on I_o is necessarily finite valued a.e. on I_o. One would not expect to be able to integrate a function which was infinite valued on a set of positive measure. Since changing a function on a set of measure 0 does not affect the value of its integral for any integral which includes that of Lebesgue, we are going to consider without any loss of generality only those functions f defined on I_o which are everywhere finite valued on I_o. The following is the definition of the Riemann Complete or $\mathcal{RC}$-integral of a finite valued function f on I_o.

$\mathcal{RC}\int f = L$ provided that, for every $\varepsilon > 0$, there is $\delta(x) > 0$ defined on I_o so that

$$|\Sigma f(w_i)\Delta x_i - L| < \varepsilon \quad \text{whenever}$$

$$a_o = x_o < x_1 < \dots < x_n = b_o \quad \text{and} \quad w_i \in [x_{i-1}, x_i]$$

$$\text{and} \quad \Delta x_i = x_i - x_{i-1} < \delta(w_i).$$

That there are always partitions of I_o satisfying $w_i \in [x_{i-1}, x_i]$ and $\Delta x_i < \delta(w_i)$ is a consequence of the compactness of I_o. The function $\delta(x)$ is called a guage. It is with some surprise that this integral of a Riemann type turns out to be equivalent to the Denjoy-Perron integral. We turn now the proof of this.

(9.16) Theorem. *For finite valued functions f defined on an interval I_o, f is Riemann-Complete integrable iff f is Denjoy-Perron integrable and then the values of the integrals agree.*

Proof. First suppose f is finite valued and Perron integrable on I_o. Given $\varepsilon > 0$, let U be a majorant and V a minorant of f on I_o so that $(U - V)(I_o) < \varepsilon$. For each $x \in I_o$, since $\underline{U}(x) \geq f(x) \neq -\infty$ and $\overline{V}(x) \leq f(x) \neq \infty$, there is $\delta(x) > 0$ so that if $x \in I$ and $|I| < \delta(x)$ then $U(I)/|I| \geq f(x) - \varepsilon$ and $V(I)/|I| \leq f(x) + \varepsilon$. This defines a $\delta(x)$ for the $\mathcal{RC}$-integral. Let $a_o = x_o < x_1 < \ldots < x_n = b_o$ and $w_i \in [x_{i-1}, x_i] = I_i$ with $\Delta x_i \leq \delta(w_i)$. Then $\Sigma(f(w_i) - \varepsilon)\Delta x_i \leq \Sigma U(I_i)$ and $\Sigma f(w_i)\Delta x_i \leq U(I_o) + \varepsilon|I_o|$. Similarly $\Sigma(f(w_i) + \varepsilon)\Delta x_i \geq \Sigma V(I_i)$ and $\Sigma f(w_i)\Delta x_i \geq V(I_o) - \varepsilon|I_o|$. Thus $|\Sigma f(w_i)\Delta x_i - \mathcal{P}\int f| \leq 2\varepsilon|I_o| + \varepsilon$. Since $\varepsilon > 0$ was arbitrary, it follows that $f \in \mathcal{RC}$ on I_o and $\mathcal{RC}\int f = \mathcal{P}\int f$. To see the converse, suppose that $f \in \mathcal{RC}$ on I_o. Given $\varepsilon > 0$, let $\delta(x)$ be given so that $|\mathcal{RC}\int f - \Sigma f(w_i)\Delta x_i| < \varepsilon$ whenever $a_o = x_o < x_1 < \ldots < x_n = b_o$ and $w_i \in [x_{i-1}, x_i]$ with $\Delta x_i < \delta(w_i)$. Let $U(x) = \sup \Sigma_x f(w_i)\Delta x_i$ where Σ_x is over the partition of $[a_o, x]$ for which $w_i \in [x_{i-1}, x_i]$ and $\Delta x_i \leq \delta(w_i)$ and the supremum is taken over all such sums. Similarly, let $V(x) = \inf \Sigma_x f(w_i)\Delta x_i$. Then U and V are finite valued functions defined on $[a_o, b_o]$; for otherwise, by adjoining to a partition of $[a_o, x]$ one of $[x, b_o]$, it would follow that f was not $\mathcal{RC}$-integrable on $[a_o, b_o]$. Now, if $x \in [a,b] = I \subset I_o$ and $b - a < \delta(x)$, then $f(x)|I| + U(a) \leq U(b)$. Hence, $f(x) \leq U(I)/|I|$ and $\underline{U}(x) \geq f(x) \neq -\infty$. Similarly, $\overline{V}(x) \leq f(x) \neq \infty$. Thus U is a majorant and V a minorant for f. However, by the definitions of U and V, $U(I_o) - V(I_o) \leq 2\varepsilon$. It follows that f is Perron integrable and its Perron integral must equal its $\mathcal{RC}$-integral.

9.1 Exercises

1. Show that i) - v) are by no means independent. For example, show v) implies each $f \in \mathcal{I}$ is measurable; show i), iii) and iv) imply $F(I) = \mathcal{I}\int f$ is an additive function of an interval.
2. Show that an integral which satisfies i) - iv) also satisfies a monotone convergence theorem; that is, if $f_n \in \mathcal{I}$ and $\{f_n\}$ is a non-decreasing sequence of functions whose limit f belongs to $\mathcal{I}$, then $\lim \mathcal{I}\int f_n = \mathcal{I}\int f$.
3. Given a closed set $E \subset I_o$ show there is a differentiable function F which is not of bounded variation in each neighborhood of each $x \in E$ but is of bounded variation on each $\bar{I} \subset E^C$.
4. Given a countable closed set E well-ordered by its natural order on the line with order type α, show that if α is sufficiently larger than β, then $F' \notin \mathcal{L}^*_\beta$ but $F' \in \mathcal{L}^*_\alpha$ where F is given by Exercise 3.
5. Give an example of $f, g \in \mathcal{L}^{H*}$ such that $f + g \notin \mathcal{L}^{H*}$.
6. Show directly that, if $f, g \in \mathcal{L}^{H*}$, then there is $\alpha < \omega_1$ so that $f + g \in \mathcal{L}^*_\alpha$.
7. If an integral satisfies i) - iv), show that the extension (C) along with the extension (H**) when carried out transfinitely yields an integral satisfying i) - iv) where (H**) is:

   $f \in \mathcal{I}^{H**}$ if $f \cdot C_P \in \mathcal{L}$ and each $f \cdot C_{I_n} \in \mathcal{I}$

   when I_n are the intervals contiguous to P

and $\Sigma\, \theta(\ell\int f\cdot c_{I_n}; I_n) < \infty$ and then $\ell^{H**}\int f\cdot c_P + \Sigma\, \ell\int f\cdot c_{I_n}$.

8. Show that if F is uniformly continuous on $E \subset I_o$ (in particular, if F is AC* on E) then there is a continuous function G defined on I_o so that G agrees with F on E (and if $F \in$ AC* on E, G can be chosen to be absolutely continuous on I_o).
9. Show that the product of two functions which are ACG* on I_o is also ACG* on I_o; show the product of two BVG* functions is BVG* on I_o.
10. Show that if F satisfies at each $x \in I_o$, "there is δ_x, M_x such that $|F(y) - F(x)| \le M_x|y - x|$ whenever $|y - x| < \delta_x$", (such functions are sometimes called locally Lipschitz) then F is ACG* on I_o.
11. Show that there is a continuous function F on $[0,1]$ so that F is BVG* but $F'(x)$ is not $\mathcal{D}$*-integrable. Hint: Construct F so that F is continuous and BVG* and $F'(x) = 1/x$ a.e. on $[0,1]$.
12. Show that if the definition of a majorant and minorant allowed $\underline{U}(x) = -\infty$ at a countable set of points or $\overline{V}(x) = +\infty$ on a countable set, the resulting integral would be the same as the Perron integral.
13. Show that if the definition of a majorant and minorant allowed $\underline{U}(x) \ge f(x)$ a.e. and $\overline{V}(x) \le f(x)$ a.e. the resulting integral would be the same as the Perron integral. (This can be done in addition to the generalization of Exercise 12.)

14. Show that there is an $f \in \mathcal{L}$ with $f(x) < \infty$ at each x so that each lower semi-continuous h with $h(x) \geq f(x)$ must take on $+\infty$ at some point. Hint: Well order the finite valued lower semi-continuous function and define f to be 0 except on a set Z of measure 0; on Z make f larger than each h at some point of Z.
15. Show how to find $\delta(x)$ for the Riemann-Complete integral for $c_G(x)$ with G open, $c_Z(x)$ when $m(Z) = 0$, $f(x)$ when $f \subset \mathcal{L}$, $f(x)$ when $f(x) = F'(x)$.

9.2 The Denjoy Integral

The four approaches to extending the Lebesgue integral that lead to the Denjoy-Perron integral which we just considered allow for a variety of additional extensions and generalizations in other contexts. Usually, when one approach produces an integral in a different setting or an extension of a given integral, the others can also be used in that setting and/or the extension can also be formulated by means of the other approaches. We consider a few of the easier generalizations here, the most noteworthy of which will be the wide-sense Denjoy integral.

It has already been mentioned that the Perron approach lends itself to extensions to $\mathbb{R}^k$. Indeed, if U and V are interpreted as additive functions of an interval defined on a k-cell $I_o \subset \mathbb{R}^k$ and if U is a majorant for a real-valued function f on I_o provided for $x \in I_o$, $-\infty \neq \underline{U}(x) \geq f(x)$ and V is a minorant if $\infty \neq \overline{V}(x) \leq f(x)$, then Theorems 9.10 and 9.11 hold and, with slight modifications, all the comments between them also hold. That is, $U - V$ is a non-negative additive function of an interval; the Perron integral is well-defined, when $\sup V(I_o) = \inf U(I_o)$, as the common value of the supremum and infimum; for each f which is Perron integrable $\mathcal{P}\int f \cdot C_{I_n}$ is a continuous additive function of an interval on I_o; the Perron integral agrees with the Lebesgue integral for functions f which are summable on I_o. We mention these readily observed facts but will not go further into the generalization of the Lebesgue integral in $\mathbb{R}^k$ or into the formulations there which use the other three approaches. We will also not be concerned with generalizations of the integral to vector-valued functions because, while these clearly exist, they usually reduce to the vector sum of the component integrals computed separately for each of the coordinate functions. Our main concern in the remainder of this chapter will be with the wide-sense Denjoy integral, or simply Denjoy integral, which is a more extensive integral than the Denjoy-Perron integral for real-valued functions of a real variable.

Having the characterization of the primitives or indefinite integrals for the Denjoy-Perron integral, the ACG* functions, it is an easy step to get to the primitives for the Denjoy integral. A function F defined on I_o is said to be <u>absolutely continuous on a set</u> E if, given $\varepsilon > 0$, there is a $\delta > 0$ so that $\Sigma|F(I_n)| < \varepsilon$ whenever $\{I_n\}$ is a sequence of pairwise non-overlapping intervals with endpoints in E and $\Sigma|I_n| < \delta$; then F is said to be AC on E. It is immediately evident that the property of being AC on E for a function F does not depend on the behavior of F on the complement of E, as does the property AC*. It is also clear that those F which are AC on a fixed set E form a vector space of functions. A function F will be said to be generalized absolutely continuous on E, or ACG on E, if F is continuous on E and E is a countable union of sets on each of which F is AC; when $E = I_o$, one simply says that F is ACG. The ACG functions are the required primitives for the Denjoy integral. We next consider some of their properties and define the class $\mathcal{D}$ of Denjoy integrable functions.

If F and G are ACG on I_o, then $I_o = \cup E_n = \cup E_k'$ where F is AC on each E_n and G is AC on each E_k'. Then both F and G are AC on each $E_{n,k} = E_n \cap E_k'$ and since $F + G$ is continuous, $F + G$ is also an ACG function on I_o. Since a constant multiple of an ACG function is clearly ACG, the generalized absolutely continuous functions form a vector space.

A problem arises with property v), which was desirable for integrals which extend the Lebesgue integral. The problem is that functions which are ACG on an interval I_o are not necessarily differentiable almost everywhere on I_o. A classic example of this fact is as follows: Let P be a nowhere dense perfect subset of $[0,1]$ with 0 and 1 in P and let $\{I_n\}$ be the sequence of intervals contiguous to P. Let d_n be the length of the largest interval in $[0,1]$ which does not meet $I_1, I_2, \ldots, I_n$ and, if $I_n = [a_n, b_n]$, let $c_n = (a_n + b_n)/2$. Now $F(x)$ is defined to be 0 if $x \in P$, $F(c_n) = |I_n| + d_n$ and F is to be linear on each interval contiguous to $P \cup \cup\{c_n\}$. Since $|I_n| \to 0$ as $n \to \infty$ and $d_n \to 0$ as $n \to \infty$, the resulting function F is continuous on P and thus is also continuous on $[0,1]$. Since F is 0 on P and linear on each $[a_n, c_n]$ and $[c_n, b_n]$, F is ACG on $[0,1]$. To see that F is not differentiable at any point of P, consider a point x in P. There is a sequence $\{I'_{n_k}\}$ of intervals contiguous to P with I_{n_k} approaching x. By the definition of F one has that $F(c_{n_k}) \geq d(x, I_{n_k}) + |I_{n_k}| \geq |c_{n_k} - x|$ and thus,

$$\frac{F(c_{n_k}) - F(x)}{|c_{n_k} - x|} \geq 1.$$

But x is a limit point of a sequence $\{x_k\}$ of points of P each distinct from x, and these points

$$\frac{F(x_k) - F(x)}{x_k - x} = 0.$$

Therefore F is not differentiable at any of the points x in P. Now, if P has positive measure, the function F is not differentiable almost everywhere.

There are two ways to resolve this problem. If one restricts one's consideration to those ACG functions which are differentiable a.e., the resulting vector space of continuous functions forms an integral which is called the <u>intermediate Denjoy integral</u>. The Denjoy integral itself is formed by taking the class of ACG functions as primitive but replacing the ordinary derivative in property v) with the approximate derivative. Thus, one of the main properties of the ACG function which we will need is that each function which is ACG on I_o is approximately derivable almost everywhere on I_o. This will be facilitated by proving somewhat more. A function F will be said to be BV or of <u>bounded variation in the wide sense on</u> $E \subset I_o$ if there is a number $M < \infty$ so that $\Sigma|F(I_n)| < M$ whenever $\{I_n\}$ is a sequence of non-overlapping intervals with endpoints in E. A function F will be said to be BVG on E or of <u>generalized bounded variation in the wide sense on</u> E if $E = \cup E_n$ and F is BV on each E_n. Such functions will be said to be BVG on E or, if $E = I_o$, to be of generalized bounded variation or simply BVG. The number

$$\sup \Sigma|F(I_n)|,$$

where the supremum is taken over all sequences of pairwise non-overlapping intervals with endpoints in E, is called the <u>variation of</u> F <u>on</u> E and is denoted by $V(F;E)$. One can also define $\overline{V}(F;E) = \sup \Sigma F(I_n)$ and $\underline{V}(F;E) = -\inf \Sigma F(I_n)$, where the supremum and infimum are each taken over all sequences of pairwise non-overlapping intervals with endpoints in E. Note that the definition of AC and BV agree with absolute continuity and bounded variation when $E = I_o$. Furthermore, the property of being BV on E does not take into account the behavior on E^c, as does BV*. Clearly, AC* implies AC and BV* implies BV on a set E and $V(F;E) \le V_*(F;E)$. Also, if F is AC (resp. BV) on E, then it is AC (resp. BV) on each subset of E. However, unlike functions which are BVG*, functions which are BVG need not be measurable; for example, $C_S(x)$ where S is a non-measurable set is BVG because it is BV on S and on S^c but it is not a measurable function.

We proceed with some theorems which illuminate these classes of functions.

(9.17) <u>Theorem</u>. <u>Each function</u> F <u>which is</u> AC <u>on</u> $E \subset I_o$ <u>is also</u> BV <u>on</u> E <u>and thus each</u> F <u>which is</u> ACG <u>on</u> E <u>is also</u> BVG <u>on</u> E.

<u>Proof</u>. If F is AC on E, then F must be bounded on E; for otherwise, a contradiction to the fact that F is AC on E can be obtained from the existence of a sequence of points $\{x_n\}$ chosen so that $\lim x_n = x_o \in I_o$ and $\lim|F(x_n)| = \infty$. Let $a = \inf E$, $b = \sup E$. Choose $\delta > 0$ so that, if $|I| < \delta$, then $V(F;I) < 1$ and let $\{a = x_o < x_1 < \ldots < x_n = b\}$ be a partition of $[a,b]$ of norm less than δ. Then

$$V(F;E) \le \sum_{i=1}^{n} V(F;E\cap[x_{i-1},x_i]) + 2M\cdot n < \infty.$$

Consequently, F is BV on E. Thus, whenever F is AC (resp. ACG) on a set E, F is also BV (resp. BVG) on E.

(9.18) <u>Theorem</u>. <u>Each function</u> F <u>which is</u> BV <u>on a set</u> $E \subset I_o$ <u>agrees on</u> E <u>with a function</u> G <u>which is of bounded variation on</u> I_o; <u>moreover, if</u> F <u>is</u> AC <u>on</u> E, <u>the function</u> G <u>can be chosen to be absolutely continuous on</u> I_o.

<u>Proof</u>. Given F which is BV on $E \subset I_o$, let $a = \inf E$, $b = \sup E$ and if $x < a$, let $V(x) = 0$ and

if $x \geq a$, let $V(x) = V(F;[a_o,x]\cap E)$. Since F is evidently bounded on E, $V(x)$ is non-decreasing on I_o. Thus the function G_1 which agrees with V on $\bar{E}$ and is linear on intervals contiguous to $\bar{E}$ with $G_1(x) = V(a) = 0$ if $x \leq a$ and $G_1(x) = V(b)$ if $x \geq b$ is non-decreasing on I_o. The function $V(x) - F(x)$ is non-decreasing on E because, for each I with endpoints in E, $V(I) \geq F(I)$ and hence $(V - F)(I) \geq 0$. Thus $V - F$ can be extended to a function G_2 which agrees with $V - F$ on E, is non-decreasing, linear on intervals contiguous to $\bar{E}$ and constant on $(-\infty,a]$ and $[b,\infty)$. Now $G_2 - G_1$ is of bounded variation on I_o and, at each $x \in E$, $(G_2 - G_1)(x) = V(x) - (V - F)(x) = F(x)$. Thus, if F is BV on $E \subset I_o$, F agrees with a function G which is of bounded variation on I_o. Now consider that F is also AC on E. In this case, it is easily seen that $V(x)$ is AC on $\bar{E}$ and hence satisfies Lusin's condition (N) on $\bar{E}$. Thus $G_1(x)$ is necessarily continuous, of bounded variation and satisfies Lusin's condition (N) and hence $G_1(x)$ is absolutely continuous on I_o. By insisting that on each interval $[a_n,b_n]$ contiguous to $\bar{E}$, $G_2(a_n) = (V - F)(a_n-)$ and $G_2(b_n) = (V - F)(b_n+)$ and that $G_2(a) = (V - F)(a+)$, $G_2(b) = (V - F)(b-)$, one obtains a function $G_2(x)$ which is continuous and non-decreasing on I_o. Since V and F are AC on E, G_2 is also AC on E and hence G_2 is AC on $\bar{E}$ and is absolutely continuous on I_o; this follows from the same argument given for the absolute continuity of G_1. Thus, when F is AC on E, the function $G = G_2 - G_1$ is absolutely continuous on I_o and at each $x \in E$, $F(x) = G(x)$.

(9.19) Corollary. If F is ACG on I_o, then F satisfies Lusin's condition (N) on I_o.

Proof. Since F is ACG on I_o, $I_o = \cup E_n$ and, for each n, there is an absolutely continuous function $G_n(x)$ defined on I_o with $G_n(x) = F(x)$ for $x \in E_n$. If $Z \subset I_o$ and $m(Z) = 0$, then $m(F(Z)) \le m(\cup F(Z \cap E_n)) = m(\cup G_n(Z \cap E_n)) = 0$. Thus F satisfies Lusin's condition (N) on I_o.

(9.20) Theorem. If E is a measurable subset of I_o and F is measurable and BVG on E (in particular, if F is ACG on a measurable set E), then F'_{ap} exists almost everywhere on E.

Proof. Assume F is measurable and BVG on the measurable set E. By Lusin's theorem, given any $\varepsilon > 0$, there is a closed set $E_\varepsilon \subset E$ with $m(E \setminus E_\varepsilon) < \varepsilon$ such that F is continuous on E_ε with respect to E_ε. Then, to show that F'_{ap} exists at almost every point of E, it suffices to show that F'_{ap} exists at almost every point of such E_ε. But, given $\varepsilon > 0$, F is BVG on E_ε and hence $E_\varepsilon = \cup E_n$ where F is BV on each set E_n. Fix n. Since F is continuous on E_n, F is BV on $\overline{E}_n \subset E_\varepsilon$ and hence F agrees with a function G on E_n where G is of bounded variation on I_o. But then, at each point $x \in \overline{E}_n$ such that x is a point of density of $\overline{E}_n$ and a point of differentiability of G, $F'_{ap}(x) = G'(x)$; that is, $F'_{ap}(x) = G'(x)$ at almost every point of E_n. Thus F'_{ap} exists at almost every point of E_ε. It follows that each BVG and measurable function F, defined on a measurable set E, is approximately derivable at almost every point of E.

The class $\mathcal{D}$ is now defined to be the collection of those functions f which agree almost everywhere with the approximate derivative of an ACG function.

In order to show that $\mathcal{D}\int f = F(b_o) - F(a_o)$ is well-defined for each $f(x) = F'_{ap}(x)$ a.e. with $F \in$ ACG on $I_o = [a_o, b_o]$, we need to show that for $F, G \in$ ACG on I_o, if $F'_{ap}(x) = G'_{ap}(x)$ a.e., then $F(x) = G(x) + C$; for then $F(b_o) - F(a_o) = G(b_o) - G(a_o) = \mathcal{D}\int f$ is well-defined. We will do this by means of what is called a "monotonicity theorem"; that is, a theorem which gives sufficient conditions for a function to be monotone non-decreasing. We have already seen one such theorem; namely, Theorem 9.10 which asserts that a function $F(x)$, satisfying $\underline{F}(x) \geq 0$ at each $x \in I_o$, is monotone non-decreasing on $\overline{I}_o$. We wish to prove that each function $H \in$ ACG with $H'_{ap}(x) \geq 0$ a.e. is monotone non-decreasing. Then $H \in$ ACG and $H'_{ap}(x) = 0$ a.e. will imply that H must be identically constant. This is because both H and $-H$ will then be monotone non-decreasing. It will follow that, if $F, G \in$ ACG and $F'_{ap}(x) = G'_{ap}(x)$ a.e., then $(F - G)'_{ap}(x) = 0$ a.e. and $F(x) = G(x) + C$.

The proof of the monotonicity theorem for the almost everywhere approximate derivative of an ACG function uses the next theorem which is a monotonicity theorem due to Zygmund.

(9.21) Theorem. *Sufficient conditions for a real-valued function F to be monotone non-decreasing on* [a,b] <u>are that at each</u> $x \in (a,b]$, $\overline{\lim}_{h\to 0^+} F(x-h) \le F(x)$; <u>and at each</u> $x \in [a,b)$, $F(x) \le \overline{\lim}_{h\to 0^+} F(x+h)$ <u>and</u> $\{y: y = F(x)$ and $\overline{F}^+(x) \le 0\}$ <u>does not contain an interval</u>. (Actually, these conditions can also be shown to be necessary.)

<u>Proof</u>. Suppose not. Then there is a function F satisfying the hypotheses of the theorem on (a,b] and $c,d \in [a,b]$ with $c < d$ and $F(c) > F(d)$. Also there must be a point y_o belonging to

$$(F(d),F(c)]\setminus\{y: y = F(x) \text{ and } \overline{F}^+(x) > 0\}.$$

Let $x_o = \sup\{x \in [c,d]: F(x) \ge y_o\}$. Since $\overline{\lim}_{h\to 0^+} F(c+h) \le F(c)$, $x_o \ne c$; since $\overline{\lim}_{h\to 0^+} F(d-h) \le F(d)$, $x_o \ne d$. It is not possible that $F(x_o) < y_o$ because then $\overline{\lim}_{h\to 0^+} F(x_o-h) \le F(x_o)$ would contradict the definition of x_o. It is also not possible that $F(x_o) > y_o$ for then $\overline{\lim}_{h\to 0^+} F(x_o+h) \ge F(x_o)$ would contradict the definition of x_o. Thus $F(x_o)$ must equal y_o. But this is also impossible because $\overline{F}^+(x_o) \le 0$ contrary to the definition of y_o. It follows that any F satisfying the hypotheses of the theorem must be monotone non-decreasing.

(9.22) <u>If</u> $F(x)$ <u>is</u> ACG <u>on</u> (a,b] <u>and</u> $F'_{ap}(x) \ge 0$ <u>at almost every point of</u> [a,b], <u>then</u> F <u>is monotone non-decreasing on</u> [a,b].

<u>Proof</u>. Fix $\varepsilon > 0$ and let $G(x) = F(x) + \varepsilon x$ on $[a,b]$; then G is ACG on $[a,b]$. Since G is continuous on $[a,b]$, $\overline{\lim}_{h\to 0^+} G(x-h) = G(x)$ at each $x \in (a,b]$ and $\overline{\lim}_{h\to 0^+} G(x+h) = G(x)$ at each $x \in [a,b)$. Also, $\bar{G}^+(x) \geq G'_{ap}(x) \geq \varepsilon > 0$ a.e. on $[a,b]$ and, if $Z = \{x: \bar{G}^+(x) \leq 0\}$, then $m(Z) = 0$ and, since G is ACG and satisfies Lusin's condition (N), $m(G(Z)) = 0$ and $G(Z)$ does not contain an interval. Thus G satisfies the conditions of the previous theorem and must be monotone non-decreasing on $[a,b]$. Then since, for each $\varepsilon > 0$, $F(x) + \varepsilon x$ is monotone non-decreasing on $[a,b]$, it easily follows that F must also be monotone non-decreasing on $[a,b]$.

(9.23) <u>Corollary</u>. <u>If</u> $F(x)$ <u>is</u> ACG* <u>on</u> $[a,b]$ <u>and</u> $F'(x) \geq 0$ a.e. <u>on</u> $[a,b]$, <u>then</u> F <u>must be monotone non-decreasing on</u> $[a,b]$.

It now follows that for $F \in$ ACG on I_o and $f(x) = F'_{ap}(x)$ a.e. on I_o that $\mathcal{D}\int f = F(b_o) - F(a_o)$ is well-defined. For functions in the set

$$\mathcal{D} = \{f: \text{there is } F \in \text{ACG on } I_o \text{ and } F'_{ap}(x) = f(x) \text{ a.e. of } I_o\},$$

we can now verify that $\mathcal{D}\int f$ satisfies properties i) - iv) of an integral.

Let $f \in \mathcal{D}$ and $[a,b] = I \subset I_o$, then there is $F \in$ ACG on I_o with $F'_{ap} = f(x)$ a.e. The function $G(x) = F(x)$ if $x \in I$, $G(x) = F(a)$ if $x < a$ and $G(x) = F(b)$ if $x > b$ is clearly ACG and $G'_{ap}(x) = f \cdot C_I$ a.e. so that $\mathcal{D}\int f \cdot C_I = G(b) - G(a) = F(b) - F(a)$.

Thus for $f \in \mathcal{D}$ and $I \subset I_o$, $f \cdot C_I \in \mathcal{D}$ and i) holds. If $f \in \mathcal{D}$ and $I = I_1 \cup I_2$ when I_1 and I_2 are pairwise non-overlapping and if $G_1, G_2 \in ACG$ with $G'_{1ap} = f \cdot C_{I_1}$ a.e. and $G'_{2ap} = f \cdot C_{I_2}$ a.e., then $G = G_1 + G_2 \in ACG$ and $G'_{ap} = G'_{1ap} + G'_{2ap} = f \cdot C_I$ a.e. Thus

$$G(I) = \mathcal{D}\int f \cdot C_I = G_1(I_o) + G_2(I_o) = \mathcal{D}\int f \cdot C_{I_1} + \mathcal{D}\int f \cdot C_{I_2}$$

and $\mathcal{D}\int f \cdot C_I$ is an additive function of an interval for each $f \in \mathcal{D}$; the continuity of $\mathcal{D}\int f \cdot C_I$ follows from the continuity of the ACG functions. Thus, ii) holds. Clearly the class $\mathcal{D}$ is a vector space of functions and by the same reasoning as above, if $f, g \in \mathcal{D}$ and $a, b \in \mathbb{R}$ and $F(I_o) = \mathcal{D}\int f$, $G(I_o) = \mathcal{D}\int g$ with $F, G \in ACG$, then $af + bg \in \mathcal{D}$ and $\mathcal{D}\int(af + bg) = (aF + bG)(I_o) = a \cdot \mathcal{D}\int f + b \cdot \mathcal{D}\int g$. To show that iii) holds, it remains to show that each $f \in \mathcal{D}$ is measurable on I_o. But if $f \in \mathcal{D}$ there is $F \in ACG$ on I_o with $F'_{ap}(x) = f(x)$ a.e. Since F is ACG on I_o, $I_o = \cup E_n$ such that for each n there is a function F_n which is absolutely continuous on I_o and thus for $x \in \overline{E}_n$, $F_n(x) = F(x)$. Then for each n, at any point of density of $\overline{E}_n$ at which $F'_n(x)$ exists, $F'_{ap}(x) = F'_n(x)$; that is, at almost every point of E_n, F'_{ap} equals F'_n. It readily follows that $f(x) = F'_{ap}(x)$ a.e. is a measurable function; indeed at almost every x, $F'_{ap}(x) = \Sigma F'_n(x) \cdot C_{E'_n}$ where $E'_1 = \overline{E}_1$ and for $n > 1$, $E'_n = \overline{E}_n \setminus \bigcup_{i=1}^{n-1} \overline{E}_i$. Finally since the absolutely continuous functions are contained in the ACG functions, $\mathcal{L} \subset \mathcal{D}$ and for $f \in \mathcal{L}$, $\mathcal{L}\int f = \mathcal{D}\int f$.

It is worth noting that if $f \in \mathcal{D}$ and $f(x) \geq 0$ a.e. on I_o, then f must be Lebesgue integrable.

This is because if $F \in ACG$ and $F'_{ap}(x) = f(x)$ a.e. on I_o, then the last theorem shows that F is monotone non-decreasing. Since F is also continuous and satisfies Lusin's condition (N), F must be absolutely continuous on I_o and $F(I_o)$ is the Lebesgue integral of f.

We turn to the analogous constructive definition of the Denjoy integral. The Cauchy (C) extension of an integral $\mathcal{I}$ will be employed along with an extension (H) analogous to (H*) which is defined as follows: If $P \subset I_o$ is a perfect set and $f \cdot C_P \in \mathcal{I}$ and on each component interval I_n of the complement of P, $f \cdot C_{I_n} \in \mathcal{I}$ and $\Sigma|\mathcal{I}\int f \cdot C_{I_n}| < \infty$ and

$$\lim_{n\to\infty} \theta(\mathcal{I}\int f \cdot C_{I_n}; I_n) = 0,$$

then the function

$$F(x) = \mathcal{I}\int f \cdot C_{P\cap[a_o,x]} + \Sigma\, \mathcal{I}\int f \cdot C_{I_n\cap[a_o,x]}$$

is defined and continuous when the improper integrals of $\mathcal{I}$ satisfy i) and ii). This is because each $\theta(\mathcal{I}\int f \cdot C_{I_n}; I_n)$ is finite and since $\Sigma|\mathcal{I}\int f \cdot C_{I_n}| < \infty$ the convergence to $F(x)$ occurs and $F(x)$ is well-defined. Moreover, $\lim \theta(\mathcal{I}\int f \cdot C_{I_n}; I_n) = 0$ guarantees the continuity of $F(x)$ on P and hence on the entire interval I_o. This defines the H-type improper integral. Then all functions f for which $S(f)$ is a perfect set and the H-type improper integral of f exists with $P = S(f)$ form the class $\mathcal{I}^H$ and $\mathcal{I}^H$ is called the (H) extension of $\mathcal{I}$. The functions in the class $\mathcal{I}^H$ are easily seen to satisfy both i) and ii) when the functions in $\mathcal{I}$ do. Again, $\mathcal{I}^H$

does not generally satisfy iii) when $\mathcal{I}$ does, but if the extension methods (C) and (H) are applied transfinitely beginning with $\mathcal{L}$, the result is the wide-sense Denjoy integral. That is, let $\mathcal{L}_o = \mathcal{L}$, $\mathcal{L}_1 = \mathcal{L}^{CH}$ and, if for each $\beta < \alpha < \omega_1$ the class $\mathcal{L}_\beta$ has been defined, let $\mathcal{L}_\alpha = (\bigcup_{\beta<\alpha} \mathcal{L}_\beta)^{CH}$. Finally $\mathcal{L}_{\omega_1} = \bigcup_{\alpha<\omega_1} \mathcal{L}_\alpha$. Again applying either (C) or (H) to $\mathcal{L}_{\omega_1}$ does not result in any new functions. We go directly from this definition of $\mathcal{L}_{\omega_1}$ to the proof that $\mathcal{L}_{\omega_1} = \mathcal{D}$ and that for each $f \in \mathcal{D}$, $\mathcal{D}\int f = \mathcal{L}_{\omega_1}\int f$.

(9.24) Theorem. The class of Denjoy integrable functions is closed under both the (C) and (H) methods of extending an integral.

Proof. The proof is similar to but simpler than that of Theorem 9.8 for the narrow-sense Denjoy integral. If $\{x_1, \ldots, x_n\} \subset I_o$ and F is ACG on each closed interval I which contains none of the points x_i and F is continuous on I_o, then (a,x_1) and (x_n,b) and each (x_{i-1},x_i) is a countable union of intervals on each of which F is ACG and since F is ACG at each point x_i, F is ACG on I_o. Now let P be a perfect subset of I_o and assume f is $\mathcal{D}$-integrable on P and on each component interval I_n to $I_o \backslash P$. Let $c_n = \mathcal{D}\int f \cdot C_{I_n}$ and assume that $\Sigma|c_n| < \infty$ and $\lim \theta(\mathcal{D}\int f \cdot C_{I_n}; I_n) = 0$. Let $F_n(x) = \mathcal{D}\int f \cdot C_{I_n \cap [a_o,x]}$

and $F(x) = \mathcal{D}\int f \cdot c_{P\cap[a_o,x]} + \sum_{n=1}^{\infty} F_n(x)$. Because $\Sigma|c_n| < \infty$, F is defined at each point of P and on each component interval of $I_o \backslash P$. Because $\lim \theta(\mathcal{D}\int f\cdot c_{I_n}; I_n) = 0$, F is continuous on P and since F is continuous on each I_n, F is continuous on I_o. To see that F is an ACG function, let $G(x) = \sum_{n=1}^{\infty} F_n(x)$ and note that G is ACG on each I_n; thus it will suffice to show that the function G is AC on P. Let $h(x) = 0$ if $x \in P$ and $h(x) = c_n/|I_n|$ if $x \in \mathring{I}_n$. Then h is Lebesgue integrable on I_o and, if $H(x) = \int_{a_o}^{x} h(t)dt$, then $H(x) = G(x)$ on P and hence G is AC on P.

That $\mathcal{L}_{\omega_1} \subset \mathcal{D}$ and the integrals agree follows by induction on the classes $\mathcal{L}_\alpha$. Clearly $\mathcal{L}_o = \mathcal{L} \subset \mathcal{D}$ and $\mathcal{L}\int f = \mathcal{D}\int f$. If for each $\beta < \alpha$, $\mathcal{L}_\beta\int f = \mathcal{D}\int f$, since $\mathcal{L}_\alpha$ is formed by applying (C) and (H) to $\bigcup_{\beta<\alpha} \mathcal{L}_\beta$, if $f \in (\bigcup_{\beta<\alpha} \mathcal{L}_\beta)^C$, $\mathcal{L}_\alpha\int f = F(I_o)$ where F is the ACG function given by applying the Cauchy process. When (H) is applied with $f\cdot c_P \in (\mathcal{L}_{\beta_o})^C$ and $f\cdot c_{I_n} \in (\mathcal{L}_{\beta_n})^C$, for each I_n which is a component interval of the complement of the perfect set P, since $\beta_n < \alpha$,

$$\begin{aligned}\mathcal{L}_\alpha\int f &= (\mathcal{L}_{\beta_o})^C\int f\cdot c_P + \Sigma(\mathcal{L}_{\beta_n})^C\int f\cdot c_{I_n} \\ &= \mathcal{D}\int f\cdot c_P + \Sigma\mathcal{D}\int f\cdot c_{I_n} = \mathcal{D}\int f.\end{aligned}$$

It now remains to show that $\mathcal{D} \subset \mathcal{L}_{\omega_1}$. For this we will need the characterization of the class ACG of Lusin; namely, a function F is ACG on I_o (or, for that matter, continuous and BVG on I_o) if each closed subset of I_o contains a portion on which F is AC (resp. continuous and BV). This follows immediately from the fact that functions which have either property AC or are continuous and BV on a set E also have the corresponding property on subsets of E and on $\bar{E}$. The characterization then follows from the proof of Theorem 9.4.

(9.25) Theorem. The class $\mathcal{D}$ is identical with $\mathcal{L}_{\omega_1}$ and for each $f \in \mathcal{L}_{\omega_1}$, $\mathcal{D}\int f = \mathcal{L}_{\omega_1}\int f$.

Proof. Because of the previous results, all that remains to prove is that $\mathcal{D} \subset \mathcal{L}_{\omega_1}$. Suppose $f \in \mathcal{D}$, $F(x) = \mathcal{D}\int_{a_o}^{x} f(t)dt$, and let $S = S(f, \mathcal{L}_{\omega_1})$ be the set of singular points of f with respect to the $\mathcal{L}_{\omega_1}$ integral. It remains to show that $S = \emptyset$. But X can not contain any isolated points because $F(x) = \mathcal{D}\int_{x_o}^{x} f(t)dt$ is a continuous function and if x_1 were the only point in $S \cap \mathring{I}$, by the Cauchy extension, $f \cdot C_I$ would belong to $(\mathcal{L}_{\omega_1})^C = \mathcal{L}_{\omega_1}$. Thus S must be a perfect set. The above argument also shows that $f \in \mathcal{L}_{\omega_1}$ on each component interval of S^C. If $S \neq \emptyset$, since F is ACG, there is a portion of S, $S \cap I$, on which $F \in AC$. But $f \in \mathcal{L}_{\omega_1}$ on each I_n contiguous to $S \cap I$ and is AC on $S \cap I$ which implies that F agrees with an absolutely continuous function G on $S \cap I$ and $G'(x) = F'_{ap}(x)$ a.e. on $S \cap I$. This implies that $F'_{ap}(x)$ is Lebesgue integrable on $S \cap I$. Since F is AC on $S \cap I$, $\Sigma|\mathcal{L}_{\omega_1}\int f \cdot C_{I_n}| = \Sigma|F(I_n)| < \infty$ and $\lim \theta(\mathcal{L}_{\omega_1}\int f \cdot C_{I_n}) = \lim \theta(F;I_n) = 0$ by the uniform continuity of F on I. Thus $f \cdot C_I \in (\mathcal{L}_{\omega_1})^H = \mathcal{L}_{\omega_1}$ and it follows that $S = \emptyset$ and that $f \in \mathcal{L}_{\omega_1}$. Consequently, $\mathcal{D} \subset \mathcal{L}_{\omega_1}$ and thus $\mathcal{D} = \mathcal{L}_{\omega_1}$, and for $f \in \mathcal{L}_{\omega_1}$, $\mathcal{D}\int f = \mathcal{L}_{\omega_1}\int f$.

We note in passing that definitions of the Denjoy integral using a Perron approach have been done as well as one which proceeds with Riemann style partitions. These will not be developed here. Instead, an additional characterization of those functions which are ACG is given in the next theorem. This will be followed by sufficient conditions for a function to be ACG.

(9.26) <u>Theorem</u>. <u>A function</u> F <u>is</u> ACG <u>on</u> I_o <u>iff</u> F <u>is</u> BVG, <u>continuous</u>, <u>and</u> <u>satisfies</u> <u>Lusin's</u> <u>condition</u> (N) <u>on</u> I_o.

<u>Proof</u>. If F is ACG on I_o, it is also BVG and continuous on I_o. Also $I_o = \cup E_n$ on each of which F

is AC. Since the function F agrees on each set E_n with a function F_n which is absolutely continuous on I_o, given a set Z with $m(Z) = 0$, $m(F(Z)) = m(\cup F(Z \cap E_n)) = m(\cup F_n(Z \cap E_n)) = 0$ and hence F also satisfies (N) on I_o. To see that the converse holds, suppose F is BVG, continuous and satisfies (N) on I_o. Then $I_o = \cup E_n$ and on each E_n, F is BV, continuous and satisfies (N). Fix n and note that the extension of F on $\overline{E}_n$ to a function which is of bounded variation and linear on intervals contiguous to $\overline{E}_n$ results in a function G which is continuous, of bounded variation and satisfies Lusin's condition (N). Thus G is absolutely continuous and since $F(x) = G(x)$ on $\overline{E}_n$, F is AC on each E_n and F is ACG on I_o.

(9.27) <u>Corollary</u>. <u>A function</u> F <u>is</u> ACG* <u>on</u> I_o <u>iff</u> F <u>is</u> BVG*, <u>continuous and satisfies Lusin's condition</u> (N) <u>on</u> I_o.

<u>Proof</u>. If $F \in$ ACG* on I_o, F is BVG*, continuous and ACG on I_o and thus satisfies (N) on I_o. For the converse, let F be BVG* on I_o. Then $I_o = \cup E_n$ on each of which F is BV*. For fixed n note that F must also be BV* on $\overline{E}_n$. Then if F is continuous, F agrees with a function G which is continuous on I_o and linear on intervals contiguous to $\overline{E}_n$. But then, when F satisfies (N), G also satisfies (N) and G must be absolutely continuous on I_o and AC on $\overline{E}_n$. Thus F, which is BV* on $\overline{E}_n$, agrees with an absolutely continuous function on $\overline{E}_n$ and hence F is AC* on $\overline{E}_n$. Since F must be

AC* on each E_n, F is ACG* on I_o whenever F is BVG*, continuous and satisfies (N) on I_o.

The Denjoy integral integrates the approximate derivatives of each continuous function which is everywhere approximately derivable. That this is the case is shown by the next theorem which gives somewhat more general conditions for a function to be ACG.

(9.28) Theorem. *If a real-valued function $F(x)$ is defined on I_o with $E \subset I_o$ and F satisfies at each $x \in E$, except for perhaps at most countably many points of E, either $-\infty < \underline{F}^+_{ap}(x) \le \overline{F}^+_{ap}(x) < \infty$ or $-\infty < \underline{F}^-_{ap}(x) \le \overline{F}^-_{ap}(x) < \infty$, then E is a countable union of sets on each of which F is AC and if F is continuous on E then F is ACG on E.*

Proof. Clearly, F is AC at each of the at most countably many points and the symmetry of the conditions makes it sufficient to show that if $-\infty < \underline{F}^+_{ap}(x) \le \overline{F}^+_{ap}(x) < \infty$ at each point of E', then E' is a countable union of sets on each of which F is AC. Given such an E', let E_n be the set of those $x \in E'$ such that $0 \le h \le 2/n$ implies

$$m^*(\{t \in [x,x+h]: |F(t) - F(x)| > n(t - x)\}) < h/4$$

Then $E' = \cup E_n$. For each integer i let $E_{i,n} = E_n \cap [(i-1)/n, i/n]$. We show that F is AC on each $E_{i,n}$. For this purpose let $x_1, x_2 \in E_{i,n}$ with $x_1 < x_2$. Consider the point x_3 satisfying $x_3 - x_2 = x_2 - x_1$. Since $x_3 - x_1 < 2/n$ and $x_2 - x_1 < 2/n$,

$$m^*(\{t \in [x_1,x_2]: |F(t)-F(x_1)| > n(t-x_1)\}) < (x_3-x_1)/4$$

and

$$m^*(\{t \in [x_2, x_3]: |F(t) - F(x_2)| > n(t - x_2)\}) < (x_2 - x_1)/4.$$
Thus the part of the union of these two sets which lies in $[x_2, x_3]$ has outer measure less than $(3/4)(x_3 - x_2)$. Thus there is a point $x_o \in [x_2, x_3]$ which belongs to neither of the two sets and then

$$|F(x_o) - F(x_1)| < n(x_0 - x_1)$$

and

$$|F(x_o) - F(x_2)| < n(x_o - x_2).$$

Hence $|F(x_1) - F(x_2)| < n(x_o - x_1) + n(x_o - x_2) \le 3n(x_2 - x_1)$. Consequently, if $\{I_k\}$ is a sequence of pairwise non-overlapping intervals with endpoints in $E_{i,n}$, then $\Sigma|F(I_k)| \le 3n \cdot \Sigma|I_k|$. It follows that F is AC on each set $E_{i,n}$. If F is also continuous on the original set E, then F is ACG on E.

A similar condition on the ordinary derivates at the points of a set E of a real-valued function F defined on I_o is sufficient for F to be ACG* on E.

(9.29) Theorem. <u>If a real-valued function F is defined on I_o with $E \subset I_o$ and F satisfies at each point of E, except for perhaps at most countably many points of E, either $-\infty < \underline{F}^+(x) \le \overline{F}^+(x) < \infty$ or $-\infty < \underline{F}^-(x) \le \overline{F}^-(x) < \infty$, then E is a countable union of sets on each of which F is AC* and if F is continuous on E then F is ACG* on E.</u>

<u>Proof</u>. As in the last theorem, it suffices to consider the case when $-\infty < \underline{F}^+(x) \le \overline{F}^+(x) < \infty$ at each point of a set $E' \subset E$. Let

$E_n = \{x \in E': |F(t) - F(x)| < n(t - x)$ when $0 < t - x \le 1/n\}$, and for each integer i, let

$$E_{i,n} = E_n \cap [(i-1)/n, i/n].$$

For $x_1, x_2 \in E_{i,n}$ with $x_1 < x_2$ and $t_1, t_2 \in [x_1, x_2]$ it follows from the definition of E_n, since $x_2 - x_1 < 1/n$, that $|F(t_1) - F(x_1)| < n(t_1 - x_1) \le n(x_2 - x_1)$ and $|F(t_2) - F(x_1)| \le n(t_2 - x_1) \le n(x_2 - x_1)$. Hence for $t_1, t_2 \in [x_1, x_2]$,

$$|F(t_2) - F(t_1)| \le 2n(x_2 - x_1)$$

and thus $\theta(F;[x_1,x_2]) \le 2n(x_2 - x_1)$. Thus, if $\{I_k\}$ is a sequence of pairwise non-overlapping intervals with endpoints in $E_{i,n}$, then $\Sigma\,\theta(F;I_k) \le 2n\cdot\Sigma|I_k|$. It follows that F is AC* on each $E_{i,n}$. If F is also continuous on E, then F is ACG* on E.

We conclude with some brief remarks concerning other integrals. As one might guess, Perron-Stieltjes integrals are possible using the derivative with respect to a function G which is monotone non-decreasing or, in general, of bounded variation. Here U is a majorant of f with respect to a non-decreasing function G if at each x, for sufficiently small intervals I with $x \in I$, $U(I) \ge f(x)\cdot G(I)$. In $\mathbb{R}^k$, U and G can be taken to be additive functions of an interval with G non-negative. A reasonably straightforward integral ensues but it will not be developed further here.

Actually, a plethora of general integrals exist which are considerably beyond the fundamental. Extensions of even the Lebesgue integral can result in two integrals which have no common extensions. Two integrals $\mathcal{I}_1$ and $\mathcal{I}_2$ are said to be <u>incompatible</u> if there is $f \in \mathcal{I}_1 \cap \mathcal{I}_2$ with $\mathcal{I}_1\!\int f \ne \mathcal{I}_2\!\int f$; incompatible extensions of the Lebesgue integral exist.

Historically, one of the purposes of the integral was that of recapturing the coefficients of a trigonometric series which converges, say to $f(x)$, by integrating the functions

$$f(x)\cdot\cos(nx) \quad \text{and} \quad f(x)\cdot\sin(nx),$$

$n = 1, 2, \ldots$ on $[-\pi,\pi]$. That the Denjoy integral is not completely adequate for this use was known to Denjoy who then developed an integral specifically for this purpose; other such integrals followed and some of these integrals turned out to be incompatible with others.

In addition, various more general derivatives have lead to integrals developed for their "anti-differentiation". General derivatives consistent with the ordinary derivative include, for example, the symmetric derivative, $\lim_{h\to 0} \frac{F(x+h) - F(x-h)}{2h}$, the approximate symmetric derivative, $\lim_{h\to 0} \text{ap} \frac{F(x+h) - F(x-h)}{2h}$, first order derivatives, $\lim_{h\to 0} \int_0^h (f(x+h) - f(x))dx/2h^2$ and n^{th} order derivatives obtained by integrating n times and taking the limit of the $(n+1)^{st}$ differences. Integrals involving these different derivatives exist; again, some are incompatible with others. It is only recently that a monotonicity theorem was developed for the approximate symmetric derivative and, naturally, a corresponding integral evolved; monotonicity theorems frequently lead to Perron type definitions.

More elaborate methods of summation have been used to produce integrals via the constructive method. In general, whenever one of the four techniques, constructive (transfinite), Perron (using majorants), descriptive (indefinite integrals), and Riemann (using partitions) results in an integral, a definition in terms of the others is usually possible.

9.2 Exercises

1. Show that a function which is continuous and of bounded variation on a set E need not agree with a continuous function on $\bar{E}$.
2. Prove that $V(F;E) = \overline{V}(F;E) + \underline{V}(F;E)$.
3. Show that the product of two BVG functions is BVG.
4. Show that the product of two ACG functions is ACG.
5. Give an example of a function which is ACG and differentiable a.e. but is not ACG*.
6. If $\overline{F}^{+}(x) < \infty$ at each $x \in E$, show that $E = \cup E_n$ such that on each E_n, $F(x) - nx$ is non-increasing.
7. If $\overline{F}_{ap}(x) < \infty$ at each $x \in E$, show that $E = \cup E_n$ such that on each E_n, $F(x) - nx$ is non-decreasing.
8. Show there is a continuous function which is not BVG on any interval.
9. Show there is a continuous function of bounded variation so that for each n, $F(x) + nx$ and $F(x) - nx$ is not monotone on any interval.

Index